RODOLFO J. AGUILAR

Professor of Civil Engineering
Louisiana State University
and
President, ADH Systems, Incorporated
Planners, Architects, Engineers, Systems Analysts

SYSTEMS ANALYSIS AND DESIGN

in Engineering, Architecture, Construction, and Planning

PRENTICE-HALL, INC.
Englewood, Cliffs, N. J.

Library of Congress Cataloging in Publication Data

Aguilar, Rodolfo J.
    Systems analysis and design in engineering,
architecture, construction, and planning.

    Includes bibliographies.
    1. Systems engineering.   2. System analysis.
3. Operations research.   I. Title.
TA168.A35   620'.72   72-10838
ISBN 0-13-881458-9

CIVIL ENGINEERING AND ENGINEERING MECHANICS SERIES

N. M. NEWMARK AND W. J. HALL, EDITORS

Printed in the United States of America

10 9 8 7 6 5 4 3 2 1

PRENTICE-HALL INTERNATIONAL, Inc., *London*
PRENTICE-HALL OF AUSTRALIA Pty. Ltd., *Sydney*
PRENTICE-HALL OF CANADA, Ltd., *Toronto*
PRENTICE-HALL OF INDIA Private Ltd., *New Delhi*
PRENTICE-HALL OF JAPAN, Inc., *Tokyo*

To Nellyn, Rudy, Richard, Robert, and Noryn ... with Love

To Frank Germano ... with Gratitude

To Jim Hand and Jules LeBlanc ... with Fraternal Friendship

Contents

PART **III**

STOCHASTIC SYSTEMS

APPENDICES

Preface

This book is an outgrowth of class notes from courses in Systems Engineering that I have taught in the Department of Civil Engineering at Louisiana State University since 1965. The notes were developed from many texts and articles in Systems Analysis, Operations Research and Planning, as well as from my publications in these fields, and have been refined and modified from semester to semester as I gained experience and insight from my students' reactions, comments, and performance. In general, students in Architecture, Landscape Architecture, Construction Technology, Geology, Industrial, Mechanical, Electrical, and Petroleum Engineering—in addition to our own in Civil Engineering—have received the information with great enthusiasm. A large number of these students have gone into graduate programs in Systems Engineering or have chosen strong minors in Systems Analysis and Operations Research.

I have attempted to use my consulting experiences to provide a measure of reality and depth to this work and to bring this experience into the classroom.

In 1968–69, I had the opportunity to conduct a series of in-house seminars for Smith, Hinchman, and Grills, Associates, Incorporated, of Detroit, Michigan, one of the largest architectural, engineering, planning firms in the United States, with over six hundred employees and with its own computer facilities. The Baton Rouge Chapter of the Association of General Contractors and the Baton Rouge Industrial Contractors Association have taken an interest in this field and have, on many occasions, aided me in organizing and conducting short courses at Louisiana State University in the management control techniques of C.P.M. and P.E.R.T. and in Construction Economics. Last year my associates and I put together five packages of lecture notes and workshops for the Louisiana Architects Association of the American Institute of Architects, on the applications of the systems approach to architectural planning, syndication, and finance. Largely because of the efforts of my friend and colleague Professor James E. Hand, former Associate Head of the Department of Architecture for Construction Technology at Louisiana State University, the system methodology is being effectively introduced into the programs of the School of Environmental Design.

This brief outline of experience should give the reader a measure of the importance that the systems approach has rapidly acquired in the study and practice of engineering, architecture, construction, and planning. The trend is clearly in this direction and, consequently, every new practitioner should equip himself with the systems tools most adaptable and valuable to his field of interest.

The book consists of fourteen chapters and is divided into three parts. *Part I*, Preliminary Considerations, consists of four chapters.

1. Introduction; the Systems Approach.
2. Economic Considerations.
3. The Differential Approach to Optimization.
4. Introduction to Matrix Analysis.

All students should read chapter 1, as it gives a general introduction to systems analysis and design. Those students with a background in engineering economy can skip over chapter 2. Chapters 3 and 4 should be easy reading to students with a strong foundation in calculus and linear algebra.
Part II deals with deterministic systems and consists of seven chapters.

5. Linear Programming; Optimization of Allocation Problems.
6. Linear Programming Applications to Planning.
7. Integer Programming.
8. Transportation, Transshipment, and Assignment Problems.
9. Management of Construction Projects.
10. Climbing Techniques.
11. Dynamic Programming; Optimization of Serial Systems.

Chapter 5 is probably the most important chapter in the book as it is the foundation for many of the topics covered later on. Chapter 6 can be skipped over by engineering students not interested in planning. Chapter 7 is optional.

Chapters 8 and 9 should be covered by most students. The subject matter of Chapter 10, Climbing Techniques, should be of special interest to students with a strong computer programming orientation. Chapter 11, Dynamic Programming, has many applications in urban and regional planning, in replacement policy studies, and in many other areas.

Part III discusses stochastic systems and consists of three chapters.

12. Stochastic Linear Programming.
13. Stochastic Dynamic Programming.
14. Introduction to the Theory of Games.

This section requires of the students a background in probability and statistics equivalent to a one-semester course in these subjects. Chapters 12, 13, and 14 should be of special interest to economic, social, and physical planners.

Although every author dreams of having his book studied in its entirety by every unsuspecting reader, I realize that most disciplines can afford only one course in this field, typically meeting 3 hours per week for one semester (14 to 16 weeks in duration). Most students should be able to cover the entire book in one full academic year (two semesters or three quarters, as the case may be).

The instructor and the student are encouraged to formulate study plans to meet their own specific requirements, nevertheless, I suggest the following basic plans of study for a one-semester first course in each of the different disciplines (in plans 2–A through 5–A the asterisk [*] indicates that in certain instances the instructor may find it desirable to treat these chapters with less mathematical rigor).

1–A For engineering students with background in calculus, linear algebra, probability, statistics, and engineering economy:
Chapters 1, 3, 4, 5, 7, 8, 9, 10, 11, 12, 13, 14

1–B For engineering students lacking one or more of the background subjects mentioned in Plan 1–A, the instructor must decide which chapters he should exclude and which he should add to the list, keeping in mind that the pace of the course would normally be slower than for Plan 1–A and that he can anticipate covering from 7 to 9 chapters, instead of the 12 suggested in Plan 1–A.

2–A For architecture students:
Chapters 1, 2, 4, 5*, 6*, 8, 9, 11*

2–B For architecture students with background in calculus, probability, and statistics:
Add chapters 12* and 13* to Plan 2–A

3–A For construction technology students:
Chapters 1, 2, 4, 5*, 8*, 9, 11*

4–A For urban and regional planning students:
Chapters 1, 2, 4, 5*, 6*, 8*, 9, 11*

5–A For landscape architecture students:

Chapters 1, 2, 4, 5*, 6*, 8*, 9

I am grateful to many of my colleagues and associates for their advice and for urging me to write this book: Dr. Frank J. Germano, Dr. B. J. Covington, and Dr. M. Al-Awady of the Civil Engineering Department, and Professor James E. Hand in Construction Technology, all at Louisiana State University.

To my students and former students Messrs. K. Movassaghi, J. T. Franques, R. S. Gonzalez, O. Argüello, J. Ristroph, T. Hebert, and W. Ketelhöhn, my gratitude for their persistent questions, suggestions, and continued interest which encouraged me to delve more deeply and more often than I would have otherwise into systems engineering.

I am also greatly indebted to Louisiana State University for supporting my writing efforts with a sabbatical leave and light teaching loads. Finally, my sincere appreciation to Miss Susan Love, Miss Norma Juban, and Miss Linda Tucker whose help and patience in typing and retyping the manuscript were an invaluable aid in completing this book.

RODOLFO J. AGUILAR

Baton Rouge, Louisiana

PART **I**

Preliminary Considerations

CHAPTER 1

Introduction: The Systems Approach

1.1 DECISION MAKING

Many contemporary problems of design and operation in engineering, architecture, construction, and urban and regional planning are of such magnitude and complexity as to require the most systematic and rational approach possible.

Generally, these problems consist of a very large number of interacting variables many of which defy quantification. Therefore, it is necessary first to classify the variables appearing in real life problems into tangible or quantifiable and intangible or unquantifiable. The purpose of the systems approach is to develop methods, mathematical or otherwise, to deal systematically and rationally with the quantifiable parameters of a problem; to increase the set of quantifiable parameters through statistical observation, testing, development of measuring techniques; and to provide a clear understanding of the situation at hand as an aid to the decision maker for subjectively evaluating the intangibles which are present in most real problems.

Physically, a system is composed of a large number of interacting components, each of which may or may not serve a different function, but all of which contribute to a common purpose. Because the components usually involve many areas of knowledge, the only way to implement the systems approach to decision making is through team action, teams involving a spectrum of specialists in seemingly unrelated fields. Systems theory is intended to provide a common basis of understanding between disciplines and, as a result, the systems approach is permeating most fields of knowledge. The physical sciences and engineering have already developed a strong vehicle for interdisciplinary teamwork through the systems approach and, recently, the life sciences as well as the humanities and the social sciences are beginning to apply to their problems the vast methodology that exists in systems analysis and design.

A complex problem can be handled with success when the systems approach is effectively applied. Cost effectiveness, which is a usual gauging device for measuring success, strives to produce a system with the lowest possible cost for a set level of effectiveness or, vice versa, the highest level of effectiveness for a set cost. Combinations of these goals are usual and trade-offs occur when such an approach is taken.

An example of the application of the systems approach to decision making, where no mathematics is needed but where the approach to problem solving recognizes the fundamental behavior of people in developing a realistic policy that can be successfully implemented, is the case of a congested urban area where two local authorities conflict in their plans for city operation and management. One of the authorities, the Planning Commission, attempts to dissuade suburbanites from bringing their personal automobiles into the Central Business District by promoting the use of the Mass Transit System, with the expectation of relieving traffic congestion and reducing the frequency and duration of traffic snarls. On the other hand, the second governmental group, the Port Authority, is charged with the planning, financing, construction, operation, and maintenance of all toll bridges and tunnels leading to the Central Business District. The Port Authority promotes the use of these facilities by offering to the public books of bridge and tunnel toll coupons which can be purchased by the frequent or routine driver at a savings of several dollars per month. A conflict immediately develops because, while one authority tries to discourage the use of personal automobiles in the Central Business District, the other encourages it by offering savings in toll charges. A behavioral solution to this sticky situation was proposed by R. L. Ackoff, Operations Researcher Professor at the University of Pennsylvania: Implement a graduated scale of toll charges as a function of the number of empty seats in a private automobile.

The policy could be developed as follows: Assume a six-passenger car arrives at the toll booth:

1. If all six places are occupied, the car passes free of charge.
2. If one seat is empty—that is, only five passengers are in the car—the charge is, say, $0.25.

3. If two seats are empty, the charge is $0.50.
4. If three seats are unoccupied, the cost now climbs to $1.00, and so on until, say,
5. The driver is the only passenger in the car, in which case the toll is $2.25.

The figures quoted may or may not be realistic but the point is well made. The day after this type of policy was implemented, a marked reduction in the number of cars visiting the Central Business District would most surely be observed. People would probably drive to the outskirts of the downtown area and then form car pools, or they would use the Mass Transit System. A suspected side effect might be an increase in sales of sports cars; with only two seats they would make an inexpensive mode of transportation. This, of course, represents at least a reduction in automobile size and can therefore be considered a desirable side effect. On the other hand, in the interest of fair play, a special tariff could be developed for nonstandard vehicles, and so on. These types of applications are primarily based on common sense, and many such decision rules can be developed without the use of mathematical techniques. However, in some of the most important applications of the systems approach, a great many complex technological questions are raised which require specialized solution techniques. This book concentrates on the development of some of the most important and useful of those techniques.

1.2 THE STRUCTURE OF SYSTEMS ANALYSIS AND DESIGN

Systems analysis is the process of separating or breaking up a whole system into its fundamental elements or component parts. It involves a detailed examination of the system in order to understand its nature and to determine its essential features. Systems design, on the other hand, is the process of selecting the components and of contriving the elements, steps, and procedures for producing a system that will optimally satisfy the stated goals. In the context of this book, systems design is used as a basis for anticipating problems and for solving them at their planning, engineering, architectural, and construction stages. Systems synthesis is akin to design because it is the process of putting together, composing, or combining parts or elements to form a whole system, completely blended to achieve its finest level of performance. Through systems synthesis, often, varied and diverse ideas, forces, or factors are combined into one coherent, consistent structure. In examining complex problems, one must recognize the following classification of elements occurring repeatedly in their solution through the systems approach:

1. *A set of decision and state variables.* The decision variables are those over which the analyst has complete control and which he can manipulate at will. The state variables are those which are dependent on the decision

variables and which, consequently, cannot be directly controlled by the decision maker. Often, the classification of variables into decisions and states is an arbitrary one. However, once they have been so stratified, their behavior follows the stated pattern. This element is basically in the analysis phase of the problem solution and the significance of each variable—that is, how sensitive the problem is to its settings—as well as whether or not the variable is quantifiable must be ascertained in this stage.

2. *An optimization model.* This solution element is necessary for understanding the problem at hand. It involves both analysis and synthesis and consists of the development of a conceptual model which is sufficiently analogous to the real problem but which, on the other hand, is simple enough to be amenable to quantitative analysis.

3. *A measure of effectiveness.* Called the objective function, this measure is formulated as a means for evaluating the degree of success or failure attained in fulfilling the problem goals. It relates various decision and state variables for the expressed purpose of ranking the outcome of the different decision sets.

4. *Generation of alternatives and optimal solution.* After the problem has been formulated quantitatively, the sets of decisions arrived at following a rational, systematic plan are evaluated by means of the objective function, and the one producing the most desirable results is selected. The different sets of decisions are the alternative plans of action and the selection of the most desirable outcome constitutes the optimization phase of the problem solution; the decision policy producing the best results is the optimal policy. Frequently, a system cannot be completely optimized. Near optimal results are often extremely valuable, especially when the objective function is not too sensitive to changes in the values of the decision and state variables near the optimum. This phase of problem solution is primarily a design phase.

5. *Policy implementation.* This step involves the carrying out of the optimal policy into the real physical situation. It constitutes, in fact, the realization of the objective and the only reason for having gone through the previous four steps. Usually, because of additional knowledge gained or because conditions change, the analyst finds it necessary to recycle the process by returning to one of the previous steps. This recycling is required in adaptive or learning processes where newly acquired data permit the system to refine itself and to adapt to a changing environment.

1.3 CLASSIFICATION OF DECISION SYSTEMS

In engineering, architecture, construction, and planning, decision systems can be classified according to size and according to predictability of behavior as follows.

I. According to size:
   1. *Simple systems* are those which involve only a relatively small number of quantifiable decision and state variables.
   2. *Complex systems* involve a large number of decision and state variables. However, the variables are, by and large, of the quantifiable type.
   3. *Exceedingly complex systems.* These consist of a large number of decision and state variables, most of which are of a non-quantifiable nature.
II. According to behavioral predictability:
   1. *Deterministic systems* are those for which every input produces a predictable response. Furthermore, the system's responses to identical inputs are themselves identical.
   2. *Stochastic systems* involve randomly determined sequences of observations, each of which is considered as a sample from a probability distribution. Stochastic variation implies system randomness in passing from one state to an adjacent state. A stochastic system's response to a specified input is not reproducible at will—that is, one cannot expect exactly the same behavior when the system is subjected to identical inputs.

Table 1–1 gives examples of different types of systems classified according to the rules previously formulated. Note that the set of exceedingly complex deterministic systems is assumed to be empty; there are no physical systems which can be classified in this category because all exceedingly complex systems possess a multitude of random parameters.

TABLE 1–1. Examples of Different Types of Systems.

Systems	Deterministic	Stochastic
Simple	Beam Deflection	Toss of Coin
Complex	Planetary System	Inventory System
Exceedingly Complex	Empty	The City

1.4 FIELDS OF SPECIALTY IN SYSTEMS ANALYSIS AND DESIGN

The systems approach to complex problem solving encompasses a broad field of work and, as such, takes on different meanings for the people involved in it, depending upon the area or areas of their specific interest and experience. Five fields of specialty can be recognized although, in most instances, they tend to overlap and run into each other as the nature of the work in systems analysis and design demands an interdisciplinary, team approach.

1. *Deterministic and stochastic optimization.* In a few words, optimization deals with the methods and procedures required to obtain the most benefit at the least possible cost. This book concentrates on the mathematical and heuristic optimization techniques applicable to deterministic and stochastic systems. In section 1.5 the general concept of optimization will be discussed further.

2. *Control theory.* Certain physical processes change so rapidly that to maintain an optimal level of performance it is necessary to adjust continuously the setting of the decision and state variables. Control theory deals with the methods and procedures needed to allow a system to adapt itself to changing conditions without upsetting it dynamically, and to keep its performance characteristics at an optimum.

Chemical and electrical engineers, for example, utilize extensively the techniques of control theory because many of the problems encountered in their fields are of the process type where continuous monitoring is essential to efficiency of operation.

3. *Information systems.* This systems area concerns itself with the structure, analysis, organization, storage, searching, and retrieval of information. The entire subject has received an increasing amount of attention in recent years, not only because simpler information handling systems are urgently needed, but also because planning and design, in whatever field, must rely on an extensive data base of high quality to be a useful tool in solving the complex problems of today and of the future. The information systems fields include elements of linguistics, mathematics, and computer programming.

4. *Numerical analysis and methods.* The computer, with its enormous speed and ability to manipulate large blocks of information, is the most useful tool available to the systems analyst and designer. Computers are of either of two basic types: (1) digital computers which are basically extremely rapid, electronic adding machines, or (2) analog computers which permit the solution of a complex problem by establishing a mathematical analogy between the problem functions and variables and a flow of electric current or a fluid. Certain problems require the coupling of digital and analog computers; this combination has come to be known as a hybrid computer. In any case, the analyst must be capable of transforming the physical problem into a mathematical one and of directing its programming for computer solution. The inclusion of the field of numerical analysis in the systems area grows out of this need. This book will present to the reader a large number of examples showing how to implement the process of going from the physical problem to its mathematical analog. This process, called modeling in the language of systems analysis, is perhaps the most difficult as well as the most important phase of problem solving.

5. *Digital, analog, and hybrid computer simulation.* This area involves the manipulation and observation of a synthetic model representing a real physical system which, for technical, economic, or other reasons, may not be suitable for experimentation. The synthetic model, usually mathematical in nature, ideally represents the essential characteristics of the physical system. In contrast with analytical models which are solved exactly or approximately, simulation models are run. The simulator observes the behavior of the model, gathers pertinent data, and draws appropriate conclusions. Digital, analog, and hybrid computers are normally used to run simulation models.

1.5 OPTIMIZATION METHODS AND APPLICATIONS

Leibniz coined the word "optimum" in 1710 as a result of his speculations about the nature of the universe and of its creation. He wrote:

> There is an infinitude of possible worlds, among which God must needs have chosen the best, since He does nothing without acting in accordance with supreme wisdom. Now this supreme wisdom, united to a goodness that is no less infinite, cannot but have chosen the best. . . . As in mathematics, when there is no maximum or minimum, everything is done equally or . . . nothing at all is done: so it may be said . . . that if there were not the best (optimum) among all possible worlds, God would not have produced any.

Those adhering to this philosophy of Leibniz came to be known as optimists when in reality they turned out to be fatalists by reasoning that if God in all of His wisdom and goodness could not make the world any better, what could a powerless human being do to change it? Optimization, with many other branches of human knowledge, had a strictly philosophical beginning. Today, however, although its philosophical implications continue to be of interest, its emphasis is on the solution of complex problems from the real world.

From the standpoint of methodology, optimization techniques can be divided into indirect and direct. Indirect techniques permit the selection of an optimal point without requiring an examination of nonoptimal points. These methods are very effective when they can be applied. This book concentrates primarily on their study. Direct methods start at an arbitrary point and proceed stepwise toward the peak by successive improvements. Chapter 10 is dedicated to the study of direct methods.

From the standpoint of the type of systems to which they are applied, optimization procedures yield various kinds of decision-making strategies:

1. *Decision making under certainty* applies to deterministic systems where the probability of occurrence of each event is one.

2. *Decision making under risk* applies to stochastic systems where the probability of occurrence of each event is known but lies somewhere between zero (nonoccurrence) and one (occurrence).
3. *Decision making under uncertainty* applies to stochastic systems with unknown probability distributions.
4. *Adaptive decision making* results when the analyst is able to conduct a sequence of experiments in order to acquire knowledge about the probability distributions involved. This category is intermediate between risk and uncertainty and in many respects can be classified as a simulation type. Chapters 12 and 13 discuss stochastic systems in some detail.
5. Finally, *decisions can be formulated individually or by groups.* In group decision making, the process may be carried out under collaboration or under competition. Chapter 14 introduces the reader to decision making under competition with risk; the subject is called "Game Theory."

REFERENCES

1. AGUILAR, R. J., "The Mathematical Formulation and Optimization of Architectural and Planning Functions," *Division of Engineering Research*, Louisiana State University, Bulletin No. 93. Baton Rouge, 1967.
2. GOSLING, W., *The Design of Engineering System.* London: Heywood & Company Ltd., 1962.
3. LEIBNIZ, G. W., *Theodicy: Essays on the Goodness of God, the Freedom of Man and the Origin of Evil* (1710), trans. E. M. Huggard, p. 128. New Haven: Yale University Press, 1952.
4. WILDE, D. J. and BEIGHTLER, C. S., *Foundations of Optimization.* Englewood Cliffs, N.J.: Prentice-Hall, Inc., 1967.

CHAPTER **2**

Economic
Considerations

2.1 INTRODUCTION

Projects of any kind become a reality only if they are technologically and
financially feasible. Technological feasibility lies in the realm of professional
know-how and ability. Financial feasibility depends upon the project's ability
to show a profit to the investor through time.

This chapter will concentrate on the development of procedures, not only
for assessing correctly the feasibility or infeasibility of a particular project,
but also for comparing alternative investment opportunities. The methods
that will be developed in this chapter will show what data are needed to
translate estimates into business decisions.

An overwhelming majority of physical projects have an "investment for
return" orientation and, for the investor, the most important question to
be answered is: Will the venture pay? Training, experience, and a great deal
of imagination are required to find the most economical or the most profitable solution to any given problem. In general, the solution must be formulated under sets of constraints; typical are a fixed budget, environmental
needs, aesthetic requirements, zoning regulations, safety standards, socio-
economic trends, technological limitations, and so on, which make arriving

at the optimal solution a very difficult task. The obvious complexity of the problem tempts one to formulate intuitive decisions. Unfortunately, these types of decisions are unreliable and dangerous and must be discarded whenever possible. Although complete optimization of a project is frequently impractical, near optimal solutions involving imperfect alternatives are often good enough to provide extremely satisfactory answers. Economic studies serve to identify clearly all of the alternative methods of solving a particular problem and of quantifying those alternatives by reducing them to a monetary basis.

2.2 THE CONCEPT OF INTEREST

Because of the existence of interest, money is worth more today than sometime in the future. Interest is the direct result of two behavioral phenomena:

1. Man has a finite life span and for this reason people are not willing to postpone enjoyment of material goods because of the inherent uncertainties of life.
2. There is a fundamental trade-off rule that appears to apply to most people: The more one has of something, the more one is willing to trade it for something else one does not have enough of. The second rule gives rise to the so-called indifference curves frequently used by economists.

If one asked somebody whether he would rather have, say, $1,000 today or $1,000 one year from today, any sane person would most assuredly respond: "Today!" Suppose that one upped the reward $100 for a delay of one year and that the question were rephrased: "$1,000 today or $1,100 one year from today?" The individual might still prefer $1,000 today. Continuing in this fashion, one would state at some point: "How about $1,000 today or $1,400 a year from today?" If, after some thought, the person responds, "I prefer to wait a year to get the $1,400," his so-called attractive annual rate of return would be 40% under this set of circumstances. That is, he is willing to wait a whole year in order to realize a 40% increase in financial return.

On this basis, *interest* can be defined as the money that must be paid for the use of borrowed money. That is, the person in the example above would be willing to lend $1,000 to a borrower in the expectation of a 40% return. The borrower would be paying 40% interest on his loan; a scandalously high rate but, nevertheless, a rate paid by many unsuspecting customers. The "truth in lending" law now requires that the true rate of interest be clearly and unequivocally stated by lenders. A review of finance company advertisements would reveal that annual interest rates in the 25% to 45% bracket are not uncommon.

A more operational definition of interest is the return derived from capital productively invested. *Interest rate* is the ratio between the amount of interest chargeable or payable at the end of a specific time period, usually one year, and the money borrowed at the beginning of the period. In general, the interest rate is "per annum" (per year), unless stated otherwise. For example, a 6% interest payable annually represents a 0.06 interest rate per year. This is equivalent to 0.015 payable quarterly, or 0.005 payable monthly. There is a slight difference between the payment of interest annually and its payment at more frequent intervals. This difference will be discussed in detail in later sections.

If the interest payable each year is computed on the total amount owed at the end of the previous year—the total amount that includes the original principal sum borrowed plus the interest that was not paid when due and that, consequently, has accumulated—the interest is said to be *compound*. *Simple* interest is that payable as a charge on the original principal sum only. This type of interest will not be discussed further because it is not generally encountered in practice and its computation is direct.

To illustrate the difference between simple and compound interest, consider the repayment of $10,000 at the end of 5 years for an 8% loan under

TABLE 2.1.   Simple and Compound Interest.

8% SIMPLE INTEREST

Year	Amount Owed at Beginning of Year	Interest Owed at End of Year	Amount Owed at End of Year
1	$10,000	$800	$10,800
2	$10,800	$800	$11,600
3	$11,600	$800	$12,400
4	$12,400	$800	$13,200
5	$13,200	$800	$14,000

8% COMPOUND INTEREST

Year	Amount Owed at Beginning of Year	Interest Owed at End of Year	Amount Owed at End of Year
1	$10,000	$800	$10,800
2	$10,800	$864	$11,664
3	$11,664	$933	$12,597
4	$12,597	$1,008	$13,605
5	$13,605	$1,088	$14,693

each plan, as given in Table 2.1. Under the 8% simple interest plan, the interest owed at the end of each year is computed on the $10,000 principal only and is a constant $800. For the 8% compound interest case, the interest is computed on the total amount (principal plus interest) owed at the end of the previous year. The amount increases from $800 at the end of the first year to $1,088 at the end of the fifth. Rounding off to the nearest dollar, the difference in cost to the borrower is $693 at the end of 5 years. This quantity is the amount the borrower pays for the use of the accumulated interest money throughout the duration of the plan.

2.3 CASH FLOW DIAGRAMS

Because of interest, money has different value through time. This results in the concept that money has "time-value." Since most investment situations involve disbursements and receipts of money through time, it is convenient to develop a semigraphical method of analysis. Cash flow diagrams provide a rational, simple way of showing how money flows through time. Figure 2.1

FIGURE 2.1.   Cash Flow Diagram.

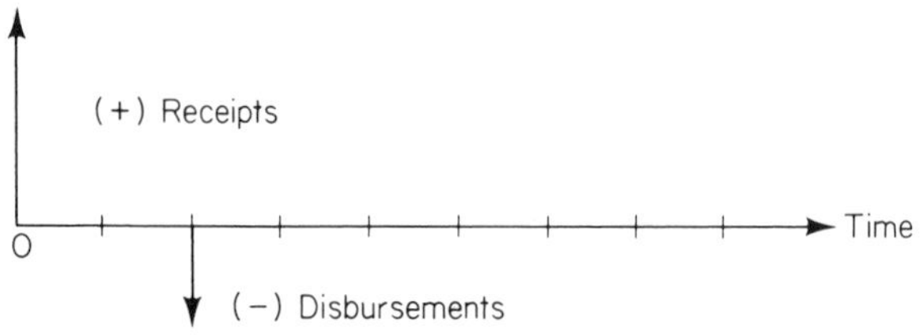

depicts a typical cash flow diagram. Plotted along a time axis are both receipts (positive cash flow) and disbursements (negative cash flow). It is irrelevant on which side of the time axis they are plotted, however; in this book receipts will be plotted above the time axis and disbursements below it.

It must be pointed out that the terms receipts and disbursements are relative as to who is receiving and paying out the money. The borrower's receipt is the lender's disbursement and vice-versa.

2.4 BASIC ASSUMPTION IN INTEREST COMPUTATIONS

For the purpose of alternative comparisons, it is well to assume that cash disbursements and receipts occur only at the end of each time period (normally one year). More frequent cash movement introduces a slight error in the computations which is of little significance in performing the types of analyses of interest to planners. For example, although repayment of principal and interest (debt service) on improved real estate property is normally done at the end of each month, with a very small error and with much simplification in the computations, the total for twelve months can be assumed to be repaid at the end of each year instead. Normally, this practice provides more than sufficient accuracy.

2.5 DECIDING BETWEEN ALTERNATIVES

It is always differences in cost or profit which are significant in selecting between alternatives. The concept of cost, or profit, must be related to specific alternatives in order to serve as a useful guide in business decision making. It must be pointed out that, although many data are irreducible to money terms, any final intelligent decision must give proper weight to the quantifiable factors as well as to the unquantifiable ones and, of course, the more one is able to quantify the elements of a problem the more reliable the decisions become. After all, the objective of an economy study is to establish a rational base for selecting between alternatives, and any information which tends to clarify the choices is of paramount importance to the goal of such a study.

2.6 THE QUESTION OF INCOME TAXES

The study of taxation is a field all to itself and no attempt will be made here to become involved in such a complex subject. Before-tax computations will be the general method used for the purposes of this book. However, the analyst must be totally aware of the tax implications of each alternative under consideration; he should seek expert advice on these and other matters which could be of vital importance to his projects.

2.7 METHODS OF ANALYSIS

Four methods of analysis of cash flow situations will be discussed in detail:

1. Present worth computations.
2. Equivalent uniform annual cash flow.
3. True rate of return computations.
4. Benefit-cost ratio.

All of these methods serve to reduce money figures through time to a common base for comparison purposes. However, before they can be studied, the reader must be totally familiar with the basic interest formulas and their use in specific situations.

2.8 INTEREST SYMBOLS AND FORMULAS

2.8.1 Symbols

Five parameters appear in interest formulas and computations. They are:

1. $i \rightarrow$ the *interest* rate per period.
2. $n \rightarrow$ the *number* of interest periods.
3. $P \rightarrow$ a *present* sum of money.
4. $F \rightarrow$ a *future* sum of money.
5. $A \rightarrow$ an *annual* payment (or a payment per period).

These symbols are mnemonic—that is, they serve as memory aids—and for this reason have been adopted for interest computations.

As will be seen later, in interest problems *four* of the *five* parameters $i, n, P, F,$ and A are always present and *three* of the *four* must be known (given as data); the object of the problem is to compute the missing fourth parameter.

An important comment must be made here; interest tables are conventionally based upon payments made at the *end* of each payment period and *not* at the beginning or anywhere in between. This characteristic of the formulas will be clearly understood when they are derived.

2.8.2 Formulas

There are six basic interest formulas belonging to two separate categories:

I. Single Payment Formulas
  { 1. Compound Amount Factor.
    2. Present Worth Factor.

II. Uniform Series of Payments
    Formulas
  { 3. Sinking Fund Factor.
    4. Compound Amount Factor.
    5. Capital Recovery Factor.
    6. Present Worth Factor.

2.8.2.1 Single Payment Formulas (I)

As the name implies, a single disbursement and receipt, through time, is involved in each of these formulas.

The cash flow diagram for the two formulas is given in Figure 2.2.

1. *Compound amount factor.* What future sum of money F can be obtained from investing P dollars today (at the end of year zero), at $i\%$ annual interest,

FIGURE 2.2.   Cash Flow Diagram for Single Payment Formulas.

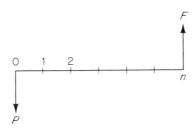

for n years? Note that the four elements involved are: P, i, n, and F. Three are known: P, i, and n. The unknown F is computed thus,

$$F = (F \longrightarrow P, i\%, n)P. \qquad (2.1)$$

Equation (2.1) is meant to be read: F equals the compound amount factor $(F \longrightarrow P, i\%, n)$, multiplied by P. Note that the symbol for the compound amount factor is functional in that it shows F is to be computed given that $\longrightarrow P$, $i\%$, and n are known. Numerical values of the compound amount factor are supplied in the tables of Appendix A. These values are tabulated under column one, headed $(F \longrightarrow P)$ for interests $i\%$ varying from 1% to 50%, and for number of interest periods, n, varying from 1 to 50 in most cases. The algebraic formula for the compound amount factor is computed as follows: If P dollars are invested at interest i, the interest for the first year is iP and the total amount at the end of the first year is

$$P + iP = P(1 + i). \qquad (2.2)$$

The second year the interest on this amount $P(1 + i)$ is $iP(1 + i)$ and the total amount accumulated at the end of the second year is

$$P(1 + i) + iP(1 + i) = P(1 + i)(1 + i)$$
$$= P(1 + i)^2. \qquad (2.3)$$

Similarly, at the end of the third year the amount is $P(1 + i)^3$; at the end of n years it is $P(1 + i)^n$. Consequently,

$$(F \longrightarrow P, i\%, n) = (1 + i)^n. \qquad (2.4)$$

Example 1:

Mr. Jones deposits $3,000 in a savings account which pays 4% annual interest. How much money would he have accumulated at the end of 5 years?

Solution:

$$P = \$3,000; \ i\% = 4\%; \ n = 5; \ F = \ ?$$
$$F = (F \longrightarrow P, 4\%, 5)3,000.$$

From Table A–9,

$$(F \longrightarrow P, 4\%, 5) = 1.2167.$$

Therefore,

$$F = 1.2167 \times 3,000 = \$3,650.$$

NOTE: For most computations, slide rule approximation provides sufficient accuracy in comparing alternative planning investments.

2. *Present worth factor.* The present worth factor is the reciprocal of the compound amount factor. How much money, P, should be invested today (at the end of year zero), at $i\%$ annual interest, for n years to accumulate a future sum of money F?

The functional relationship for present worth is

$$P = (P \longrightarrow F, i\%, n)F. \tag{2.5}$$

The algebraic equivalence of the functional formula for present worth is easily computed from the compound amount relationship.

From equations (2.1) and (2.4),

$$F = (1 + i)^n P. \tag{2.6}$$

Therefore,

$$P = \left[\frac{1}{(1 + i)^n} \right] F. \tag{2.7}$$

Numerical values for single payment present worth factors are given in column two of each table in Appendix A.

Example 2:

Mr. Smith determines that he will need \$4,000, 10 years from today, to help defray the cost of his eldest son's college education. How much money should he deposit now in a savings account paying 5% annual interest to accumulate the needed funds in 10 years?

Solution:

$$F = \$4,000; \ i\% = 5\%; \ n = 10; \ P = \ ?$$
$$P = (P \longrightarrow F, 5\%, 10)4\ 000.$$

From Table A–11,

$$(P \longrightarrow F, 5\%, 10) = 0.6139.$$

Therefore,

$$P = 0.6139 \times 4,000 = \$2,456.$$

2.8.2.2   Uniform Series of Payments Formulas (II)

These formulas involve series of equal payments.

3. *Sinking fund factor.* How much money A should be invested at the end of each year, at $i\%$ for n years, to accumulate a stipulated future sum of money F? The cash flow diagram for this situation is given in Figure 2.3.

FIGURE 2.3. Cash Flow for a Sinking Fund.

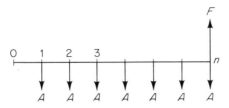

The functional relationship is

$$A = (A \longrightarrow F, i\%, n)F. \tag{2.8}$$

The numerical values of the sinking fund factor $(A \longrightarrow F, i\%, n)$ are given in column three of the tables in Appendix A. The algebraic formula for the sinking fund factor is obtained as follows:

If A is invested at the end of each year for n years, the total amount F accumulated at the end of n years will obviously equal the sum of the compound amounts of the individual investments. The amount A invested at the end of the first year will earn interest for $n - 1$ years; therefore, its compound amount will be $A(1 + i)^{n-1}$. The second year's investment will amount to $A(1 + 1)^{n-2}$ at the end of the $n - 2$ years left. The third year's investment will be $A(1 + i)^{n-3}$ and so on, until the last year (the end of year n) in which the investment earns no interest and, consequently, amounts to only A.

The total amount F is, therefore,

$$F = A(1 + i)^{n-1} + A(1 + i)^{n-2} + A(1 + i)^{n-3} + \cdots + A. \tag{2.9}$$

This can be rewritten,

$$F = A[1 + (1 + i) + (1 + i)^2 + \cdots + (1 + i)^{n-2} + (1 + i)^{n-1}]. \tag{2.10}$$

Multiplying both sides by $(1 + i)$, obtain

$$F(1 + i) = A[(1 + i) + (1 + i)^2 + (1 + i)^3 + \cdots + (1 + i)^{n-1}$$
$$+ (1 + i)^n]. \tag{2.11}$$

Subtract equation (2.10) from equation (2.11), thus

$$iF = A[(1 + i)^n - 1].$$

Consequently,

$$A = \left[\frac{i}{(1 + i)^n - 1} \right] F. \tag{2.12}$$

Therefore,

$$(A \longrightarrow F, i\%, n) = \frac{i}{(1 + i)^n - 1}. \tag{2.13}$$

Example 3:

Suppose that Mr. Smith in Example 2 cannot afford to deposit $2,456 today at 5% to accumulate $4,000 at the end of 10 years. Instead, he wishes to determine how much money he should deposit at the end of each year for 10 years to generate the same $4,000 savings.

Solution:

$$F = \$4,000; \; i\% = 5\%, \; n = 10; \; A = \;?$$
$$A = (A \longrightarrow F, 5\%, 10)4,000.$$

From Table A–11,

$$(A \longrightarrow F, 5\%, 10) = 0.07950.$$

Therefore,

$$A = 0.07950 \times 4,000 = \$318 \text{ per year.}$$

4. *Compound amount factor.* If A dollars are invested at $i\%$ for n years, how much money F will be accumulated during that period?

$$F = (F \longrightarrow A, i\%, n)A. \tag{2.14}$$

The algebraic value of the compound amount factor for a uniform series of payments is obtained directly from equation (2.12).
    Since

$$A = \left[\frac{i}{(1 + i)^n - 1} \right] F,$$

then

$$F = \left[\frac{(1 + i)^n - 1}{i} \right] A,$$

and, therefore,

$$(F \longrightarrow A, i\%, n) = \frac{(1 + i)^n - 1}{i}. \tag{2.15}$$

The compound amount factor for a uniform series of payments is the reciprocal of the sinking fund factor. Its numerical values are tabulated in column five of the tables in Appendix A.

Example 4:

What sum of money will be generated by $1,000 invested at the end of each year for 8 years in an investment yielding 6% return?

Solution:

$$A = \$1,000; \; i\% = 6\%; \; n = 8; \; F = \;?$$
$$F = (F \longrightarrow A, 6\%, 8)1,000.$$

From Table A–8,

$$(F \longrightarrow A, 6\%, 8) = 9.897.$$

Therefore,

$$F = 9.897 \times 1,000 = \$9.897.$$

5. *Capital recovery factor.* If P dollars are invested today at $i\%$, how many dollars, A, can be secured at the end of each year for n years such that the initial investment, P, is depleted? The cash flow diagram for this plan is given in Figure 2.4.

FIGURE 2.4.   Cash Flow for Capital Recovery.

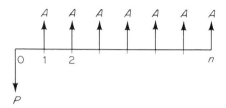

Note that capital recovery, as its name implies, refers to the number of future equal end-of-the-year payments, A, needed to recover an initial investment, P, in n years at $i\%$. The functional relationship is,

$$A = (A \longrightarrow P, i\%, n)P.$$

The algebraic formula for the capital recovery factor $(A \longrightarrow P, i\%, n)$ can be computed from the sinking fund relationship, equation (2.8), as follows:

$$A = (A \longrightarrow F, i\%, n)F. \tag{2.16}$$

To reduce F to P, use the compound amount relation, equation (2.1), and substitute it in the expression given above to obtain

$$A = (A \longrightarrow F, i\%, n)(F \longrightarrow P, i\%, n)P. \tag{2.17}$$

Consequently, from equations (2.16) and (2.17),

$$(A \longrightarrow P, i\%, n) = (A \longrightarrow F, i\%, n)(F \longrightarrow P, i\%, n). \tag{2.18}$$

The algebraic equivalence is derived from equations (2.4) and (2.13)

$$(A \longrightarrow P, i\%, n) = \left[\frac{i}{(1 + 1)^n - 1}\right](1 + i)^n,$$

or

$$(A \longrightarrow P, i\%, n) = \frac{i(1 + i)^n}{(1 + i)^n - 1}. \tag{2.19}$$

This equation can also be expressed as

$$(A \longrightarrow P, i\%, n) = \frac{i + i(1 + i)^n - i}{(1 + i)^n - 1},$$

$$= \frac{i[1 + (1 + i)^n - 1]}{(1 + i)^n - 1},$$

$$(A \longrightarrow P, i\%, n) = \frac{i}{(1 + i)^n - 1} + i. \tag{2.20}$$

Expression (2.20) shows that the capital recovery factor equals the sinking fund factor, equation (2.13) plus the interest rate. Thus,

$$(A \longrightarrow P, i\%, n) = (A \longrightarrow F, i\%, n) + i. \tag{2.21}$$

Numerical values of the capital recovery factor are given in column four of the tables in Appendix A.

Example 5:

What should be the annual return from an investment of $10,000 at 4.5% for 15 years?

Solution:

$$P = \$10,000; \; i\% = 4.5\%; \; n = 15; \; A = \; ?$$
$$A = (A \longrightarrow P, 4.5\%, 15)10,000.$$

From Table A–10,

$$(A \longrightarrow P, 4.5\%, 15) = 0.09311.$$

Therefore,

$$A = 0.09311 \times 10,000 = \$931 \text{ per year.}$$

6. *Present worth factor.* How much money P should be invested today at $i\%$ to recover A dollars at the end of each year for n years?

$$P = (P \longrightarrow A, i\%, n)A. \tag{2.22}$$

The present worth factor $(P \longrightarrow A, i\%, n)$ is the reciprocal of the capital recovery factor. Therefore, its algebraic formula is

$$(P \longrightarrow A, i\%, n) = \frac{(1 + i)^n - 1}{i(1 + i)^n}. \tag{2.23}$$

Equation (2.23) is the reciprocal of equation (2.19). Numerical values of $(P \longrightarrow A, i\%, n)$ are given in column six of the tables in Appendix A.

Example 6:

What amount P should be invested at 8% to generate a return of $1,000 at the end of each year for the next 12 years?

Solution:

$$A = \$1,000; \; i\% = 8\%; \; n = 12; \; P = \; ?$$
$$P = (P \longrightarrow A, 8\%, 12)1,000.$$

From Table A–15,

$$(P \longrightarrow A, 8\%, 12) = 7.536.$$

Therefore,

$$P = 7.536 \times 1,000 = \$7,536.$$

2.8.3 Summary of Interest Factors

1. Single Payment Compound Amount Factor

$$(F \longrightarrow P, i\%, n) = (1 + i)^n. \tag{2.24}$$

2. Single Payment Present Worth Factor

$$(P \longrightarrow F, i\%, n) = \frac{1}{(1 + i)^n}. \tag{2.25}$$

3. Series of Payments Sinking Fund Factor

$$(A \longrightarrow F, i\%, n) = \frac{i}{(1 + i)^n - 1}. \tag{2.26}$$

4. Series of Payments Compound Amount Factor

$$(F \longrightarrow A, i\%, n) = \frac{(1 + i)^n - 1}{i}. \tag{2.27}$$

5. Series of Payments Capital Recovery Factor

$$(A \longrightarrow P, i\%, n) = \frac{i(1 + i)^n}{(1 + i)^n - 1} = \frac{i}{(1 + i)^n - 1} + i. \tag{2.28}$$

6. Series of Payments Present Worth Factor

$$(P \longrightarrow A, i\%, n) = \frac{(1 + i)^n - 1}{i(1 + i)^n}. \tag{2.29}$$

2.8.4 Relationship Between Interest Factors

1. $(F \longrightarrow P, i\%, n) = \dfrac{1}{(P \longrightarrow F, i\%, n)}.$ $\qquad$ (2.30)

2. $(A \longrightarrow F, i\%, n) = \dfrac{1}{(F \longrightarrow A, i\%, n)}.$ $\qquad$ (2.31)

3. $(A \longrightarrow P, i\%, n) = \dfrac{1}{(P \longrightarrow A, i\%, n)}.$ $\qquad$ (2.32)

4. $(A \longrightarrow P, i\%, n) = (A \longrightarrow F, i\%, n) + i.$ $\qquad$ (2.33)

2.8.5 Other Interest Formulas

Although other interest formulas can be just as easily derived, they are not as important as the six basic formulas already given. Formulas for gradient series, continuous compounding, and so on, as well as numerical tables for many of them, can be found in the references.

2.9 NOMINAL AND EFFECTIVE RATES OF INTEREST

Suppose that interest is compounded more frequently than once per year. Many banks, for instance, compound interest quarterly or even more frequently on their savings accounts. The more frequent the compounding, the higher the yield of a specified interest rate.

Let interest be compounded m times per year at a rate i/m per compounding period.

The nominal rate per year is

$$m\left(\frac{i}{m}\right) = i. \tag{2.34}$$

The effective rate per year is

$$\left(1 + \frac{i}{m}\right)^m - 1. \tag{2.35}$$

Formula (2.35) simply states that the effective interest per unit (dollar) is the compound amount at the end of m periods per year, minus the initial investment of one unit (dollar).

For example, a nominal rate $i = 0.12$ (12% annual interest) has the following effective rates:

1. Compounded annually,

$$\text{Effective Rate} = 0.12$$

2. Compounded semiannually,

$$\text{Effective Rate} = \left(1 + \frac{0.12}{2}\right)^2 - 1 = 0.1236$$

3. Compounded monthly,

$$\text{Effective Rate} = \left(1 + \frac{0.12}{12}\right)^{12} - 1 = 0.1268$$

2.10 PROCEDURE FOR FINDING AN UNKNOWN RATE OF INTEREST

Because interest formulas are nonlinear, the most efficient method for finding an unknown rate of interest is by trial and error, using the tables of Appendix A and linearly interpolating between values. The interpolation produces an error which is negligible for most applications.

Example 7:

A company invested $10,000 in a piece of property 15 years ago. Today the company sold the property for $22,000. What is the return on the $10,000 investment?

Solution:

$$P = \$10,000; \ F = \$22,000; \ n = 15; \ i\% = ?$$

By interpolation:

$$F = (F \longrightarrow P, \ i\%, \ 15)P,$$

or,

$$22,000 = (F \longrightarrow P, \ i\%, \ 15)10,000.$$

Consequently,

$$(F \longrightarrow P, \ i\%, \ 15) = \frac{22,000}{10,000} = 2.2000.$$

Looking through the tables one finds:

$$(F \longrightarrow P, \ 5\%, \ 15) = 2.0789, \text{ from Table A.11.}$$

$$(F \longrightarrow P, \ 5.5\%, \ 15) = 2.2325, \text{ from Table A.12.}$$

Consequently, by interpolation,

$$i\% = 5\% + \left(\frac{2.2000 - 2.0789}{2.2325 - 2.0789}\right)0.5\%,$$

$$i\% = 5\% + 0.395\% = 5.39\% \text{ return.}$$

Exact Solution:

$$(F \longrightarrow P, \ i\%, \ 15) = (1 + i)^{15} = 2.2000.$$

Therefore,

$$1 + i = \sqrt[15]{2.2},$$

$$i = \sqrt[15]{2.2} - 1.$$

Using logarithms, find

$$i\% = 5.397\%.$$

The error is, in fact, negligible.

2.11 ADDITIONAL EXAMPLES OF APPLICATION

It is well to repeat here three important observations made previously:

1. In each problem *four* of the five elements P, F, A, i, and n appear. Three of the four must be known (must be given as data).
2. Interest formulas and their associated tables are based upon payments made at the *end* of each payment period.
3. Slide rule accuracy is sufficient here for most applications of interest.

Example 8:

If $4,000 is invested now, $5,000 two years hence, and $10,000 four years hence, all at 5% interest, what will be the total accumulated amount 12 years from today?

Solution:

The cash flow diagram for this problem is

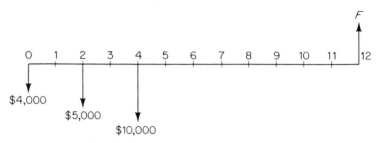

To compute F, note that the compound amount factor must be applied to the different deposits for different numbers of years. Thus,

$$F = (F \longrightarrow P, 5\%, 12)4,000 + (F \longrightarrow P, 5\%, 10)5,000$$
$$+ (F \longrightarrow P, 5\%, 8)10,000.$$

From Table A–11.

$$F = 1.7959 \times 4,000 + 1.6289 \times 5,000 + 1.4775 \times 10,000,$$
$$F = \$30,100.$$

Example 9:

What is the compound amount of $6,000 for 7 years with interest at 8% compounded quarterly?

Solution:

Because the interest is compounded quarterly (four times per year), the effective interest for a nominal 8% per year is equivalent to 2% per period (quarter) for 28 periods (7 years). Therefore,

$$P = \$6,000; \ i\% = 2\%; \ n = 28; \ F = \ ?$$
$$F = (F \longrightarrow P, \ 2\%, \ 28)6,000 = 1.7410 \times 6,000 = \$10,446.$$

Example 10:

Compute the present worth of $2,000 at the end of each 6-year period for the next 18 years if interest is at 6%.

Solution:

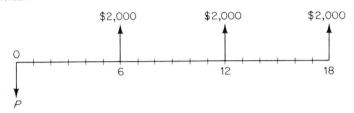

$$P = 2,000[(P \longrightarrow F, \ 6\%, \ 6) + (P \longrightarrow F, \ 6\%, \ 12) + (P \longrightarrow F, \ 6\%, \ 18)]$$
$$= 2,000[0.7050 + 0.4970 + 0.3503] = \$3,100.$$

Example 11:

How long will it take for money to double itself with interest at 6%?

Solution:

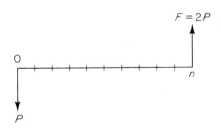

From the 6% interest table, find that for $n = 11$,

$$(F \longrightarrow P, \ 6\%, \ 11) = 1.8983.$$

For $n = 12$,

$$(F \longrightarrow P, \ 6\%, \ 12) = 2.0122.$$

Therefore, it takes, approximately,

$$n \simeq 11 + \left(\frac{2.0000 - 1.8983}{2.0122 - 1.8983}\right)1 = 11.89 \text{ years}$$

for money to double itself at 6% interest.

Example 12:

How much must be paid each year during 10 years to repay a loan of $15,000 if interest is 8%?

Solution:

$$P = \$15,000; \; i\% = 8\%; \; n = 10; \; A = ?$$
$$A = (A \longrightarrow P, 8\%, 10) \; 15,000$$
$$= 0.14903 \times 15,000 = \$2,240 \text{ per year.}$$

Example 13:

How much would still be owed after the sixth payment has been made on the loan of Example 12?

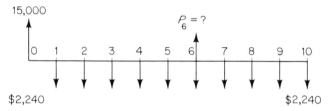

Solution:

Immediately after the sixth payment has been made, there are still four more payments to make. The question could be rephrased: What is the present worth (at the end of year 6), P_6, of four annual payments of $2,240 at 8% interest?

$$P_6 = (P \longrightarrow A, 8\%, 4) \, 2,240$$
$$= 3.312 \times 2240 = \$7,400.$$

This shows that during the first six years only $7,600 was amortized. The rest of the money went to pay the interest.

Example 14:

An investment of $100,000 produces a return of $14,000 per year for 16 years. What is this investment's approximate rate of return?

Solution:

$$P = \$100,000; \; A = \$14,000; \; n = 16; \; i\% = ?$$
$$100,000 = (P \longrightarrow A, i\%, 16) \, 14,000.$$

Therefore,

$$(P \longrightarrow A, i\%, 16) = \frac{100,000}{14,000} = 7.14.$$

Scan the tables in Appendix A for a series of payments present worth factor of 7.14 in 16 years.

$$\text{For } i = 10\% \ (P \longrightarrow A, \ 10\%, \ 16) = 7.824.$$

$$\text{For } i = 12\% \ (P \longrightarrow A, \ 12\%, \ 16) = 6.974.$$

By interpolation,

$$i\% \simeq 10\% + \left(\frac{7.82 - 7.14}{7.82 - 6.97}\right)2\% = 11.6\%.$$

2.12 PRESENT WORTH COMPUTATIONS

Proposed investments are feasible only if they can be recovered with interest. The interest rate should be at least as large as that which is attractive to the investor under the particular set of circumstances. In what follows this interest will be called the attractive rate of return.

The first method to be discussed for comparing alternative investment situations consists of reducing series of estimated money receipts and disbursements to their present worth as of the zero date of the series being compared.

Because first costs are already at zero date, no interest factors need be applied to them. In a cost situation, present worth of salvage values must be subtracted to obtain the present worth of the net disbursements. Present worth computations also serve to place a value on prospective net money receipts and, as will be shown later, are used to compute true rates of return.

In accounting circles, reduction to present worth is called discounting.

Example 15:

An architect is faced with the following alternatives for providing on-site power generation to a hospital:

Alternative A
1. Purchase equipment to supply an expanded 500-bed hospital facility at a cost of $250,000.
2. The expected life of the equipment is 25 years with zero salvage value.
3. Maintenance and operating costs are estimated at $30,000 per year.

Alternative B
1. Purchase equipment package No. 1 today at a cost of $150,000 to supply the needs of a 250-bed facility.
2. Ten years hence, buy equipment package No. 2 at an estimated cost of $200,000 to supply the hospital facility expanded to 500 beds.
3. The expected life of each equipment package is 25 years. However, equipment 15 years old can be sold at an estimated price of $50,000.

4. Maintenance and operating costs are estimated at $18,000 per year for each equipment package.

The architect wishes to select the least costly alternative assuming a minimum attractive rate of return of 8% per annum.

Solution:

The cash flow diagrams for each alternative are given below:

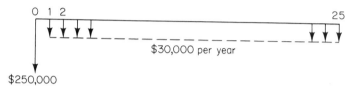

ALTERNATIVE A

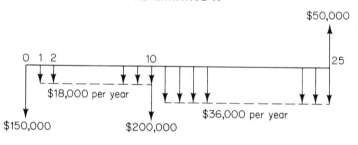

ALTERNATIVE B

The computations of present worth (P.W.) are carried out as follows:

Alternative A

Initial Investment	= $250,000
P.W. of Annual Disb., years 1 to 25	
= $30,000(P \longrightarrow A$, 8%, 25) = $30,000(10.675)	= $320,000
Total P.W.	= $570,000

Alternative B

Initial Investment	= $150,000
P.W. of Annual Disb. of $18,000, years 1 to 10	
= $18,000(P \longrightarrow A$, 8%, 10) = $18,000(6.710)	= $121,000
P.W. of $200,000, 10 years hence	
= $200,000(P \longrightarrow F$, 8%, 10) = $200,000(0.4632)	= $ 92,600
P.W. of Annual Disb. of $36,000, years 11 to 25	
= $36,000(P \longrightarrow A$, 8%, 15)$(P \longrightarrow F$, 8%, 10)	
= $36,000(8.559)(0.4632)	= $143,000
Total P.W.	= $506,600
Less P.W. of salvage value	
= $50,000(P \longrightarrow F$, 8%, 25) = $50,000(0.1460)	= $ 7,300
P.W. of *Net* Disbursements	= $499,300

Consequently, Alternative B is more economical.

Comment on Example 15:

1. It was previously stated that in computing present worth it is necessary to establish the attractive rate of return. This is the rate at which money could be profitably invested in an alternate venture. Naturally, this rate should not be less than the cost of borrowed money and, in general, should be higher.
2. Under Plan B, the reduction to present worth of the $36,000 annual disbursement from years 11 to 25 was accomplished as follows:
   a. First, the series was reduced to P.W. at year 10

   $$= 36{,}000(P \longrightarrow A, 8\%, 15) = P_{10}.$$

   b. Then this sum of money at year 10 was reduced to present worth at year zero

   $$= P_{10}(P \longrightarrow F, 8\%, 10) = \$143{,}000.$$

   c. The calculation was performed in one operation, thus

   $$36{,}000(P \longrightarrow A, 8\%, 15)(P \longrightarrow F, 8\%, 10)$$
   $$= 36{,}000(8.559)(0.4632) = \$143{,}000.$$

2.12.1   Alternatives with Perpetual Lives

There are cases when a present worth study must be done for an alternative with a long service life (50 years or more) or for one with a perpetual period of service. Economy studies are, in general, not sensitive to cash flow in the distant future, unless the interest rate is very low. For example, the present worth of one dollar 50 years hence at 4% interest is only fourteen cents. At 8% interest, it is only two cents. Present worth computed for a perpetual period is called *capitalized* costs. Capitalized costs are simply annual costs divided by the interest rate.

Example 16:

What is the present worth of an annual receipt of $450, forever, if the attractive rate of return is 8%?

Solution:

The question can be rephrased as follows: How much money P should be invested at 8% to yield annual receipts of $450 without ever touching the principal (that is, forever)?
The solution is obvious.

$$0.08\,P = 450.$$

That is, 8% of P must equal $450 each year. Thus,

$$P = \frac{450}{0.08} = \$5{,}620.$$

If $5,620 is deposited in a savings account paying 8% annual interest,

$450 can be withdrawn every year, forever, without depleting the original amount. Consequently, the present worth of $450 annually, in perpetuity, is $5,620 at 8%.

Example 17:

Determine the difference in present worth of the following two plans.

Plan C

Initial Cost	= $50,000
Annual Cost	= $ 2,000
Service Life	= 25 years
Interest Rate	= 0.08

Plan D

Assume that Plan C will continue in perpetuity—that is, every 25 years there will be an investment of $50,000 and the annual costs will remain at $2,000, forever.

Solution:

Plan C

Initial Cost	= $50,000
P.W. of Annual Disb., years 1 to 25	
$= 2,000(P \longrightarrow A, 8\%, 25) = 2,000(10.675)$	= $21,400
Total P.W.	= $71,400

Plan D

Initial Cost	= $50,000
P.W. of Annual Disb., forever	
$= 2,000 \div 0.08$	= $25,000
P.W. of 50,000 every 25 years, forever	
$= 50,000(A \longrightarrow F, 8\%, 25) \div 0.08$	
$= 50,000(0.01368) \div 0.08$	= $ 8,550
Total P.W.	$83,550

Comment on Example 17:

1. The present worth of $50,000 every 25 years, forever, is computed as follows
   a.

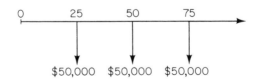

First, reduce the $50,000 to an equivalent annual cost by multiplying it by the sinking fund factor for each 25-year period,

$A = \$50,000 \, (A \longrightarrow F, 8\%, 25).$

b.

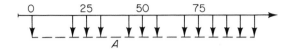

Now that the equivalent annual cost to a $50,000 disbursement every 25 years has been computed, divide that annual cost by the interest rate, as in Example 16, to obtain

P.W. $\$50,000 \, (A \longrightarrow F, 8\%, 25) \div 0.08 = \$8,550$

2. Plan D is, in fact, a perpetually cycling Plan C. The difference in present worth between the two plans is $12,150—an increase of only 17% between 25 years of life and perpetual service.

2.12.2 Alternatives with Different Lives

In order to provide an unbiased comparison between alternatives with different lives, it is necessary to transform them, somehow, into alternative plans supplying the needed service for the *same* number of years.

Because distant disbursements and receipts have little effect upon present worth computations at normal interest charges (4% or higher), a convenient simple assumption to make is that replacement assets will repeat the costs forecasted for the initial assets.

Example 18:

Compare the following plans at 6% interest.

Plan E

Initial Equipment Cost	= $50,000
Annual Cost of Oper. & Maint.	= $ 3,000
Service Life	= 25 years
Salvage Value	= $10,000

Plan F

Initial Equipment Cost	= $75,000
Annual Cost of Oper. & Maint.	= $ 2,500
Service Life	= 50 years
Salvage Value	= 0

Solution:

The first step is to make the service life of Plan E 50 years by assuming that the costs will be repeated during the second 25-year cycle. Thus the

cash flow diagram for Plan E becomes

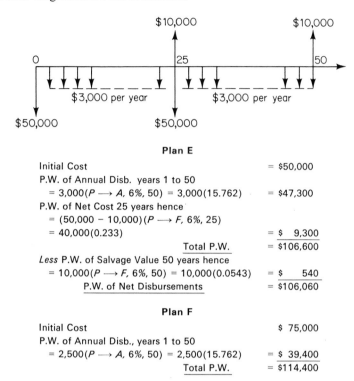

Plan E

Initial Cost	= $50,000
P.W. of Annual Disb. years 1 to 50	
= 3,000($P \longrightarrow A$, 6%, 50) = 3,000(15.762)	= $47,300
P.W. of Net Cost 25 years hence	
= (50,000 − 10,000)($P \longrightarrow F$, 6%, 25)	
= 40,000(0.233)	= $ 9,300
Total P.W.	= $106,600
Less P.W. of Salvage Value 50 years hence	
= 10,000($P \longrightarrow F$, 6%, 50) = 10,000(0.0543)	= $ 540
P.W. of Net Disbursements	= $106,060

Plan F

Initial Cost	$ 75,000
P.W. of Annual Disb., years 1 to 50	
= 2,500($P \longrightarrow A$, 6%, 50) = 2,500(15.762)	= $ 39,400
Total P.W.	= $114,400

On the surface, Plan E is more economical; however, there are additional considerations:

1. The cost of the equipment 25 years hence for Plan E may be much higher than the $50,000 assumed.
2. On the other hand, technological breakthroughs may make the present equipment less desirable or obsolete and, consequently, the shorter life of Plan E may turn out to be a blessing in disguise.

Considerations such as these fall in the realm of unquantifiable factors. The function of management is to make the right decision after having looked into the hard figures as well as the imponderables. Long-range planning results in economical decisions when the forecasts for company growth and other factors are essentially correct. Being a good manager (decision maker) is a very difficult, demanding task.

2.12.3 A Warning About Present Worth Computations

Although municipal engineers and other professionals involved in public works projects have used capitalized costs for many years (long-lived structures), a common error in present worth computations is that of assuming a perpetual life when the situation is otherwise. Also, interest rates used in present worth computations are frequently too low and do not reflect the true cost of money.

2.12.4 Establishing Bond Valuations

Suppose that a 4%, $15,000 government bond is due after 20 years. The bond calls for a semiannual payment of $300 and of $15,000 at the end of the 20-year period. If the bond is to be valued to yield a nominal 6% compounded *semiannually*, what should its purchase price be? The value of the bond is the present worth of the interest payments of $300 every six months plus the principal payment of $15,000, 20 years from today, at 3% interest per six-month period (6% annual interest).

P.W. of semiannual payments, periods 1 to 40
  = $300(P \longrightarrow A, 3\%, 40) = 300(23.115)$ = \$ 6,934.50
P.W. of principal payment
  = $15,000(P \longrightarrow F, 3\%, 40) = 15,000(0.3066)$ = \$ 4,599.00
       Value of bond to yield a nominal   = \$11,533.50
       6% compounded semiannually

The effect of bond ownership on the payment of income taxes should be carefully studied. Published tables of bond values give the relationship between price and yield for bonds with several coupon rates and years to maturity.

2.13 EQUIVALENT UNIFORM ANNUAL CASH FLOW

The second procedure to be discussed for comparing cash flow through time is the so-called equivalent uniform annual cash flow method. Nonuniform series of money receipts and disbursements are transformed into equivalent uniform series by applying the proper compound interest factors.

Once the present worth of an investment plan has been computed, its equivalent annual cash flow value can be obtained immediately by multiplying the present worth by the capital recovery factor. By the same token, equivalent uniform annual cash flow can be converted to present worth by multiplying the present worth by the capital recovery factor. Whichever alternative is favored by one of the methods must also be favored by the other, and in the same proportion.

Example 19:

What are the equivalent uniform annual cash flows of the plans given in Example 18?

Solution:

The equivalent uniform annual costs can be computed from the present worths.

Plan E

$i = 0.06$, $n = 50$ years
P.W. of Plan E $= \$106,060.$
Equivalent Uniform Annual Cost
$= 106,060(A \longrightarrow P, 6\%, 50)$
$= 106,060(0.06344)$ $= \$6,720$ per year.

Plan F

$i = 0.06$, $n = 50$ years
P.W. of Plan F $= \$114,400.$
Equivalent Uniform Annual Cost
$= 114,400(A \longrightarrow P, 6\%, 50)$
$= 114,400(0.06344)$ $= \$7,250$ per year.

The equivalent uniform annual costs can also be computed directly from the data. The different service lives of Plan E (25 years) and Plan F (50 years) no longer have to be made the same when using the equivalent annual cash flow method. However, the assumption is implied that the annual cost of service for the shorter lived alternative will continue the same for the additional number of years of service provided by the longer lived alternative.

Plan E

Equivalent Annual Cost of Initial Cost
$= 50,000(A \longrightarrow P, 6\%, 25) = 50,000(0.07823)$ $= \$3,900$
Annual Disbursements, years 1 to 25 $= \$3,000$
Less Equivalent Annual Cost of Salvage Value, 25 years hence
$= 10,000(A \longrightarrow F, 6\%, 25) = 10,000(0.01823)$ $= \$-180$
Equivalent Uniform Annual Cost $= \$6,720$

Plan F

Equivalent Annual Cost of Initial Cost
$= 75,000(A \longrightarrow P, 6\%, 50) = 75,000(0.6344)$ $= \$4,750$
Annual Disbursements, years 1 to 50 $= \$2,500$
Equivalent Uniform Annual Cost $\$7,250$

Comment on Example 19:

1. The equivalent uniform annual costs computed both ways are the same, as should be expected.
2. Only capital costs require conversion to uniform series by the appropriate compound interest factors.
3. In transforming the salvage value to equivalent uniform annual

receipts for Plan E, the sinking fund factor was used because the salvage value occurs in the future. Several formulas for reducing salvage values can be derived; however, the method used in Example 19 is adequate and is the one adopted for future computations.

2.13.1 Difference in Annual Cost Between Long Life and Perpetual Life

As the number of years, n, increases, the capital recovery factor $(A \rightarrow P, i\%, n)$ approaches the interest rate i.

From Equation (2.20),

$$(A \longrightarrow P, i\%, n) = \frac{i}{(1 + i)^n - 1} + i.$$

Therefore,

$$\lim_{n \to \infty}(A \longrightarrow P, i\%, n) = i \tag{2.36}$$

because the denominator of the fraction increases without bound, thus making the fraction vanish in the limit.

For example,

$$(A \longrightarrow P, 6\%, 100) = 0.06018 \simeq 0.06 = i.$$

This concludes the discussion of the second method presented for comparing alternative cash flows.

2.14 TRUE RATE OF RETURN COMPUTATIONS

The true rate of return of an investment is the interest rate at which the present worth of the *net* cash flow is zero. It is determined by using trial-and-error computations. Straight line interpolation between tabulated rates of interest is considered to give enough accuracy for most applications of interest.

Cash flow analyses of the type being discussed in this book make management aware of the economic value of time; money today is more valuable than money in the distant future. It also makes management cognizant of the fact that in business situations cash flow is what matters and that income taxes, their amount and timing, play an important role in business decisions for they have a very marked effect upon cash flow.

Example 20:

A parcel of land was bought 20 years ago for $20,000 and sold today for $55,000. Through the years the average property tax paid annually amounts to $600. What is the true rate of return yielded by this investment?

Solution:

The cash flow diagram is

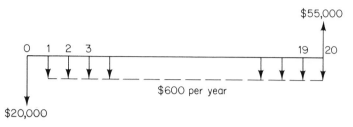

$55,000

0   1   2   3   19 20

$600 per year

$20,000

The true rate of return is determined by trial and error, as follows:

For i = 0.03 (3% interest)

Initial Investment	= $-20,000
P.W. of $600 Annual Disb., years 1 to 20	
= 600(P ⟶ A, 3%, 20) = 600(14.877)	= $ -8,930
P.W. of Disbursements	= $-28,930
P.W. of $55,000 receipt 20 years hence	
= 55,000(P ⟶ F, 3%, 20) = 55,000(0.5537)	= $+30,400
P.W. of Receipts	= $+30,400
P.W. of Net Cash Flow = +30,400 − 28,930	= $ +1,470

Because the P.W. of the net cash flow turned out to be positive the interest rate must be higher.

Try i = 0.035 (3.5% interest)

Initial Investment	= $-20,000
P.W. of $600 Annual Disb., years 1 to 20	
= 600(P ⟶ A, 3.5%, 20) = 600(14.212)	= $ -8,540
P.W. of Disbursements	= $-28,540
P.W. of $55,000 receipt 20 years hence	
= 55,000(P ⟶ F, 3.5%, 20) = 55,000(0.5026)	= $+27,600
P.W. of Receipts	= $+27,600
P.W. of Net Cash Flow = +27,600 − 28,540	= $ -940

The P.W. of the net cash flow is now negative. This indicates that the true rate of return is between 3% and 3.5%. From a linear interpolation (See Figure 2.5), find

$$i\% = 3\% + \left(\frac{1,470}{1,470 + 940}\right)0.5\% = 3.3\%.$$

Thus, the true rate of return from this land investment is a modest 3.3%. The linear interpolation is usually valid to the nearest tenth of a percent.

FIGURE 2.5   Linear Interpolation to Determine True Rate of Return.

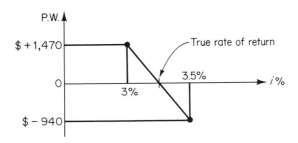

2.14.1   Equivalent Uniform Annual Cash Flow Method

Equivalent uniform annual cash flow can be used to compute the true rate of return. The definition which applies to this method is: True rate of return is the interest rate at which the equivalent uniform annual *net* cash flow is zero. Whichever method is used, the results will be identical (except for round-off error).

The rate of return computation is given different names by different analysts: discounted cash flow method, investors method, interest rate or return, profitability index, and so on. All of them follow the same basic procedure. However, the reader is warned to look out for many incorrect—sometimes deceivingly so—methods of computing so-called rates of return.

2.14.2   Alternatives with Different Lives

As in the case of present worth computations, the lives of the competing alternatives must be equalized, as explained previously, before computing their true rates of return.

Example 21:

Table 2–2 shows receipts and disbursements on income property from the

TABLE 2.2.   Cash Flow of Income Property.

Year	Disbursements	Receipts	Net Cash Flow
0	$-19,000	0	$-19,000
1	-2,000	$ +2,500	+500
2	-1,500	+2,500	+1,000
3	-1,200	+2,800	+1,600
4	-1,800	+2,800	+1,000
5	-800	+3,000	+2,200
6	-1,200	+3,000	+1,800
7	-1,700	+3,200	+1,500
8	-2,000	+27,000	+25,000
			$+15,600

time it was bought to its sale 8 years later. Compute the rate of return on the investment.

Solution:

Note that the sum of column four in Table 2.2 is $15,600. This is the present worth of the net cash flow at 0% interest. Only if this column yields a positive value has the investment been profitable. In fact, if the net cash flow column summed to zero the investment would have yielded a zero rate of return. To determine the true rate of return, set up the following table for the trial-and-error computations:

Year	Net Cash Flow	$(P \longrightarrow F, 8\%, n)$	P.W. at 8%	$(P \longrightarrow F, 10\%, n)$	P.W. at 10%
0	$-19,000	1.0000	$-19,000	1.0000	$-19,000
1	+500	0.9259	+460	0.9091	+450
2	+1,000	0.8573	+860	0.8264	+830
3	+1,600	0.7938	+1,270	0.7513	+1,200
4	+1,000	0.7350	+740	0.6830	+680
5	+2,200	0.6808	+1,500	0.6209	+1,370
6	+1,800	0.6302	+1,130	0.5645	+1,020
7	+1,500	0.5835	+880	0.5132	+770
8	+25,000	0.5403	+13,500	0.4665	+11,700
	$+15,600		$ +1,340		$ −980

Therefore, the rate of return $= 8\% + \left(\dfrac{1,340}{1,340 + 980}\right)2\% = 9.1\%.$

2.15 BENEFIT-COST RATIO

The fourth and final method of alternative evaluation discussed in this book is the benefit-cost ratio. This approach was born out of the Flood Control Act of 1936 which stipulates that "benefits to whom-soever they may accrue" should exceed "estimated costs." Identical conclusions concerning the worthiness of alternative investment plans are reached with the benefit-cost ratio as with present worth, equivalent uniform annual cash flow, or rate of return computations.

2.15.1 Calculation of Benefit Costs

Let $B =$ Benefits,
$\quad D =$ Disbenefits,
and $C =$ Costs.

Then the stipulation that $B - (C + D) > 0$ is also expressible $B - D/C > 1$. However, in the ratio computation it makes a great deal of difference whether an item is classified as a disbenefit or as a cost. In the first case, the item is subtracted from the benefit figure on the numerator; in the second, it is added to the cost figure in the denominator. This same consideration does not affect the difference approach to benefit cost. The benefit-cost computations yield the same results whether they are obtained from present worth or from equivalent uniform annual values.

Example 22:

Consider a public works project with estimated benefits of $1,600,000 and costs of $650,000. In addition, assume that there exists an adverse item valued at $200,000. Then,

$$B - (C + D) = 1,600,000 - (650,000 + 200,000)$$
$$= \$750,000$$

whether or not the adverse item is considered a cost or a disbenefit. On the other hand, for the ratio computation:

1. Assume that the $200,000 is a cost, then

$$\frac{B}{C} = \frac{1,600,000}{850,000} = 1.88.$$

2. If the $200,000 is assumed to be a disbenefit, obtain

$$\frac{B - D}{C} = \frac{1,600,000 - 200,000}{650,000} = 2.15.$$

The project appears to have different merits according to which decision is made. The benefit-cost ratio can also lead to incorrect, even catastrophic, conclusions when used improperly or in situations where it is not relevant, such as in warfare, countless social problems, and many business situations.

Example 23:

A municipality is studying the worthiness of constructing a large airport facility. The estimated cost of all improvements is $35,000,000, to be financed with sales taxes and through landing fees and other income generated by the airport. The municipality plans to float a $35,000,000, 5% bond issue payable, with interest, during the life of the facility. Benefits to the community are in the form of increased tourism, industrial expansion, business upturn, and other intangibles which have been estimated to generate an increase in income of approximately $4,000,000 per year, in perpetuity. The decision makers decide to use the benefit-cost ratio, based on present worth computations at 5% interest, to evaluate the proposed plan.

P.W. of Benefits, in perpetuity
= 4,000,000 ÷ 0.05   = $80,000,000
P.W. of Costs   = $35,000,000

$$\text{Benefit-Cost Ratio} = \frac{80,000,000}{35,000,000} = 2.28$$

This ratio is excellent provided that the municipal government does not overextend itself to the detriment of other programs such as education, transportation, law enforcement, sanitation, and other vital services.

The references cited at the end of the chapter are excellent sources of additional information to the reader interested in pursuing further the topics discussed.

EXERCISES

2-1. What is the effective rate of a nominal 8% annual interest paid
   a. semiannually?
   b. quarterly?
   c. monthly?

2-2. If \$2,500 is deposited in a savings account on December 31, 1971, how much money would have accumulated by December 31, 1985, if interest is 5% compounded semiannually?

2-3. How much would have to be invested at 8% today to recover \$20,000 six years hence?

2-4. What should the quarterly notes be on a loan of \$8,000 to be repaid in 5 years at a nominal 8% annual interest?

2-5. Mr. Smith wants to subscribe to a savings plan returning 6% annual interest. He wishes to prepare to meet the expenses of his son's college education for which he anticipates the need of \$15,000 ten years from today. How much should Mr. Smith deposit every year in the savings account?

2-6. How much money should a wealthy industrialist donate to set up in perpetuity a \$30,000 per year university chair if the money can be safely invested at 4%?

2-7. What is the maximum amount of money that can be withdrawn in equal amounts at the end of 1985, '86, '87, and '88 from a savings account paying 6% interest if deposits of \$1,500 each were made at the end of 1971, '72, and '73?

2-8. What is the present worth in 1971 of a uniform series of \$2,000 receipts from 1990 through '99 (10 years) if interest is 3.5%?

2-9. How much should be invested today at 10% to withdraw \$5,000,

5 years hence; $10,000, 8 years from today; and $9,000, 10 years from today and completely deplete the investment?

2-10. Use present worth computations at 6% annual interest to determine the most economical of the plans given below.

Item	Plan A	Plan B	Plan C
First Cost	$25,000	$50,000	$35,000
Annual Oper. & Maint.	$ 3,000	$ 2,000	$ 2,500
Salvage Value at End of Life	$ 5,000	$ 7,000	$ 6,000
Life	10 years	30 years	15 years

2-11. Mr. Jones wishes to determine the approximate value of income property based on a present worth computation at 8% interest (his minimum attractive rate of return). Annual receipts (including the effect of depreciation on income taxes) are, on the average, $20,000. Disbursements amount to $5,000 per year. Jones wishes to hold the property seven years, at which time he expects to sell it for a net $150,000 after the payment of long-term capital gains. What is the maximum purchase price Mr. Jones can afford to pay to meet his investment objectives?

2-12. Determine the better of the following two plans using present worth computations at 5% interest.

Year(s)	Receipts	Disbursements	Receipts	Disbursements
0	0	$-150,000	0	$-300,000
1-20	$ +15,000	-3,000	$ +25,000	-4,000
21	+15,000	-4.000	+25,000	-4,000
22	+15,000	-5,000	+25,000	-4,000
23	+15,000	-5,000	+25,000	-4,000
24	+15,000	-6,000	+25,000	-4,000
25	+15,000	-186,000	+25,000	-54,000
26-49	+20,000	-2,000	+30,000	-5,000
50	+40,000	-2,000	+60,000	-5,000
	$+895,000	$-466,000	$+1,405,000	$-575,000

2-13. Use the equivalent uniform annual cost method at 7% interest to solve problem 2-10.

2-14. Use the equivalent uniform annual cash flow method at 4% interest to solve problem 2-12.

2-15. Compute the true rate of return of Plan D in problem 2-12.

2-16. Compute the true rate of return of Plan E in problem 2-12.

2-17. Assume a $100,000 purchase price for the property in problem 2-11. What is Mr. Jones's true rate of return on this seven-year investment venture?

2–18. The Department of Public Works is considering a project with a cost of $2,000,000; resulting benefits valued at $5,500,000; and a $75,000 adverse item.
  a. Compute the benefit-cost ratio for the project with the adverse item considered as a cost.
  b. Compute the benefit-cost ratio with the adverse item taken as a disbenefit.

2–19. Compare plans D and E in problem 2–12 using the benefit-cost ratio computed on present worth at 3.5% interest.

REFERENCES

1. GRANT, E. L. and IRESON, W. G., *Principles of Engineering Economy*, 5th ed. New York: The Ronald Press Company, 1970.

2. KAHN, S. A., CASE, F. E., and SCHIMMEL, A., *Real Estate Appraisal and Investment*. New York: The Ronald Press Company, 1963.

3. MARSTON, A., WINFREY, R., and HEMPSTEAD, J. C., *Engineering Valuation and Depreciation*, 4th printing. Ames, Iowa: Iowa State University Press, 1968.

4. SEILER, K., III, *Introduction to Systems Cost Effectiveness*. New York: John Wiley & Sons, Inc., 1969.

3

The Differential Approach to Optimization

3.1 PRELIMINARY CONCEPTS

The typical problem in optimization is expressed by an objective function of n independent variables

$$y(x_1, x_2, \ldots, x_n) \tag{3.1}$$

that is to be maximized or minimized—in general, optimized—subject to a set of constraints written in the form of equations and inequalities,

$$f_i(x_1, \ldots, x_n) = 0 \quad \text{for } i = 1, \ldots, m;$$
$$f_i(x_1, \ldots, x_n) \leq 0 \quad \text{for } i = m + 1, \ldots, r; \tag{3.2}$$
$$f_i(x_1, \ldots, x_n) \geq 0 \quad \text{for } i = r + 1, \ldots, p.$$

Points satisfying the constraints collectively form the problem's feasible region F.

If equalities are possible for all constraints, F is said to be closed; otherwise, F is open.

For reasons that will become apparent in Chapter 5, problems must be formulated with closed regions whenever possible.

As an example, consider

$$\text{Max } y = 3x_1^2 + 2x_2^3 + x_3^{-1},$$

$$\text{subject to } x_1 + x_2 \leq 4,$$

$$x_3 \geq 10.$$

Because equalities are possible for both constraints, the feasible region, F, is closed. To simplify the notation, let

$$y(x_1, x_2, \ldots, x_n) = y(\bar{x}), \tag{3.3}$$

where, as will be discussed in the next chapter, $\bar{x}$ is a row vector of the independent variables x_i, $i = 1, \ldots, n$. Such a vector, $\bar{x}$, represents a point in n dimensional space. This point is a feasible point if all of its components x_i, taken simultaneously, satisfy the set of constraints. This is expressed symbolically as follows,

$$\bar{x} \epsilon F, \tag{3.4}$$

where ϵ is read "is a member of" or "belongs to" the feasible region F, in this case.

Consider a point $\bar{x}^*$ such that

$$y(\bar{x}^*) < y(\bar{x}) \tag{3.5}$$

for all $\bar{x} \neq \bar{x}^*$ in F. Then, $y(\bar{x}^*)$ is said to be the global minimum of y and $\bar{x}^*$ is the global minimizing policy (the symbol $\neq$ is read "not equal to"). If $y(\bar{x}^*) \leq y(\bar{x})$ for all $\bar{x} \neq \bar{x}^*$ in F, there is more than one global minimal policy but only one global minimum value. One can write

$$y(\bar{x}^*) = y^* = \min_{x \epsilon F} [y(\bar{x})].$$

If $\bar{x}^*$ is on the boundary of F, y^* is a boundary global minimum, otherwise y^* is an interior global minimum. The same statements hold true for global maximum points with the inequalities reversed. To cover optimization in general, talk about the global optimum, y^*, and the global optimal policy or policies, $\bar{x}^*$, such that

$$y^* = y(\bar{x}^*) = \underset{x \epsilon F}{\text{opt}} [y(x)]. \tag{3.6}$$

Consider $\bar{x}^*$ and a feasible neighborhood of $\bar{x}^*$. A feasible neighborhood of $\bar{x}^*$ is comprised by all points $\bar{x}$ satisfying $\bar{x} \epsilon F$ and also

$$0 < |\bar{x} - \bar{x}^*| < \delta, \tag{3.7}$$

where

$$|\bar{x} - \bar{x}^*| = \left[\sum_{j=1}^{n} (\bar{x}_j - \bar{x}_j^*)^2 \right]^{\frac{1}{2}}. \tag{3.8}$$

Consequently, δ is the radius of an n-dimensional spherical, open region centered at $\bar{x}^*$ and containing the points $\bar{x}$ and $\bar{x}^*$. Let the feasible neighborhood be denoted by η.

If

$$y(\bar{x}^*) < y(\bar{x}) \text{ for } \bar{x}\epsilon\eta, \tag{3.9}$$

then $y(\bar{x}^*)$ is a local minimum and $\bar{x}^*$ is a locally minimizing policy. This is expressed

$$y^* = y(\bar{x}^*) = l \min_{x\epsilon\eta} [y(\bar{x})]. \tag{3.10}$$

The notation "l min" signifies that the points $\bar{x}$ are restricted to a feasible neighborhood of $\bar{x}^*$. A similar argument holds true for local maxima. If $y(\bar{x})$ has a unique optimum in its feasible region, $y(\bar{x})$ is said to be unimodal in the region.

3.2 IMPORTANT ELEMENTARY RELATIONSHIPS

1. If $\bar{x}^*$ minimizes $y(\bar{x})$ it also minimizes $a + by(\bar{x})$ where a is an arbitrary constant and b is any positive constant.
2. If

$$a + by(\bar{x}^*) = \min_{\bar{x}\epsilon F} [a + by(\bar{x})], \tag{3.11}$$

then it is also true that

$$a - by(\bar{x}^*) = \max_{\bar{x}\epsilon F} [a - by(\bar{x})]. \tag{3.12}$$

In other words, the policy $\bar{x}^*$ that minimizes the objective function $y(\bar{x})$ also maximizes the negative objective $[-y(\bar{x})]$.

3.3 THE CALCULUS METHOD

The differential calculus provides classical methods for determining maxima, minima, or critical points of functions of real variables. These methods are applicable to decision problems in which the objective function $y(\bar{x})$ is at least twice differentiable within the feasible region, and the constraints are either nonexistent or consist of equations only. If no constraint equations are present, the optimization consists of nothing more than direct differentiation of the objective function; otherwise, constraint equations may be solved first for pertinent variables which are substituted into the objective function before differentiation is performed, or Lagrange multipliers may be used in the optimization scheme. The Langrange multiplier method will be discussed later in this chapter. Three simple examples with unimodal objective functions are given as illustrations.

3.3.1 The Economics of Simple Bridges

A very instructive example problem attributed to T. Au and T. E. Stelson, (see References) concerns the determination of the most economical bridge span length for a series of equal span, simple trusses, or girders supported by piers assumed to be constructed at about the same cost per pier anywhere at the bridge crossing. Figure 3.1 is a diagrammatic representation of the bridge structure under consideration.

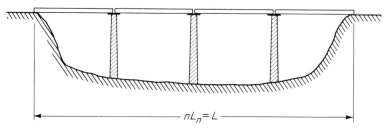

FIGURE 3.1.

Let C, the total cost of the bridge structure, be expressed in terms of the following parameters:

L = total length of the crossing; assumed fixed.
L_n = span length of individual truss or girder.
n = number of spans = L/L_n.
A = cost of the two abutments (end piers).
P = cost of one pier.
w = weight of steel, in pounds per foot of span, for trusses or girders, including bracing.
c = cost per pound of steel for trusses or girders, including bracing.
m = cost per foot of span for floor system, roadway, and miscellaneous items such as railings and sidewalks.

The weight of steel per foot of span is a function of the span length L_n, and can safely be assumed to be proportional to L_n—that is, $w = kL_n$ where k is a constant for a given type of truss or girder. The numerical value of k may or may not be known. However, even if it is unknown, it will be shown that a decision rule can still be formulated and used in establishing economic criteria for this type of bridge structure.

The total bridge cost can be expressed as the sum of the cost of the two abutments, the cost of the piers, of the steel superstructure, and of the floor system. Thus,

$$C = A + (n - 1)P + cwL + mL. \tag{3.13}$$

Because $n = L/L_n$ and $w = kL_n$, the objective function may be expressed in the form

$$\text{Min } C = A + \left(\frac{L}{L_n} - 1\right)P + ckL_nL + mL. \tag{3.14}$$

This is a function of one variable only, L_n. In order to minimize the span length, set $dC/dL_n = 0$.

$$\frac{dC}{dL_n} = -\frac{LP}{L_n^2} + ckL = 0, \tag{3.15}$$

from which

$$L_n^2 = \frac{P}{ck}, \quad \text{or} \quad L_n = \sqrt{\frac{P}{ck}}. \tag{3.16}$$

The second derivative of C with respect to L_n gives

$$\frac{d^2C}{dL_n^2} = \frac{2LP}{L_n^3} > 0. \tag{3.17}$$

Therefore, the total bridge cost C is a minimum for the span length L_n given by equation (3.17).

The result of the optimization is useful even if one does not know the value of the proportionality constant k. Note that

$$L_n^2 = \frac{P}{ck}$$

can be written as

$$P = ckL_n^2 = c(kL_n)L_n. \tag{3.18}$$

But, by assumption,

$$w = kL_n.$$

Therefore, equation (3.18) can also be expressed as

$$P = cwL_n. \tag{3.19}$$

This implies that the cost of one pier must equal the cost of one span of trusses or girders. This is a simple decision rule for determining the optimum span length at a long crossing.

It should be emphasized that the mathematical analysis given is only as good as the assumption that the piers can be constructed anywhere at the crossing at about the same cost per pier. If, for example, part of the crossing is shallow and part of it is over a depression, the above formulation is no longer correct and the simple conclusion derived is not applicable.

3.3.2 A River Transportation Problem

As a second example, consider the following problem: 400 cubic yards of building material are to be transported across a river in an open box of length x_1, width x_2, and height x_3. It costs \$10 per square yard to build the sides and bottom of the box, and \$20 per square yard to build the ends. Each round trip across the river costs \$0.10. Find the dimensions x_1, x_2, and x_3, and the number of round trips x_4 that will result in the minimum total cost of transportation for the material.

The objective function to be minimized is

$$C = 10(x_1 x_2 + 2x_1 x_3) + 20(2x_2 x_3) + 0.10x_4 \qquad (3.20)$$

where $10(x_1 x_2 + 2x_1 x_3)$ is the cost of the bottom plus two sides of the box, $20(2x_2 x_3)$ is the cost of the two ends, and $0.10x_4$ is the cost of the total number of round trips of the ferry. In this case there exists one constraint, namely, that the total number of round trips multiplied by the volume of the box must equal the volume of material to be shipped, thus

$$x_1 x_2 x_3 x_4 = 400. \qquad (3.21)$$

Consequently, the mathematical model of this problem is

$$\text{Min } C = 10x_1 x_2 + 20x_1 x_3 + 40x_2 x_3 + 0.10x_4,$$

subject to

$$x_1 x_2 x_3 x_4 = 400.$$

In this case it is possible to solve for x_4 in the constraint equation and to substitute its value in the objective function (this procedure is obviously not always feasible). Thus, after substitution, the unconstrained objective function is

$$\text{Min } C = 10x_1 x_2 + 20x_1 x_3 + 40x_2 x_3 + 40x_1^{-1} x_2^{-1} x_3^{-1}. \qquad (3.22)$$

Differentiate partially with respect to each of the independent variables and set each derivate equal to zero to obtain

$$\frac{\partial C}{\partial x_1} = 10x_2 + 20x_3 - 40x_1^{-2} x_2^{-1} x_3^{-1} = 0,$$

$$\frac{\partial C}{\partial x_2} = 10x_1 + 40x_3 - 40x_1^{-1} x_2^{-2} x_3^{-1} = 0,$$

$$\frac{\partial C}{\partial x_3} = 20x_1 + 40x_2 - 40x_1^{-1} x_2^{-1} x_3^{-2} = 0.$$

Simultaneous solution of these nonlinear equations yields the following critical point,

$$x_1^* = 2; \ x_2^* = 1; \ x_3^* = \tfrac{1}{2};$$

from which

$$x_4^* = \frac{400}{x_1^* x_2^* x_3^*} = 400.$$

Thus, the most economical design condition is to build a box 2 yards long, by 1 yard wide, by 1/2 yard high, or 1 cubic yard in volume, and to make 400 round trip crossings of the river. This is a result of the low cost of each round trip: only \$0.10. The minimum total cost of transportation is $C^* = \$100$. The simple second derivative test which was applied in the first example problem to verify the minimum is no longer applicable because the objective function (3.22) has more than one independent variable. This condition will be discussed later in this chapter. However, for the given problem it is intuitive that the objective function is unimodal (one minimum value) and, consequently, the computed optimum is the global minimum.

3.3.3 An Inventory Problem

As a final example, consider the following production situation also described by T. Au and T. E. Stelson. The manufacturer of a certain product has an agreement to supply a customer at a constant rate of R units per day. If the manufacturer starts a production run, a constant number of Q units must be produced over a period of t_0 days averaging K units per day. A fixed cost C_0 is involved in setting up each production run. If the product is not shipped out on the same day, the storage cost is C_1 per unit per day. What is the optimal size of each run to minimize the average production cost per day, and how often should the runs be made? This is an example of a deterministic, linear inventory problem.

The fluctuations of inventory I with time t are represented by the series of solid line triangles in Figure 3.2. Let t_1 be the time interval between production runs. Because the volume Q produced in t_0 days is exhausted in t_1 days, one has that $Q = Kt_0 = Rt_1$. During the production period t_0, the inventory is accumlated at a rate of $(K - R)$ units per day since R units are depleted while K units are produced every day. The inventory reaches $(K - R)t_0$ units when the production stops. This amount will be depleted further at a rate of R units per day until it is completely exhausted. The area of each triangle in Figure 3.2 represents the number of units that must be stored in each cycle of duration t_1. Let the average production cost per day be C. Then the objective function is

$$\text{Min } C = \frac{1}{t_1}\left[C_0 + \frac{C_1(K - R)t_0 t_1}{2}\right]. \tag{3.23}$$

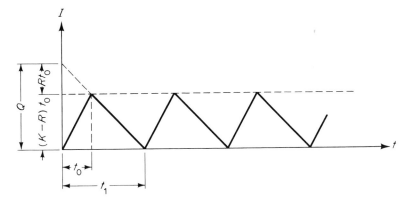

FIGURE 3.2.

Since $Q = Kt_0 = Rt_1$, the objective function can be written,

$$\text{Min } C = \frac{R}{Q}\left[C_0 + \frac{C_1(K - R)Q^2}{2KR}\right]$$

$$= \frac{C_0 R}{Q} + \frac{C_1}{2}\left(1 - \frac{R}{K}\right)Q. \tag{3.24}$$

Equation (3.24) is a function of only one variable, Q.

In order to obtain the optimum size of each production run, let the first derivative of C with respect to Q vanish. Thus,

$$\frac{dC}{dQ} = -\frac{C_0 R}{Q^2} + \frac{C_1}{2}\left(1 - \frac{R}{K}\right) = 0, \tag{3.25}$$

from which

$$Q = \sqrt{\frac{2C_0 R}{C_1(1 - R/K)}}. \tag{3.26}$$

Note that the second derivative of C is positive,

$$\frac{d^2C}{dQ^2} = \frac{2C_0 R}{Q^3} > 0. \tag{3.27}$$

Hence, the optimum value of Q leads to a minimum cost. The optimum time interval between production runs can be obtained by substituting the optimum value of Q into the relation $t_1 = Q/R$.

If the production time for each run is negligible, such as in many reordering situations, then $t_0 = 0$ and $K = \infty$, and the optimum value of Q becomes

$$Q = \sqrt{\frac{2C_0 R}{C_1}}. \tag{3.28}$$

This equation is generally known as the economic lot size formula for holding inventory of merchandise.

3.4   THE TAYLOR SERIES EXPANSION

The Taylor Series Expansion of a function about a point is covered in detail in most calculus books. The concept is reviewed in this section because of its important applications to optimization theory.

3.4.1   Taylor Expansion of One-Variable Functions

Consider a function of one variable, $f(x)$, which one wishes to expand into an infinite geometric series of the form

$$f(x) = a_0 + a_1(x - x_0) + a_2(x - x_0)^2 + \cdots$$

$$\cdots + a_n(x - x_0)^n + \cdots \tag{3.29}$$

This series is called the MacLaurin Expansion of $f(x)$. It can also be written more compactly as follows,

$$f(x) = \sum_{i=0}^{\infty} \frac{f^i(0)}{i!} x^i. \tag{3.39}$$

3.4.3 Taylor Expansion of Functions with Several Variables

Let $f(x, y)$ and its partial derivatives of all orders be continuous in the neighborhood of (x_0, y_0) contained in the x, y plane.

Introduce the real parameters α and β, such that

$$x = x_0 + \alpha t,$$

and

$$y = y_0 + \beta t. \tag{3.40}$$

Then,

$$x - x_0 = \alpha t,$$
$$y - y_0 = \beta t, \tag{3.41}$$

and

$$t = \frac{x - x_0}{\alpha} = \frac{y - y_0}{\beta}.$$

Consequently,

$$f(x, y) = f(x_0 + \alpha t, y_0 + \beta t) = F(t). \tag{3.42}$$

That is, the function of two variables $f(x, y)$ can be replaced by a function $F(t)$ of only one variable: the parameter t. The MacLaurin expansion of $F(t)$ yields

$$F(t) = F(0) + F'(0)t + \frac{F''(0)}{2!}t^2 + \frac{F'''(0)}{3!}t^3 + \dots \tag{3.43}$$

The coefficients $F(0)$, $F'(0)$, $F''(0)$, $F'''(0)$, ..., can be evaluated as follows:

1. From equation (3.42)

$$F(0) = f(x_0, y_0). \tag{3.44}$$

2.
$$F'(t) = \frac{d}{dt}[F(t)] = \frac{\partial f}{\partial x}\frac{dx}{dt} + \frac{\partial f}{\partial y}\frac{dy}{dt}. \tag{3.45}$$

But, from equation (3.40),

$$\frac{dx}{dt} = \alpha \quad \text{and} \quad \frac{dy}{dt} = \beta. \tag{3.46}$$

Consequently,

$$F'(t) = \left(\alpha\frac{\partial f}{\partial x} + \beta\frac{\partial f}{\partial y}\right). \tag{3.47}$$

This expression can be written in operator form as follows:

$$F'(t) = \left(\alpha\frac{\partial}{\partial x} + \beta\frac{\partial}{\partial y}\right)f, \tag{3.48}$$

where the operater $[\alpha(\partial/\partial x) + \beta(\partial/\partial y)]$ is defined to operate on the function f so as to yield equation (3.47).
Therefore,

$$F'(0) = \left(\alpha\frac{\partial}{\partial x} + \beta\frac{\partial}{\partial y}\right)f(x_0, y_0), \tag{3.49}$$

with the understanding that the operator must operate on f before the expression is evaluated at the point (x_0, y_0).

3.　$F''(t) = \dfrac{d}{dt}[F'(t)] = \dfrac{d}{dt}\left[\alpha\dfrac{\partial f}{\partial x} + \beta\dfrac{\partial f}{\partial y}\right]$

$$= \alpha\frac{d}{dt}\left[\frac{\partial f}{\partial x}\right] + \beta\frac{d}{dt}\left[\frac{\partial f}{\partial y}\right]$$

$$= \alpha\left[\frac{\partial^2 f}{\partial x^2}\frac{dx}{dt} + \frac{\partial^2 f}{\partial x\partial y}\frac{dy}{dt}\right] + \beta\left[\frac{\partial^2 f}{\partial y\partial x}\frac{dx}{dt} + \frac{\partial^2 f}{\partial y^2}\frac{dy}{dt}\right]$$

$$= \alpha\left[\frac{\partial^2 f}{\partial x_2}\alpha + \frac{\partial^2 f}{\partial x\partial y}\beta\right] + \beta\left[\frac{\partial^2 f}{\partial x\partial y}\alpha + \frac{\partial^2 f}{\partial y^2}\beta\right]$$

$$= \left(\alpha^2\frac{\partial^2 f}{\partial x^2} + 2\alpha\beta\frac{\partial^2 f}{\partial x\partial y} + \beta^2\frac{\partial^2 f}{\partial y^2}\right). \tag{3.50}$$

$F''(t)$ can be written in operator form,

$$F''(t) = \left(\alpha\frac{\partial}{\partial x} + \beta\frac{\partial}{\partial y}\right)^2 f, \tag{3.51}$$

with the definition

$$\left(\alpha\frac{\partial}{\partial x} + \beta\frac{\partial}{\partial y}\right)^2 = \alpha^2\frac{\partial^2}{\partial x^2} + 2\alpha\beta\frac{\partial^2}{\partial x\partial y} + \beta^2\frac{\partial^2}{\partial y^2}. \tag{3.52}$$

That is, the operator is raised to a power as a binomial expansion with mixed partial derivatives.
　Now, again,

$$F''(0) = \left(\alpha\frac{\partial}{\partial x} + \beta\frac{\partial}{\partial y}\right)^2 f(x_0, y_0), \tag{3.53}$$

with the same stipulation that the expression be evaluated at (x_0, y_0) only after the operator has operated on f.

　4. Following the same procedure, one finds that

$$F'''(t) = \left(\alpha\frac{\partial}{\partial x} + \beta\frac{\partial}{\partial y}\right)^3 f,$$

and

$$F'''(0) = \left(\alpha\frac{\partial}{\partial x} + \beta\frac{\partial}{\partial y}\right)^3 f(x_0, y_0). \tag{3.54}$$

Thus, the operator

$$\left(\alpha \frac{\partial}{\partial x} + \beta \frac{\partial}{\partial y}\right) \tag{3.55}$$

serves to express, compactly, the series (3.43).

$$F(t) = f(x, y) = f(x_0, y_0) + \left(\alpha \frac{\partial}{\partial x} + \beta \frac{\partial}{\partial y}\right) f(x_0, y_0)t$$

$$+ \frac{1}{2!}\left(\alpha \frac{\partial}{\partial x} + \beta \frac{\partial}{\partial y}\right)^2 f(x_0, y_0)t^2$$

$$+ \frac{1}{3!}\left(\alpha \frac{\partial}{\partial x} + \beta \frac{\partial}{\partial y}\right)^3 f(x_0, y_0)t^3$$

$$+ \cdots, \tag{3.56}$$

or more compactly, still,

$$F(t) = f(x, y) = \sum_{i=0}^{\infty} \frac{1}{i!}\left(\alpha \frac{\partial}{\partial x} + \beta \frac{\partial}{\partial y}\right)^i f(x_0, y_0)t^i. \tag{3.57}$$

This is the Taylor Series Expansion of $f(x, y)$. From expression (3.41), one finds that

$$t = \frac{x - x_0}{\alpha} = \frac{y - y_0}{\beta}.$$

Consequently, the general term in (3.57)

$$\frac{1}{i!}\left(\alpha \frac{\partial}{\partial x} + \beta \frac{\partial}{\partial y}\right)^i f(x_0, y_0)t^i \text{ can be expressed as}$$

$$\frac{1}{i!}\left[(x - x_0)\frac{\partial}{\partial x} + (y - y_0)\frac{\partial}{\partial y}\right]^i f(x_0, y_0), \tag{3.58}$$

and equation (3.57) can be rewritten,

$$F(t) = f(x, y) = \sum_{i=0}^{\infty} \frac{1}{i!}\left[(x - x_0)\frac{\partial}{\partial x} + (y - y_0)\frac{\partial}{\partial y}\right]^i f(x_0, y_0). \tag{3.59}$$

As in section 3.4.2, replace x by $x_0 + \Delta x$ and y by $y_0 + \Delta y$. Equation (3.59) becomes

$$f(x_0 + \Delta x, y_0 + \Delta y) = \sum_{i=0}^{\infty} \frac{1}{i!}\left[\frac{\partial}{\partial x}\Delta x + \frac{\partial}{\partial y}\Delta y\right]^i f(x_0, y_0). \tag{3.60}$$

It can also be written

$$f(x_0 + \Delta x, y_0 + \Delta y) = f(x_0, y_0) + \sum_{i=1}^{\infty} \frac{1}{i!}\left[\frac{\partial}{\partial x}\Delta x + \frac{\partial}{\partial y}\Delta y\right]^i f(x_0, y_0), \tag{3.61}$$

from which,

$$\Delta f(x_0, y_0) = f(x_0 + \Delta x, y_0 + \Delta y) - f(x_0, y_0)$$

$$= \sum_{i=1}^{\infty} \frac{1}{i!}\left[\frac{\partial}{\partial x}\Delta x + \frac{\partial}{\partial y}\Delta y\right]^i f(x_0, y_0). \tag{3.62}$$

In the limit as $\Delta x \to 0$ and $\Delta y \to 0$, $\Delta f(x_0, y_0) \to df(x_0, y_0)$ and equation (3.62) becomes,

$$df(x_0, y_0) = \sum_{i=1}^{\infty} \frac{1}{i!} \left[\frac{\partial}{\partial x} dx + \frac{\partial}{\partial y} dy \right]^i f(x_0, y_0). \tag{3.63}$$

This is the Taylor Series Expansion of the change in $f(x, y)$ evaluated at (x_0, y_0).

To generalize, the Taylor Series Expansion of $f(x_1, x_2, \ldots, x_n)$ about a point $(x_{10}, x_{20}, \ldots, x_{n0})$ is given by

$$f(x_1, x_2, \ldots, x_n) = \sum_{i=0}^{\infty} \frac{1}{i!} \Delta^i f(x_{10}, x_{20}, \ldots, x_{n0}), \tag{3.64}$$

where

$$\Delta = \left[(x_1 - x_{10}) \frac{\partial}{\partial x_1} + (x_2 - x_{20}) \frac{\partial}{\partial x_2} + \ldots + (x_n - x_{n0}) \frac{\partial}{\partial x_n} \right] \tag{3.65}$$

is the operator for the general function of n variables.

Similarly,

$$df(x_{10}, x_{20}, \ldots, x_{n0}) = \sum_{i=1}^{\infty} \frac{1}{i!} D^i f(x_{10}, x_{20}, \ldots, x_{n0}), \tag{3.66}$$

where

$$D = \left[\frac{\partial}{\partial x_1} dx_1 + \frac{\partial}{\partial x_2} dx_2 + \ldots + \frac{\partial}{\partial x_n} dx_n \right]. \tag{3.67}$$

This is a very convenient and compact representation of all the terms of the Taylor Series Expansion of $f(\bar{x})$ and of the change in $f(\bar{x})$ about the point $(x_{10}, x_{20}, \ldots, x_{n0})$.

3.5 CRITICAL POINTS OF FUNCTIONS

MacLaurin formulated rules for critical (stationary) points of functions of one variable. Calculus texts treat them in detail:

1. The coordinates of critical points are established by setting the first derivative of the function to zero and solving the resulting equation for the variable values.
2. The point thus obtained is a minimum if and only if the lowest order nonvanishing derivative is positive and of even order, and a maximum, if negative and of even order. These rules were applied in Examples 3.3.1 and 3.3.3.

To extend the theory to functions of more than one variable—in general, n variables—consider the Taylor Series Expansion of the change in $y(x_1,$

$x_2, \ldots, x_n)$ about the point $(x_1^*, x_2^*, \ldots, x_n^*)$ assumed to be a minimum. From equations (3.66) and (3.67), this is

$$dy^* = \sum_{i=1}^{\infty} \frac{1}{i!} D^i y^*,$$

where

$$y^* = y(x_1^*, x_2^*, \ldots, x_n^*)$$

and

$$D = \left(\frac{\partial}{\partial x_1} dx_1 + \frac{\partial}{\partial x_2} dx_2 + \cdots + \frac{\partial}{\partial x_n} dx_n \right).$$

The expression for the change in y evaluated at the n dimensional point $\bar{x}^*$ can be written,

$$dy^* = \sum_{j=1}^{\infty} \left(\frac{\partial y}{\partial x_j} \right)^* dx_j + \tfrac{1}{2} \sum_{j=1}^{n} \sum_{k=1}^{n} \left(\frac{\partial^2 y}{\partial x_j \partial x_k} \right)^* dx_j dx_k + 0(dx_j^3), \qquad (3.68)$$

where $0(dx_j^3)$ represents all terms of order three and of higher orders. At a critical point

$$\sum_{j=1}^{n} \left(\frac{\partial y}{\partial x_j} \right)^* dx_j = 0 \qquad (3.69)$$

because the first partial derivatives must vanish for all possible perturbations dx_j. This implies

$$\left(\frac{\partial y}{\partial x_j} \right)^* = 0 \qquad \text{for } j = 1, 2, \ldots, n. \qquad (3.70)$$

If, for example, $(\partial y/\partial x_j)^* \neq 0$ for any one j, a perturbation dx_j with sign opposite from that of the corresponding $(\partial y/\partial x_j)^*$ would result in

$$\left(\frac{\partial y}{\partial x_j} \right)^* \partial x_j < 0. \qquad (3.71)$$

Assuming that all terms of order two and higher in equation (3.68) approach zero more rapidly than the first-order terms, that is,

$$\lim_{dx_j \to 0} \left[\frac{0(dx_j^2)}{\left(\frac{\partial y_j}{\partial x_j} \right)^* dx_j} \right] = 0, \qquad (3.72)$$

then, equation (3.71) indicates that if all other perturbations are held at zero,

$$dy^* < 0 \qquad (3.73)$$

and a better point would be generated in every feasible neighborhood of y^*, contradicting the assumption that y^* was a minimum. Therefore, all first partial derivatives must vanish at a minimum. The same argument holds true for maxima and for other critical points such as ridge, valley, and saddle points.

Consequently, at a minimum, equation (3.68) becomes

$$dy^* = \frac{1}{2} \sum_{j=1}^{n} \sum_{k=1}^{n} \left(\frac{\partial^2 y}{\partial x_j \partial x_k}\right)^* dx_j dx_k + O(dx_j^3). \tag{3.74}$$

Again, assuming that the terms of order three and higher approach zero more rapidly than the second-order terms as each dx_j goes to zero, the sign of dy^* depends upon the sign of the term

$$\frac{1}{2} \sum_{j=1}^{n} \sum_{k=1}^{n} \left(\frac{\partial^2 y}{\partial x_j \partial x_k}\right) dx_j dx_k. \tag{3.75}$$

This scalar expression is called a *differential quadratic form*. If this quadratic form is positive for all $dx_j \neq 0$, it is then said to be positive-definite; this is a sufficient condition for $\bar{x}^*$ to be a local minimum. At a local maximum the differential quadratic form is negative-definite. In valleys the form is non-negative, vanishing for perturbations along the valley, in which case it is said to be positive-semidefinite. At ridges it is negative-semidefinite—that is, nonpositive—vanishing for perturbations along the ridge. At saddle points the differential quadratic form can be positive, negative, or zero, depending on the value of $dx_j, j = 1, 2, \ldots, n$, chosen, in which case it is called indefinite. Table 3–1 summarizes these conclusions.

It must be stressed that the conditions given in Table 3–1 are *only* necessary. They are also sufficient in special cases which will not be discussed further in this text.

TABLE 3.1.

Sign	Type	Classification of Critical Point
D.Q.F.* > 0	Positive-Definite	Minimum
D.Q.F.* < 0	Negative-Definite	Maximum
D.Q.F.* ≥ 0	Positive-Semidefinite	Valley
D.Q.F.* ≤ 0	Negative-Semidefinite	Ridge
D.Q.F.* ≶ 0	Indefinite	Saddle

$$* \text{ D.Q.F.} = \text{Differential Quadratic Form} = \frac{1}{2} \sum_{j=1}^{n} \sum_{k=1}^{n} \left(\frac{\partial^2 y}{\partial x_j \partial x_k}\right)^* dx_j\, dx_k$$

3.5.1 Example

In section 3.3.2 a critical point for the unconstrained objective (3.22) was found at

$$x^* = 2; \ x^* = 1; \ x^* = \tfrac{1}{2}.$$

The intuitive conclusion that this is a minimum can be verified by analyzing the differential quadratic form. For this problem,

$$
\begin{aligned}
\text{D.Q.F} &= \tfrac{1}{2} \sum_{j=1}^{3} \sum_{k=1}^{3} \left(\frac{\partial^2 y}{\partial x_j \, \partial x_k} \right)^* dx_j dx_k \\
&= \tfrac{1}{2} \left[\frac{\partial^2 y}{\partial x_1^2} dx_1^2 + \frac{\partial^2 y}{dx_2^2} dx_2^2 + \frac{\partial^2 y}{dx_3^2} dx_3^2 \right. \\
&\quad + 2 \frac{\partial^2 y}{\partial x_1 \, \partial x_2} dx_1 dx_2 + 2 \frac{\partial^2 y}{\partial x_1 \, \partial x_3} dx_1 dx_3 \\
&\quad \left. + 2 \frac{\partial^2 y}{\partial x_2 \, \partial x_3} dx_2 dx_3 \right]^*,
\end{aligned}
$$

where

$$ y = C = 10x_1x_2 + 20x_1x_3 + 40x_2x_3 + 40x_1^{-1}x_2^{-1}x_3^{-1}. $$

The second partial derivatives evaluated at $\bar{x}^*$ are:

$$ \left(\frac{\partial^2 y}{\partial x_1^2} \right)^* = (80x_1^{-3}x_2^{-1}x_3^{-1})^* = 20, $$

$$ \left(\frac{\partial^2 y}{\partial x_2^2} \right)^* = (80x_1^{-1}x_2^{-3}x_3^{-1})^* = 80, $$

$$ \left(\frac{\partial^2 y}{\partial x_3^2} \right)^* = (80x_1^{-1}x_2^{-1}x_3^{-3})^* = 320, $$

$$ \left(\frac{\partial^2 y}{\partial x_1 \partial x_2} \right)^* = (10 + 40x_1^{-2}x_2^{-2}x_3^{-1})^* = 30, $$

$$ \left(\frac{\partial^2 y}{\partial x_1 \partial x_2} \right)^* = (20 + 40x_1^{-2}x_2^{-1}x_3^{-2})^* = 60, $$

$$ \left(\frac{\partial^2 y}{\partial x_2 \partial x_3} \right)^* = (40 + 40x_1^{-1}x_2^{-2}x_3^{-2})^* = 120. $$

Therefore,

$$
\begin{aligned}
\text{D.Q.F.} &= \tfrac{1}{2}[20dx_1^2 + 80dx_2^2 + 320dx_3^2 + 60dx_1dx_2 + 120dx_1dx_3 \\
&\quad + 240dx_2dx_3].
\end{aligned}
$$

In order to make a definite statement about the sign of the D.Q.F., it is necessary to write it in terms of perfect squares, as follows:

$$ \text{D.Q.F.} = 10dx_1^2 + 40dx_2^2 + 160dx_3^2 + 30dx_1dx_2 + 60dx_1dx_3 + 120dx_2dx_3 $$

or

$$
\begin{aligned}
\text{D.Q.F.} &= 10(dx_1^2 + 3dx_1dx_2 + 6dx_1dx_3) \\
&\quad + 40dx_2^2 + 160dx_3^2 + 120dx_2dx_3, \\
\text{D.Q.F.} &= 10[(dx_1 + 1.5dx_2 + 3dx_3)^2 - 2.25dx_2^2 - 9dx_3^2 - 9dx_2dx_3] \\
&\quad + 40dx_2^2 + 160dx_3^2 + 120dx_2dx_3,
\end{aligned}
$$

D.Q.F. $= 10[dx_1 + 1.5dx_2 + 3dx_3]^2 + 17.5dx_2^2$
$+ 70dx_3^2 + 30dx_2dx_3.$

To this point, all terms containing dx_1 have been written as perfect squares. Now, proceed to write as perfect squares all terms containing dx_2. Thus,

$17.5dx_2^2 + 70dx_3^2 + 30dx_2dx_3$

$= 17.5[dx_2^2 + 1.72dx_2dx_3] + 70dx_3^2$

$= 17.5[(dx_2 + 0.86dx_3)^2 - 0.74dx_3^2] + 70dx_3^2$

$= 17.5[dx_2 + 0.86dx_3]^2 + 57dx_3^2.$

The result is that all terms are now in the form of perfect squares. Thus,

D.Q.F. $= 10[dx_1 + 1.5dx_2 + 3dx_3]^2$
$+ 17.5[dx_2 + 0.86dx_3]^2 + 57[dx_3]^2 > 0$

for any arbitrary values of dx_1, dx_2, and dx_3. Consequently, the differential quadratic form is positive-definite and the critical point is, in fact, a minimum.

3.6 LAGRANGE MULTIPLIERS

The method of Lagrange multipliers can be used when a function $y(x_1, x_2, \ldots, x_n)$ is to be optimized subject to equality constraints of the form

$$f_i(x_1, x_2, \ldots, x_n) = 0; \; i = 1, 2, \ldots, m \qquad (3.76)$$

where the f_i are not functionally dependent, so that the constraints are neither equivalent nor incompatible.

Suppose that the function y and f_i, $i = 1, 2, \ldots, m$, have first partial derivatives everywhere in a region which includes the critical points. Because the f_i are all equal to zero their total differentials must vanish. Specifically, they must vanish at each critical point $\bar{x}^* = (x_1^*, x_2^*, \ldots, x_n^*)$. Thus,

$$df_i = \sum_{j=1}^{n} \left(\frac{\partial f_i}{\partial x_j}\right)^* dx_j = 0 \qquad \text{for } i = 1, 2, \ldots, m. \qquad (3.77)$$

At the critical points, the linear terms of the Taylor Series Expansion of the change in y must be zero. This condition yields

$$\sum_{j=1}^{n} \left(\frac{\partial y}{\partial x_j}\right)^* dx_j = 0. \qquad (3.78)$$

One may multiply equation (3.77) by the arbitrary constants λ_i, $i = 1, 2, \ldots, m$, respectively, and subtract the results from equation (3.78) to yield the requirement

$$\sum_{j=1}^{n} \left[\left(\frac{\partial y}{\partial x_j}\right)^* - \sum_{i=1}^{m} \lambda_i \left(\frac{\partial f_i}{\partial x_j}\right)^*\right] dx_j = 0. \qquad (3.79)$$

Because each dx_j is an arbitrary perturbation, equation (3.79) implies that

$$\left(\frac{\partial y}{\partial x_1}\right)^* - \sum_{i=1}^{m} \lambda_i \left(\frac{\partial f_i}{\partial x_1}\right)^* = 0,$$

$$\left(\frac{\partial y}{\partial x_2}\right)^* - \sum_{i=1}^{m} \lambda_i \left(\frac{\partial f_i}{\partial x_2}\right)^* = 0, \tag{3.80}$$

and

$$\left(\frac{\partial y}{\partial x_n}\right)^* - \sum_{i=1}^{m} \lambda_i \left(\frac{\partial f_i}{\partial x_n}\right)^* = 0,$$

which together with the conditions

$$f_i(x_1, x_2, \ldots, x_n) = 0 \qquad \text{for } i = 1, 2, \ldots, m \tag{3.81}$$

at $\bar{x}^* = (x_1^*, x_2^*, \ldots, x_n^*)$ provide $n + m$ equations in n unknowns x_j^* plus m unknown λ_i. These equations are, in general, nonlinear and, consequently, may be extremely difficult to solve. The λ_i coefficients are called Lagrange multipliers and the method described can be easily remembered by noticing that equations (3.80) and (3.81) can be generated by differentiating partially with respect to each x_j and λ_i, and setting equal to zero the Lagrangian function,

$$L(x_1, x_2, \ldots, x_n, \lambda_1, \lambda_2, \ldots, \lambda_m) = y - \sum_{i=1}^{m} \lambda_i f_i. \tag{3.82}$$

Hence, at the critical points

$$\left(\frac{\partial L}{\partial x_1}\right)^* = \left(\frac{\partial y}{\partial x_1}\right)^* - \sum_{i=1}^{m} \lambda_i \left(\frac{\partial f_i}{\partial x_1}\right)^* = 0,$$

$$\left(\frac{\partial L}{\partial x_2}\right)^* = \left(\frac{\partial y}{\partial x_2}\right)^* - \sum_{i=1}^{m} \lambda_i \left(\frac{\partial f_i}{\partial x_2}\right)^* = 0,$$

$$\vdots \qquad \qquad \vdots$$

$$\left(\frac{\partial L}{\partial x_n}\right)^* = \left(\frac{\partial y}{\partial x_n}\right)^* - \sum_{i=1}^{m} \lambda_i \left(\frac{\partial f_i}{\partial x_n}\right)^* = 0,$$

$$\left(\frac{\partial L}{\partial \lambda_1}\right)^* = -f_1(x_1^*, x_2^*, \ldots, x_n^*) = 0, \tag{3.83}$$

$$\left(\frac{\partial L}{\partial \lambda_2}\right)^* = -f_2(x_1^*, x_2^*, \ldots, x_n^*) = 0,$$

$$\vdots \qquad \qquad \vdots$$

and

$$\left(\frac{\partial L}{\partial \lambda_m}\right)^* = -f_m(x_1^*, x_2^*, \ldots, x_n^*) = 0.$$

This procedure is equivalent to that of obtaining the critical points of the Lagrangian function $L(x_1, x_2, \ldots, x_n, \lambda_1, \lambda_2, \ldots, \lambda_m)$ with no constraints.

3.6.1 Example

Consider the solution of Example 3.3.2, the transportation problem, with the method of Lagrange multipliers. The situation is

$$\text{Min } y = C = 10x_1x_2 + 20x_1x_3 + 40x_2x_3 + 0.10x_4,$$

$$\text{subject to } x_1x_2x_3x_4 = 400.$$

The Lagrangian function is

$$L(x_1, x_2, x_3, x_4, \lambda_1) = 10x_1x_2 + 20x_1x_3 + 40x_2x_3$$
$$+ 0.10x_4 - \lambda_1(x_1x_2x_3x_4 - 400).$$

Note that the constraint has to be written in the standard form

$$f_1 = x_1x_2x_3x_4 - 400 = 0$$

before introducing it into the Lagrangian function. Differentiating partially with respect to each x_i and λ_1 and setting the partial derivatives to zero, obtain the system of nonlinear equations,

$$\left(\frac{\partial L}{\partial x_1}\right)^* = 10x_2^* + 20x_3^* - \lambda_1 x_2^* x_3^* x_4^* = 0,$$

$$\left(\frac{\partial L}{\partial x_2}\right)^* = 10x_1^* + 40x_3^* - \lambda_1 x_1^* x_3^* x_4^* = 0,$$

$$\left(\frac{\partial L}{\partial x_3}\right)^* = 20x_1^* + 40x_1^* - \lambda_1 x_1^* x_2^* x_4^* = 0,$$

$$\left(\frac{\partial L}{\partial x_4}\right)^* = 0.10 - \lambda_1 x_1^* x_2^* x_3^* = 0,$$

and

$$\left(\frac{\partial L}{\partial \lambda_1}\right)^* = -x_1^* x_2^* x_3^* x_4^* + 400 = 0.$$

This is a system of five nonlinear equations in five unknowns. Simultaneous solution shows that the point

$$x_1^* = 2; \ x_2^* = 1; \ x_3^* = \tfrac{1}{2}; \ x_4^* = 400;$$

and $\lambda_1 = 0.10$, is, in fact, a critical point. Substitution of the numerical values of the point coordinates into the system of equations serves to verify the solution. The differential quadratic form, checked in section 3.5.1, is positive-definite, showing that the point is a minimum.

EXERCISES

3–1. A silo has burned down but its foundation is undamaged. The insurance company settles with the owner for $1,500, and because the foundation can be reused, only the cylindrical sides and flat top of the silo must be rebuilt. The building costs are $1.25 per sq. ft. of surface area and the total volume of the silo is specified at 4,000 cu. ft. Can the owner afford to rebuild the structure for $1,500?

3–2. A vertical cylindrical tank is open at the top. Its sides and bottom are made of the same material, but the bottom is twice the thickness of the sides. Compute its optimal diameter for any given volume.

3–3. Determine the minimum cost of transportation in Problem 3.3.2 if the cost of each round trip for the ferry has gone up to $1.00.

3–4. Plot a graph to show the relationship between the volume of the box and the cost of each round trip in Problem 3.3.2.

3–5. A crate manufacturer has in inventory a large quantity of steel bars used to reinforce their frames. Each bar is L feet long and must be cut into three sections, x_1, x_2, and x_3, representing the crate's length, width, and depth, respectively. Determine the lengths x_1, x_2, and x_3 that maximize the volume of each crate.

3–6. Determine the span lengths x_1, x_2, and x_3 that minimize the cost of the bridge in Figure 3.3, given the following information on costs:

Cost of abutments = $1,500/ft. of length.

Cost of piers = $1,200/ft. of length.

Weight of steel for trusses = $100x_i$ pounds/ft. of span for $i = 1, 2, 3$.

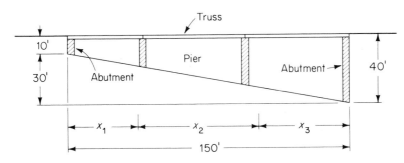

FIGURE 3.3.

NOTE: Truss unit weight is assumed proportional to the corresponding span length.

Cost of truss steel = $0.50 per pound.

All miscellaneous costs such as railings and sidewalks can be ignored in the optimization study.

3–7. To plot a set of five points from an experiment, it appears that the following function will adequately represent the variation in x and y:

$$y = ax^2 + bx + c.$$

Given that the coordinates (x, y) of the five points are (1, 12), (2, 9),

(3, 7), (4, 6), and (5, 5), determine the constants a, b, and c such that the sum of the squares of the errors in y at these five points will be a minimum.

3–8. Write the MacLaurin Series Expansions of the following functions:
   a. e^{-x}.
   b. $\sin x$.
   c. $\cos x$.

3–9. Write the linear and the second-degree terms of the MacLaurin Expansions of the following functions:
   a. $e^{(x+y)}$.
   b. $\sin x \cos y$.

3–10. Obtain critical points of the function

$$f(x, y) = x^2 + y^2 - xy$$

constrained by $xy = 4$.

3–11. Obtain stationary points of

$$y = 3x_1^2 + 2x_1x_2^2 - 3x_2x_3 + x_3^2$$

subject to

$$x_1 + x_2 + x_3 = 3,$$

and

$$x_1^2 - x_2 = 0.$$

3–12. Use Lagrange multipliers to compute the stationary points of

$$y = 6x_1^2 + 2x_1x_2 - 3x_2x_3 + x_3^2$$

subject to

$$x_1 + x_2 + x_3 = 1,$$

and

$$x_1 - 3x_2 = 5.$$

3–13. Solve Problem 3–1 using the method of Lagrange multipliers.

REFERENCES

1. Au, T. and T. E. Stelson, *Introduction to Systems Engineering, Deterministic Models*. Reading, Mass.: Addison-Wesley, 1969.

2. Bolsa, O., *Lectures on the Calculus of Variations*. New York: Dover Publications, Inc., 1961.

3. Hildebrand, F. B. *Advanced Calculus for Applications*. Englewood Cliffs, N.J.: Prentice-Hall, Inc., 1962.

4. Kaplan, W., *Advanced Calculus*. Reading, Mass.: Addison-Wesley, 1952.

5. Taylor, A. E., *Advanced Calculus*. Boston: Ginn and Company, 1955.

CHAPTER **4**

Introduction to Matrix Analysis

4.1 INTRODUCTION

Because much of the work in optimization requires a basic understanding of matrices, the fundamentals of matrix theory will be presented in this chapter. The matrix concepts discussed in the following sections will be sufficient to develop the theory in this book. However, the reader is warned that, to conserve time and space, only the most basic notions of matrix analysis are given and, consequently, he is encouraged to expand his knowledge by consulting the references listed at the end of the chapter.

4.2 MATRICES

A matrix is a rectangular array of elements of the form:

$$\bar{A} = \begin{pmatrix} a_{11} & a_{12} & \cdots & a_{1n} \\ a_{21} & a_{22} & \cdots & a_{2n} \\ \hdotsfor{4} \\ a_{m1} & a_{m2} & \cdots & a_{mn} \end{pmatrix}. \tag{4.1}$$

In this text a matrix will be represented by an upper or a lower case letter with a bar across the top. The elements a_{ij} of the matrix are enclosed by parentheses. Each element in the array is subscripted with two indices. The first indicates the row position of the element; the second, its column position. Thus, a_{ij} represents an element located at the intersection of the ith row and the jth column.

The matrix given in equation (4.1) has m rows and n columns and its dimensions are said to be specified by these parameters. Consequently, the array of equation (4.1) is an $m \times n$ matrix.

The individual elements a_{ij} can take several forms, and in some applications they can be vectors, functions, and other expressions. In the discussion that follows, the elements will represent only real numbers—that is, zero plus all positive and negative integers, positive and negative fractions, and positive and negative irrational numbers.

Although a matrix does not have a value, one can speak of a matrix being physically larger than another. For example, if $\bar{A}$ is an $m \times n$ matrix and $\bar{B}$ is a $p \times q$ matrix and if $m > p$ and $n > q$, then one would say that $\bar{A}$ is physically larger than $\bar{B}$. ($\bar{A}$ has more rows and more columns than $\bar{B}$.)

The following sections will give examples of how matrices can be generated; however, a very important application of matrices (and the one with which this text is primarily concerned) is in the solution of systems of linear equations such as

$$a_{11}x_1 + a_{12}x_2 + \cdots + a_{1n}x_n = b_1,$$
$$a_{21}x_1 + a_{22}x_2 + \cdots + a_{2n}x_n = b_2, \qquad (4.2)$$
$$.$$
$$.$$
$$.$$
$$a_{m1}x_1 + a_{m2}x_2 + \cdots + a_{mn}x_n = b_m.$$

It is primarily because of the interrelationship that exists between systems of simultaneous equations and matrices that it was necessary to include this chapter in the book.

4.3   MATRIX DEFINITIONS

For the purposes of this text, the following matrix definitions must be stated:

1. Matrix $\bar{A}$ is said to be square if it has the same number of rows and columns.

2. A matrix with m rows and 1 column is called a column vector. Example:

$$\bar{a} = \begin{pmatrix} a_1 \\ a_2 \\ \cdot \\ \cdot \\ \cdot \\ a_m \end{pmatrix}. \tag{4.3}$$

3. A matrix with n columns and 1 row is called a row vector. Example:

$$\bar{b} = (b_1, b_2, \ldots, b_n). \tag{4.4}$$

4. If for matrix $\bar{A}$, $a_{ij} = a_{ji}$ for $i \neq j$, the matrix is called symmetric. The following is a numerical example of a symmetric matrix

$$\bar{A} = \begin{pmatrix} 3 & 4 & 2 \\ 4 & 5 & 0 \\ 2 & 0 & 7 \end{pmatrix}. \tag{4.5}$$

5. If matrix $\bar{A}$ is square and $a_{ij} = 0$ for $i \neq j$, then $\bar{A}$ is called a diagonal matrix. Example:

$$\bar{A} = \begin{pmatrix} a_{11} & 0 & \cdots & 0 \\ 0 & a_{22} & \cdots & 0 \\ \cdots & \cdots & \cdots & \cdots \\ 0 & 0 & \cdots & a_{nn} \end{pmatrix}. \tag{4.6}$$

6. If all the elements a_{ii} along the main diagonal of a diagonal matrix $\bar{A}$ are equal to the same constant k, the matrix is called scalar. An example of a scalar matrix is

$$\bar{A} = \begin{pmatrix} 5 & 0 & \cdots & 0 \\ 0 & 5 & \cdots & 0 \\ \cdots & \cdots & \cdots & \cdots \\ 0 & 0 & \cdots & 5 \end{pmatrix}. \tag{4.7}$$

7. If for matrix $\bar{A}$, $a_{ij} = -a_{ji}$ for $i \neq j$ and $a_{ii} = 0$, then matrix $\bar{A}$ is called skew-symmetric. Example:

$$\bar{A} = \begin{pmatrix} 0 & -1 & 3 & 4 \\ 1 & 0 & 5 & -7 \\ -3 & -5 & 0 & 8 \\ -4 & 7 & -8 & 0 \end{pmatrix}. \tag{4.8}$$

8. The unit matrix $\bar{I}$ is a square matrix with $a_{ij} = 0$ for $i \neq j$ and $a_{ii} = 1$.

$$\bar{I} = \begin{pmatrix} 1 & 0 & \cdots & 0 \\ 0 & 1 & \cdots & 0 \\ \cdots\cdots\cdots\cdots \\ 0 & 0 & \cdots & 1 \end{pmatrix}. \tag{4.9}$$

9. A matrix in which all elements are zero is called a null matrix.

10. Two matrices $\bar{A}$ and $\bar{B}$ are equal if and only if they are identical—that is, if they have the same dimensions and if $a_{ij} = b_{ij}$ for all values of i and j.

11. If matrix B is written on the right side of A simply as a matrix operation, and if B has the same number of rows of A, then the matrix

$$(\bar{A} \,|\, \bar{B}) \tag{4.10}$$

is said to be the matrix $\bar{A}$ _augmented_ by $\bar{B}$. For example, if

$$\bar{A} = \begin{pmatrix} 1 & 3 & 5 \\ 2 & 1 & 2 \\ 1 & 1 & 3 \end{pmatrix},$$

and

$$\bar{B} = \begin{pmatrix} 1 \\ 2 \\ 3 \end{pmatrix},$$

then $\bar{A}$ augmented by $\bar{B}$ is

$$(\bar{A} \,|\, \bar{B}) = \begin{pmatrix} 1 & 3 & 5 & | & 1 \\ 2 & 1 & 2 & | & 2 \\ 1 & 1 & 3 & | & 3 \end{pmatrix}.$$

4.4 MATRIX OPERATIONS

1. Addition of two matrices $\bar{A}$ and $\bar{B}$ is defined only if the matrices have the same dimensions, in which case they are said to be conformable for addition. The sum of $\bar{A}$ plus $\bar{B}$ is another matrix $\bar{C}$, such that each of its elements is defined by

$$c_{ij} = a_{ij} + b_{ij} \tag{4.11}$$

for all i and all j.

Example:

$$\begin{pmatrix} 3 & 5 & 2 \\ 1 & 1 & 4 \end{pmatrix} + \begin{pmatrix} 0 & 1 & 5 \\ 2 & 1 & 0 \end{pmatrix} = \begin{pmatrix} 3 & 6 & 7 \\ 3 & 2 & 4 \end{pmatrix}.$$

2. Subtraction is defined under the same conditions as addition. Thus

$$\bar{C} = \bar{A} - \bar{B}$$

implies that $\bar{A}$ and $\bar{B}$ have the same dimensions and that each element of $\bar{C}$ is computed

$$c_{ij} = a_{ij} - b_{ij}$$

for all i and all j.
Example:

$$\begin{pmatrix} 3 & 5 & 2 \\ 1 & 1 & 4 \end{pmatrix} - \begin{pmatrix} 0 & 1 & 5 \\ 2 & 1 & 0 \end{pmatrix} = \begin{pmatrix} 3 & 4 & -3 \\ -1 & 0 & 4 \end{pmatrix}.$$

3. Multiplication of a matrix $\bar{A}$ by a scalar k results in a matrix $\bar{P}$, such that each element of $\bar{P}$ is given by

$$p_{ij} = ka_{ij} \tag{4.12}$$

for all i and all j.
Example:

$$10\begin{pmatrix} 3 & 5 & 2 \\ 1 & 1 & 4 \end{pmatrix} = \begin{pmatrix} 30 & 50 & 20 \\ 10 & 10 & 40 \end{pmatrix}.$$

4. Two matrices $\bar{A}$ and $\bar{B}$ are conformable for multiplication if and only if the number of columns in $\bar{A}$ is equal to the number of rows in $\bar{B}$, in which case the product $\bar{A} \times \bar{B}$ is a matrix $\bar{C}$, such that each of its elements is computed as follows:

$$c_{ik} = \sum_{j=1}^{n} a_{ij}b_{jk} \tag{4.13}$$

for all i and all k.

According to the definition of matrix multiplication, if $\bar{A}$ has m rows and n columns (symbolized $\bar{A}_{mn}$) and $\bar{B}$ has n rows and p columns (symbolized $\bar{B}_{np}$), then the product matrix $\bar{C}$ must have m rows and p columns:

$$\bar{C}_{mp} = \bar{A}_{mn}\bar{B}_{np}. \tag{4.14}$$

Note that the two central indices (n) in the product must match. This is an easy check of conformability for multiplication. The following numerical example will serve to clarify ideas:

$$\begin{pmatrix} 3 & 4 & 5 \\ 1 & 2 & 3 \end{pmatrix}_{2\times 3} \begin{pmatrix} 1 & 1 \\ 0 & 1 \\ 2 & 0 \end{pmatrix}_{3\times 2} = \begin{pmatrix} 13 & 7 \\ 7 & 3 \end{pmatrix}_{2\times 2}$$

where

$$c_{11} = 13 = 3(1) + 4(0) + 5(2),$$
$$c_{12} = 7 = 3(1) + 4(1) + 5(0),$$
$$c_{21} = 7 = 1(1) + 2(0) + 3(2),$$
$$c_{22} = 3 = 1(1) + 2(1) + 3(0).$$

It is also of interest to note that c_{ik} is obtained by multiplying the ith row of $\bar{A}$ with the kth column of $\bar{B}$, such that

$$c_{ik} = (a_{i1}, a_{i2}, \ldots, a_{in})_{1 \times n} \begin{pmatrix} b_{1k} \\ b_{2k} \\ \vdots \\ b_{nk} \end{pmatrix}_{n \times 1}, \qquad (4.15)$$

or

$$c_{ik} = \sum_{j=1}^{n} a_{ij} b_{jk}.$$

The product of a row vector on the left by a column vector on the right results in a scalar. This operation is also called dot product or scalar product of two vectors. Notice that if the relative position of the vectors is reversed—that is, the column vector on the left is multiplied by the row vector on the right—the product is no longer a scalar c_{ij} but, rather, an n-dimensional square matrix. This observation reveals that multiplication of matrices is, in general, *not* commutative.

The set of linear equations mentioned in section 4.2,

$$a_{11}x_1 + a_{12}x_2 + \cdots + a_{1n}x_n = b_1,$$
$$a_{21}x_1 + a_{22}x_2 + \cdots + a_{2n}x_n = b_2,$$
$$\vdots$$
$$a_{m1}x_1 + a_{m2}x_2 + \cdots + a_{mn}x_n = b_m,$$

can now be represented in matrix form as follows:

$$\bar{A}_{mn} \bar{X}_n = \bar{B}_m$$

where

$$\bar{A}_{mn} = \begin{pmatrix} a_{11} & a_{12} & \cdots & a_{1n} \\ a_{21} & a_{22} & \cdots & a_{2n} \\ \cdots\cdots\cdots\cdots\cdots \\ a_{m1} & a_{m2} & \cdots & a_{mn} \end{pmatrix},$$

$$\bar{X}_n = \begin{pmatrix} x_1 \\ x_2 \\ \cdot \\ \cdot \\ \cdot \\ x_n \end{pmatrix},$$

and

$$\bar{B}_m = \begin{pmatrix} b_1 \\ b_2 \\ \cdot \\ \cdot \\ \cdot \\ b_m \end{pmatrix}.$$

5. The transpose of a matrix $\bar{A}$ is another matrix $\bar{A}^t$ such that each element a_{ij} in $\bar{A}^t$ is equal to the element a_{ji} in $\bar{A}$. For example, let

$$\bar{A} = \begin{pmatrix} 3 & 2 & 1 & 4 \\ 3 & 2 & 0 & 2 \\ 1 & 1 & 1 & 3 \end{pmatrix},$$

then,

$$\bar{A}^t = \begin{pmatrix} 3 & 3 & 1 \\ 2 & 2 & 1 \\ 1 & 0 & 1 \\ 4 & 2 & 3 \end{pmatrix}.$$

6. Division is not explicitly defined in matrix algebra. However, under certain conditions which will be stated later, for a square matrix $\bar{A}$ there may exist another matrix $\bar{A}^{-1}$, called the inverse of $\bar{A}$, such that

$$\bar{A}\bar{A}^{-1} = \bar{A}^{-1}\bar{A} = \bar{I}. \tag{4.16}$$

That is, the product of $\bar{A}$ with its inverse $\bar{A}^{-1}$, on the left or on the right, is the unit matrix $\bar{I}$. This is one of the few cases where matrix multiplication is commutative. If a square matrix $\bar{A}$ has an inverse, it is said to be nonsingular, otherwise $\bar{A}$ is singular. The inverse is not defined for nonsquare (rectangular) matrices.

4.5 MATRIX PROPERTIES

1. Matrices are commutative under addition, that is,

$$\bar{A} + \bar{B} = \bar{B} + \bar{A}. \tag{4.17}$$

2. Matrices are associative under addition,

$$\bar{A} + (\bar{B} + \bar{C}) = (\bar{A} + \bar{B}) + \bar{C}. \tag{4.18}$$

3. Matrix multiplication is associative,

$$\bar{A}(\bar{B} \cdot \bar{C}) = (\bar{A} \cdot \bar{B})\bar{C}. \tag{4.19}$$

4. Matrices are distributive under multiplication. Thus,

$$\bar{A}(\bar{B} + \bar{C}) = \bar{A}\bar{B} + \bar{A}\bar{C} \tag{4.20}$$

5. In general, matrix multiplication is not commutative,

$$\bar{A}\bar{B} \neq \bar{B}\bar{A}, \tag{4.21}$$

even when the commuted operation is defined (in the case of square matrices, for example).
Exceptions are:
a. Multiplication by the inverse,

$$\bar{A}\bar{A}^{-1} = \bar{A}^{-1}\bar{A} = \bar{I}. \tag{4.22}$$

b. Multiplication of a matrix by itself (raising to the second power),

$$\bar{A}\bar{A} = \bar{A}\bar{A} = \bar{A}^2. \tag{4.23}$$

c. Multiplication of a square matrix by its conformable unit matrix,

$$\bar{A}\bar{I} = \bar{I}\bar{A} = \bar{A}. \tag{4.24}$$

d. Multiplication of a square matrix by is conformable null (or zero) matrix,

$$\bar{A}\bar{O} = \bar{O}\bar{A} = \bar{O}. \tag{4.25}$$

6. It must be pointed out that the null matrix is the additive identity, that is,

$$\bar{A} + \bar{O} = \bar{A}, \tag{4.26}$$

and that the unit matrix is the multiplicative identity, that is,

$$\bar{A}\bar{I} = \bar{A}. \tag{4.27}$$

4.6 PARTITIONING OF MATRICES

A matrix can be partitioned into submatrices simply by grouping its elements into rectangular arrays. For example,

$$\bar{A} = \begin{pmatrix} 1 & 3 & \vdots & 5 \\ 7 & 2 & \vdots & 1 \\ \cdots & \cdots & & \cdots \\ 4 & 3 & \vdots & 2 \end{pmatrix}$$

can be partitioned as shown by the indicated lines, such that one is able to write

$$\bar{A} = \begin{pmatrix} \bar{A}_{11} & \bar{A}_{12} \\ \bar{A}_{21} & \bar{A}_{22} \end{pmatrix},$$

where

$$\bar{A}_{11} = \begin{pmatrix} 1 & 3 \\ 7 & 2 \end{pmatrix},$$

$$\bar{A}_{12} = \begin{pmatrix} 5 \\ 1 \end{pmatrix},$$

$$\bar{A}_{21} = (4, 3),$$

$$\bar{A}_{22} = (2).$$

Specifically, a matrix can be partitioned by rows or by columns. Thus,

$$\bar{A} = \begin{pmatrix} \bar{R}_1 \\ \bar{R}_2 \\ \bar{R}_3 \end{pmatrix}$$

where

$$\bar{R}_1 = (1, 3, 5),$$
$$\bar{R}_2 = (7, 2, 1),$$

and

$$\bar{R}_3 = (4, 3, 2).$$

Also,

$$\bar{A} = (\bar{C}_1, \bar{C}_2, \bar{C}_3)$$

where

$$\bar{C}_1 = \begin{pmatrix} 1 \\ 7 \\ 4 \end{pmatrix},$$

$$\bar{C}_2 = \begin{pmatrix} 3 \\ 2 \\ 3 \end{pmatrix},$$

and

$$\bar{C}_3 = \begin{pmatrix} 5 \\ 1 \\ 2 \end{pmatrix}.$$

Partitioning is sometimes necessary when operating with large matrices to conserve computer core space. When a product of two matrices is being formed, the matrices can be partitioned; however, one must be careful to partition the rows of the right matrix as the columns of the left in order to maintain conformability for multiplication. For example, the product

$$\begin{pmatrix} 3 & 5 & \vdots & 3 \\ 2 & 1 & \vdots & 0 \\ 4 & 2 & \vdots & 1 \end{pmatrix} \begin{pmatrix} 1 & 0 \\ 2 & 1 \\ \text{--} & \text{--} \\ 4 & 2 \end{pmatrix}$$

can be partitioned as shown. Notice that the first two rows of the right matrix were partitioned as were the first two columns of the left. This maintained conformability of the submatrices, for now

$$\begin{pmatrix} 3 & 5 & \vdots & 3 \\ 2 & 1 & \vdots & 0 \\ 4 & 2 & \vdots & 1 \end{pmatrix} \begin{pmatrix} 1 & 0 \\ 2 & 1 \\ \text{--} & \text{--} \\ 4 & 2 \end{pmatrix} = \left[\begin{pmatrix} 3 & 5 \\ 2 & 1 \\ 4 & 2 \end{pmatrix} \begin{pmatrix} 1 & 0 \\ 2 & 1 \end{pmatrix}, \begin{pmatrix} 3 \\ 0 \\ 1 \end{pmatrix}(4, 2) \right]$$

and the products of the submatrices can still be formed.

4.7 THEOREM NO. 1

Let $\bar{C}_{mp} = \bar{A}_{mn}\bar{B}_{np}$, then $\bar{C}^t_{pm} = \bar{B}^t_{pn}\bar{A}^t_{nm}$. That is, the transpose of the product of two matrices is equal to the product of the transpose of each matrix, taken in reverse order.

Proof:

First observe that all matrix dimensions after taking the transpose are in agreement with conformability for multiplication.
Now, form the product $\bar{C} = \bar{A}\bar{B}$,

$$\bar{A} \cdot \bar{B} = \begin{pmatrix} a_{11} & \cdots & a_{1n} \\ \cdots\cdots\cdots\cdots \\ a_{m1} & \cdots & a_{mn} \end{pmatrix} \begin{pmatrix} b_{11} & \cdots & b_{1p} \\ \cdots\cdots\cdots\cdots \\ b_{n1} & \cdots & b_{np} \end{pmatrix}$$

$$= \begin{pmatrix} \sum_{j=1}^{n} a_{1j}b_{j1} & \cdots & \sum_{j=1}^{n} a_{1j}b_{jp} \\ \vdots & & \vdots \\ \sum_{j=1}^{n} a_{mj}b_{j1} & \cdots & \sum_{j=1}^{n} a_{mj}b_{jp} \end{pmatrix} = \bar{C}. \qquad (4.28)$$

Therefore,

$$\bar{C}^t = \begin{pmatrix} \sum_{j=1}^{n} a_{1j}b_{j1} & \cdots & \sum_{j=1}^{n} a_{mj}b_{j1} \\ \cdots\cdots\cdots\cdots\cdots\cdots \\ \sum_{j=1}^{n} a_{1j}b_{jp} & \cdots & \sum_{j=1}^{n} a_{mj}b_{jp} \end{pmatrix}. \qquad (4.29)$$

Also, obtain the matrix product $\bar{B}^t \bar{A}^t$,

$$\bar{B}^t \bar{A}^t = \begin{pmatrix} b_{11} & \cdots & b_{n1} \\ \cdots\cdots\cdots\cdots \\ b_{1p} & \cdots & b_{np} \end{pmatrix} \begin{pmatrix} a_{11} & \cdots & a_{m1} \\ \cdots\cdots\cdots\cdots \\ a_{1n} & \cdots & a_{mn} \end{pmatrix}$$

$$= \begin{pmatrix} \displaystyle\sum_{j=1}^{n} b_{j1} a_{1j} & \cdots & \displaystyle\sum_{j=1}^{n} b_{j1} a_{mj} \\ \cdots\cdots\cdots\cdots\cdots\cdots \\ \displaystyle\sum_{j=1}^{n} b_{jp} a_{1j} & \cdots & \displaystyle\sum_{j=1}^{n} b_{jp} a_{jn} \end{pmatrix}. \tag{4.30}$$

Since the matrix elements are scalars, matrix (4.30) can be rewritten,

$$\bar{B}^t \bar{A}^t = \begin{pmatrix} \displaystyle\sum_{j=1}^{n} a_{1j} b_{j1} & \cdots & \displaystyle\sum_{j=1}^{n} a_{mj} b_{j1} \\ \cdots\cdots\cdots\cdots\cdots\cdots \\ \displaystyle\sum_{j=1}^{n} a_{1j} b_{jp} & \cdots & \displaystyle\sum_{j=1}^{n} a_{mj} b_{jp} \end{pmatrix}. \tag{4.31}$$

A comparison of matrix (4.29) with matrix (4.31) shows that they are equal term by term, thus

$$\bar{C}^t = (\bar{A}\bar{B})^t = \bar{B}^t \bar{A}^t \tag{4.32}$$

and the theorem is proved.

4.7.1 Corollary

By induction,

$$(\bar{A}\bar{B}\bar{C})^t = [(\bar{A}\bar{B})\bar{C}]^t = \bar{C}^t(\bar{A}\bar{B})^t = \bar{C}^t\bar{B}^t\bar{A}^t. \tag{4.33}$$

The same argument can be extended to the transpose of the product of any number of matrices.

4.8 THEOREM NO. 2

Let $\bar{C} = \bar{A}\bar{B}$, then $\bar{C}^{-1} = \bar{B}^{-1}\bar{A}^{-1}$. That is, the inverse of the product of two matrices is equal to the product of the inverse of each matrix, taken in reverse order.

Proof:

Consider

$$\bar{C} = \bar{A}\bar{B}, \text{ then } \bar{A}^{-1}\bar{C} = \bar{A}^{-1}\bar{A}\bar{B} = \bar{I}\bar{B} = \bar{B}.$$

Also,

$$\bar{B}^{-1}\bar{A}^{-1}\bar{C} = \bar{B}^{-1}\bar{B} = \bar{I}.$$

By the associative law,

$$(\bar{B}^{-1}\bar{A}^{-1})\bar{C} = \bar{I}.$$

Consequently,

$$\bar{C}^{-1} = \bar{B}^{-1}\bar{A}^{-1}.$$

Hence,

$$(\bar{A}\bar{B})^{-1} = \bar{B}^{-1}\bar{A}^{-1}, \tag{4.34}$$

and the theorem is proved.

4.8.1 Corollary

By induction,

$$(\bar{A}\bar{B}\bar{C})^{-1} = [(\bar{A}\bar{B})\bar{C}]^{-1} = \bar{C}^{-1}(\bar{A}\bar{B})^{-1} = \bar{C}^{-1}\bar{B}^{-1}\bar{A}^{-1}. \tag{4.35}$$

The same argument can be extended to the inverse of the product of any number of matrices.

4.9 Elementary Transformations

The work of this section is preliminary to the development of the method of counters for the computation of inverses, which is, after all, the central point in the study of the material presented in this chapter.

There exist a total of six elementary matrix transformations: three elementary row transformations and three elementary column transformations. For the rows they are:

1R. Interchange two rows.
2R. Multiply a row by a scalar.
3R. Add the multiple of one row to another row.

Similarly, for the columns:

1C. Interchange two columns.
2C. Multiply a column by a scalar.
3C. Add the multiple of one column to another column.

These operations are simple enough and can be performed directly on any matrix. However, for reasons that will become clear later, it is necessary to define elementary transformation matrices.

Suppose any one, or any combination of the elementary row transformations, is to be performed on a matrix $\bar{A}$. To accomplish this, multiply, on the *left*, the matrix $\bar{A}$ by the elementary row transformation matrix $\bar{E}_R$ which is formed by taking the unit matrix, conformable to $\bar{A}$ on the left, and performing on it the desired elementary row transformation, or set of row transformations. The elementary column transformation matrix $\bar{E}_C$ is defined in an analogous manner, but the multiplication is performed on the *right* of the matrix $\bar{A}$, and thus $\bar{E}_C$ must be conformable to $\bar{A}$, on the right.

4.9.1 Examples

$$\text{Let matrix } \bar{A} = \begin{pmatrix} 1 & 3 & 5 \\ 2 & 1 & 4 \end{pmatrix}. \tag{4.36}$$

As a first case, suppose that one wishes to multiply the first row by the scalar 2 and add it to the second row, leaving the original first row intact. These are elementary row transformations and the corresponding row transformation matrix, $\bar{E}_R$, will multiply and be conformable to $\bar{A}$ on the left. Thus, $\bar{E}_R$ is constructed by performing the stated row transformations on a 2×2 unit matrix,

$$\bar{E}_R = \begin{pmatrix} 1 & 0 \\ 2 & 1 \end{pmatrix}. \tag{4.37}$$

Twice, the first row was added to the second, and the first row was left undisturbed. Now, to perform the stated transformations on A,

$$\bar{E}_R \bar{A} = \begin{pmatrix} 1 & 0 \\ 2 & 1 \end{pmatrix}\begin{pmatrix} 1 & 3 & 5 \\ 2 & 1 & 4 \end{pmatrix} = \begin{pmatrix} 1 & 3 & 5 \\ 4 & 7 & 14 \end{pmatrix}, \tag{4.38}$$

which accomplishes the intended row operations.

Assume now that one wishes to interchange the first column with the third column of the original matrix $\bar{A}$ and, further, to add three times the original first column to the second. Since these are elementary column transformations, $\bar{E}_C$ is formed by operating on a 3×3 unit matrix for multiplication on the right of $\bar{A}$. Thus,

$$\bar{E}_C = \begin{pmatrix} 0 & 3 & 1 \\ 0 & 1 & 0 \\ 1 & 0 & 0 \end{pmatrix}. \tag{4.39}$$

Now, to perform the column transformations on $\bar{A}$, take

$$\bar{A}\bar{E}_C = \begin{pmatrix} 1 & 3 & 5 \\ 2 & 1 & 4 \end{pmatrix}\begin{pmatrix} 0 & 3 & 1 \\ 0 & 1 & 0 \\ 1 & 0 & 0 \end{pmatrix}$$

$$= \begin{pmatrix} 5 & 6 & 1 \\ 4 & 7 & 2 \end{pmatrix}, \tag{4.40}$$

which results in a matrix with the specified column operations on $\bar{A}$.

These concepts will be of the utmost importance in the development of computational techniques for obtaining inverses of nonsingular (square) matrices.

4.10 THE CANONICAL OR REDUCED FORM OF A MATRIX

Using elementary transformations alone, it is possible to reduce any matrix to a matrix having zeros on all off-diagonal elements and ones or zeros on its diagonal elements. This form of a matrix is called its canonical or reduced form and the number of ones on its principal diagonal is called the *rank* of the matrix. A square matrix of order n is nonsingular if and only if its rank is equal to n. The matrix is singular if its rank is smaller than n.

4.10.1 Examples

1. First consider a rectangular matrix,

$$\bar{A} = \begin{pmatrix} 2 & 3 & 1 \\ 0 & 1 & 2 \end{pmatrix}.$$

To reduce it to its canonical form, perform elementary row or column transformations, or both, until there are at most two ones along its principal diagonal with all other elements reduced to zeros.
Thus,

$$\bar{A} = \begin{pmatrix} 2 & 3 & 1 \\ 0 & 1 & 2 \end{pmatrix} \sim \begin{pmatrix} 1 & \frac{3}{2} & \frac{1}{2} \\ 0 & 1 & 2 \end{pmatrix} \sim \begin{pmatrix} 1 & 0 & -\frac{5}{2} \\ 0 & 1 & 2 \end{pmatrix}$$

$$\sim \begin{pmatrix} 1 & 0 & 0 \\ 0 & 1 & 2 \end{pmatrix} \sim \begin{pmatrix} 1 & 0 & 0 \\ 0 & 1 & 0 \end{pmatrix}.$$

The symbol ($\sim$) indicates that elementary transformations have been performed on the preceding matrix. The original matrix was transformed into the second by dividing the first row by 2. The third matrix was obtained from the second by multiplying the second row by $(-\frac{3}{2})$ and adding it to the first row. The fourth matrix was obtained from the third by multiplying the first column by $(\frac{5}{2})$ and adding it to the third. Finally, the canonical form was obtained by multiplying the second column by (-2) in the previous matrix and adding it to the third column. In general, the operations are much easier to perform than to describe. The canonical form of the matrix $\bar{A}$ is a rectangular matrix, its inverse is not defined, and the question of singularity does not come up.

2. As a second example, obtain the reduced form of the following square matrix:

$$\bar{B} = \begin{pmatrix} 2 & 3 & 1 \\ 0 & 1 & 2 \\ 4 & 2 & 0 \end{pmatrix} \sim \begin{pmatrix} 1 & \frac{3}{2} & \frac{1}{2} \\ 0 & 1 & 2 \\ 0 & -4 & -2 \end{pmatrix} \sim \begin{pmatrix} 1 & 0 & -\frac{5}{2} \\ 0 & 1 & 2 \\ 0 & 0 & 6 \end{pmatrix}$$

$$\sim \begin{pmatrix} 1 & 0 & -\frac{5}{2} \\ 0 & 1 & 2 \\ 0 & 0 & 1 \end{pmatrix} \sim \begin{pmatrix} 1 & 0 & 0 \\ 0 & 1 & 0 \\ 0 & 0 & 1 \end{pmatrix}.$$

The canonical form of $\bar{B}$ shows that it is a nonsingular matrix because its rank is 3; the same as its dimensional order.

4.11 THEOREM NO. 3

If $\bar{A}$ is a square, nonsingular matrix, it is possible to obtain its inverse, $\bar{A}^{-1}$, by row transformations alone.

Proof:

Since $\bar{A}$ was assumed to be nonsingular, its canonical form is the unit matrix I with the same dimensions of $\bar{A}$. $\bar{I}$ can be obtained by performing row transformations alone, as follows:

$$\bar{A}' = \bar{E}_{R_1}\bar{A},$$
$$\bar{A}'' = \bar{E}_{R_2}\bar{E}_{R_1}\bar{A},$$
$$\cdot$$
$$\cdot$$
$$\cdot$$
$$\bar{I} = \bar{E}_{R_p}\bar{E}_{R_{p-1}} \quad \cdots \quad \bar{E}_{R_2}\bar{E}_{R_1}\bar{A}, \tag{4.41}$$

where the $\bar{E}_{R_i}$, $i = 1, \ldots, p$ are elementary row transformation matrices. Since matrix multiplication is associative, one can rewrite (4.41),

$$\bar{I} = (\bar{E}_{R_p}\bar{E}_{R_{p-1}} \quad \cdots \quad \bar{E}_{R_2}\bar{E}_{R_1})\bar{A}. \tag{4.42}$$

Now, the product in parenthesis is in fact a matrix which when multiplied by $\bar{A}$ yields $\bar{I}$. But this is the definition of the inverse of $\bar{A}$. Consequently,

$$\bar{E}_{R_p}\bar{E}_{R_{p-1}} \quad \cdots \quad \bar{E}_{R_2}\bar{E}_{R_1} = \bar{A}^{-1}, \tag{4.43}$$

thus proving the theorem.

4.11.1 Corollary

$\bar{A}^{-1}$ is unique.

Proof:

One has that $\bar{A}\bar{A}^{-1} = \bar{A}^{-1}\bar{A} = \bar{I}$. Assume there exists another matrix $\bar{B}$ such that

$$\bar{A}\bar{B} = \bar{I}.$$

Then,

$$\bar{A}^{-1}A\bar{B} = \bar{A}^{-1}\bar{I},$$

or

$$\bar{I}\bar{B} = \bar{A}^{-1}.$$

Therefore,

$$\bar{B} = \bar{A}^{-1}.$$

And the corollary is proven since $\bar{B}$ turned out to be $\bar{A}^{-1}$ itself.

4.12 THE METHOD OF COUNTERS: A COMPUTATIONAL SCHEME FOR OBTAINING THE INVERSE

At this point it may be well to ask, Why the inverse? Suppose one must solve the following system of equations:

$$a_{11}x_1 + a_{12}x_2 + \cdots + a_{1n}x_n = b_1,$$
$$a_{21}x_1 + a_{22}x_2 + \cdots + a_{2n}x_n = b_2,$$
$$\vdots$$
$$a_{n1}x_1 + a_{n2}x_2 + \cdots + a_{nn}x_n = b_n.$$

This system can be written,

$$\begin{pmatrix} a_{11} & a_{12} & \cdots & a_{1n} \\ a_{21} & a_{22} & \cdots & a_{2n} \\ \cdots\cdots\cdots\cdots\cdots \\ a_{n1} & a_{n2} & \cdots & a_{nn} \end{pmatrix} \begin{pmatrix} x_1 \\ x_2 \\ \cdot \\ \cdot \\ x_n \end{pmatrix} = \begin{pmatrix} b_1 \\ b_2 \\ \cdot \\ \cdot \\ b_n \end{pmatrix},$$

or

$$\bar{A}\bar{X} = \bar{B}.$$

If $\bar{A}$ is nonsingular, $\bar{A}^{-1}$ exists. Therefore,

$$\bar{A}^{-1}\bar{A}\bar{X} = \bar{A}^{-1}\bar{B},$$
$$\bar{I}\bar{X} = \bar{A}^{-1}\bar{B},$$
$$\bar{X} = \bar{A}^{-1}\bar{B},$$

and the system is solved. Thus, if one were able to compute the inverse of the matrix of coefficients, one could immediately solve the system of equations simply by multiplying the inverse by the vector of constants (or stipulations).

Furthermore, the system could be loaded with different vectors of constants and the solutions for each would be obtained with equal ease; no new inversions would be required.

The following computational technique for obtaining the inverse, called the method of *counters*, is extremely simple and easy to carry out even by hand operation with relatively small or large, but sparse, matrices:

1. Augment the nonsingular square matrix $\bar{A}$, the inverse of which is desired, with the unit matrix $\bar{I}$ of the same dimensional order,

$$(\bar{A} \,|\, \bar{I}).$$

2. Perform elementary row transformations on $\bar{A}$, and perform the same transformation on $\bar{I}$, until $\bar{A}$ is reduced to its canonical form (unit matrix, $\bar{I}$).

$$(\bar{E}_{R_p} \cdots \bar{E}_{R_1} \bar{A} \,|\, \bar{E}_{R_p} \cdots \bar{E}_{R_1} \bar{I}) = (\bar{I} \,|\, \bar{E}_{R_p} \cdots \bar{E}_{R_1} \bar{I}).$$

3. When this has been done, the matrix product of the row transformations on $\bar{I}$ must, by Theorem No. 3, be the inverse of $\bar{A}$.
4. In summary, the operation proceeds as follows:

$$(\bar{A} \,|\, \bar{I}) \sim (\bar{A}^{-1} \bar{A} \,|\, \bar{A}^{-1} \bar{I}) = (\bar{I} \,|\, \bar{A}^{-1}). \tag{4.44}$$

Example:

Suppose one wished to obtain the inverse of

$$\bar{A} = \begin{pmatrix} 2 & 3 & 1 \\ 0 & 1 & 2 \\ 4 & 2 & 0 \end{pmatrix}.$$

This matrix has already been shown to be nonsingular in Example 2, section 4.10.1. Nevertheless, even if the matrix were singular, this condition would show up during the course of the computations. Proceed as follows:

$$\begin{pmatrix} 2 & 3 & 1 & | & 1 & 0 & 0 \\ 0 & 1 & 2 & | & 0 & 1 & 0 \\ 4 & 2 & 0 & | & 0 & 0 & 1 \end{pmatrix} \sim \begin{pmatrix} 1 & \frac{3}{2} & \frac{1}{2} & | & \frac{1}{2} & 0 & 0 \\ 0 & 1 & 2 & | & 0 & 1 & 0 \\ 4 & 2 & 0 & | & 0 & 0 & 1 \end{pmatrix}$$

$$\sim \begin{pmatrix} 1 & \frac{3}{2} & \frac{1}{2} & | & \frac{1}{2} & 0 & 0 \\ 0 & 1 & 2 & | & 0 & 1 & 0 \\ 0 & -4 & -2 & | & -2 & 0 & 1 \end{pmatrix} \sim \begin{pmatrix} 1 & 0 & -\frac{5}{2} & | & \frac{1}{2} & -\frac{3}{2} & 0 \\ 0 & 1 & 2 & | & 0 & 1 & 0 \\ 0 & 0 & 6 & | & -2 & 4 & 1 \end{pmatrix}$$

$$\sim \begin{pmatrix} 1 & 0 & 0 & | & -\frac{1}{3} & \frac{1}{6} & \frac{5}{12} \\ 0 & 1 & 0 & | & \frac{2}{3} & -\frac{1}{3} & -\frac{1}{3} \\ 0 & 0 & 1 & | & -\frac{1}{3} & \frac{2}{3} & \frac{1}{6} \end{pmatrix}.$$

Matrix $\bar{A}$ has been reduced to $\bar{I}$. Consequently,

$$\bar{A}^{-1} = \begin{pmatrix} -\frac{1}{3} & \frac{1}{6} & \frac{5}{12} \\ \frac{2}{3} & -\frac{1}{3} & -\frac{1}{3} \\ -\frac{1}{3} & \frac{2}{3} & \frac{1}{6} \end{pmatrix} = \frac{1}{12}\begin{pmatrix} -4 & 2 & 5 \\ 8 & -4 & -4 \\ -4 & 8 & 2 \end{pmatrix}.$$

The correctness of this computation can be verified simply by forming the product

$$\bar{A}^{-1}\bar{A} = \bar{I},$$

and, in fact,

$$\begin{pmatrix} 2 & 3 & 1 \\ 0 & 1 & 2 \\ 4 & 2 & 0 \end{pmatrix}\begin{pmatrix} -4 & 2 & 5 \\ 8 & -4 & -4 \\ -4 & 8 & 2 \end{pmatrix}\frac{1}{12} = \frac{1}{12}\begin{pmatrix} 12 & 0 & 0 \\ 0 & 12 & 0 \\ 0 & 0 & 12 \end{pmatrix}$$

$$= \begin{pmatrix} 1 & 0 & 0 \\ 0 & 1 & 0 \\ 0 & 0 & 1 \end{pmatrix}.$$

4.13   EQUIVALENT MATRICES

If a matrix $\bar{A}$ can be transformed into another matrix $\bar{B}$ by using elementary transformations alone, $\bar{A}$ and $\bar{B}$ are said to be equivalent matrices.

4.13.1   Theorem No. 4

If $\bar{A}$ and $\bar{B}$ are equivalent matrices, they have the same dimensions and the same rank.

Proof:

Since $\bar{A}$ and $\bar{B}$ are equivalent, in going from $\bar{A}$ to $\bar{B}$ with elementary transformations, one can reduce each of them to the same canonical form, thus

$$\bar{A} \sim \text{canonical form} \sim \bar{B}, \tag{4.45}$$

and, consequently, they must have not only the same dimensions, but also the same rank.

4.14   SYSTEMS OF SIMULTANEOUS EQUATIONS

Assume that the system

$$\bar{A}\bar{X} = \bar{B}, \tag{4.46}$$

possesses the solution

$$\bar{X} = \bar{A}^{-1}\bar{B}. \tag{4.47}$$

The best computational scheme for finding its solution parallels the method of counters of section 4.13, and is implemented as follows:

$$(\bar{A}\,|\,\bar{B}) \sim (\bar{A}^{-1}\bar{A}\,|\,\bar{A}^{-1}\bar{B}) = (\bar{I}\,|\,\bar{X}). \tag{4.48}$$

If one wishes to find $\bar{A}^{-1}$ also, use

$$(\bar{A}\,|\,\bar{I}\,|\,\bar{B}) \sim (\bar{A}^{-1}\bar{A}\,|\,\bar{A}^{-1}\bar{I}\,|\,\bar{A}^{-1}\bar{B}) = (\bar{I}\,|\,\bar{A}^{-1}\,|\,\bar{X}). \tag{4.49}$$

Furthermore, one can with similar ease solve the system when loaded with several stipulation or constant vectors by augmenting the matrix of coefficients with these vectors, simultaneously, before performing the elementary row transformations. Thus,

$$(\bar{A}\,|\,\bar{B}_1\,|\,\bar{B}_2\,|\,\ldots\,|\,\bar{B}_p) \sim (\bar{A}^{-1}\bar{A}\,|\,\bar{A}^{-1}\bar{B}_1\,|\,\bar{A}^{-1}\bar{B}_2\,|\,\ldots\,|\,\bar{A}^{-1}\bar{B}_p)$$
$$= (\bar{I}\,|\,\bar{X}_1\,|\,\bar{X}_2\,|\,\ldots\,|\,\bar{X}_p). \tag{4.50}$$

Example:

Solve the following system of equations loaded with two different vectors of constants,

$$\begin{pmatrix} 2 & 3 & 1 \\ 0 & 1 & 2 \\ 4 & 2 & 0 \end{pmatrix}\begin{pmatrix} x_1 \\ x_2 \\ x_3 \end{pmatrix} = \begin{pmatrix} 4 \\ 6 \\ 12 \end{pmatrix} \quad \text{and} \quad \begin{pmatrix} 4 \\ 0 \\ 8 \end{pmatrix}.$$

In the solution process the inverse will also be computed:

$$(\bar{A}\,|\,\bar{I}\,|\,\bar{B}_1\,|\,\bar{B}_2) = \begin{pmatrix} 2 & 3 & 1 & | & 1 & 0 & 0 & | & 4 & | & 4 \\ 0 & 1 & 2 & | & 0 & 1 & 0 & | & 6 & | & 0 \\ 4 & 2 & 0 & | & 0 & 0 & 1 & | & 12 & | & 8 \end{pmatrix}$$

$$\sim \begin{pmatrix} 1 & \frac{3}{2} & \frac{1}{2} & | & \frac{1}{2} & 0 & 0 & | & 2 & | & 2 \\ 0 & 1 & 2 & | & 0 & 1 & 0 & | & 6 & | & 0 \\ 4 & 2 & 0 & | & 0 & 0 & 1 & | & 12 & | & 8 \end{pmatrix}$$

$$\sim \begin{pmatrix} 1 & \frac{3}{2} & \frac{1}{2} & | & \frac{1}{2} & 0 & 0 & | & 2 & | & 2 \\ 0 & 1 & 2 & | & 0 & 1 & 0 & | & 6 & | & 0 \\ 0 & -4 & -2 & | & -2 & 0 & 1 & | & 4 & | & 0 \end{pmatrix}$$

$$\sim \begin{pmatrix} 1 & 0 & -\frac{5}{2} & | & \frac{1}{2} & -\frac{3}{2} & 0 & | & -7 & | & 2 \\ 0 & 1 & 2 & | & 0 & 1 & 0 & | & 6 & | & 0 \\ 0 & 0 & 6 & | & -2 & 4 & 1 & | & 28 & | & 0 \end{pmatrix}$$

$$\sim \begin{pmatrix} 1 & 0 & 0 & | & -\frac{1}{3} & \frac{1}{5} & \frac{5}{12} & | & \frac{14}{3} & | & 2 \\ 0 & 1 & 0 & | & \frac{2}{3} & -\frac{1}{3} & -\frac{1}{3} & | & -\frac{10}{3} & | & 0 \\ 0 & 0 & 1 & | & -\frac{1}{3} & \frac{2}{3} & \frac{1}{6} & | & \frac{14}{3} & | & 0 \end{pmatrix} = (\bar{I}\,|\,\bar{A}^{-1}\,|\,\bar{X}_1\,|\,\bar{X}_2).$$

Consequently,

$$\bar{A}^{-1} = \begin{pmatrix} -\frac{1}{3} & \frac{1}{6} & \frac{5}{12} \\ \frac{2}{3} & -\frac{1}{3} & -\frac{1}{3} \\ -\frac{1}{3} & \frac{2}{3} & \frac{1}{6} \end{pmatrix} \text{ as before.}$$

For the first solution,

$$\begin{pmatrix} x_1 \\ x_2 \\ x_3 \end{pmatrix} = \begin{pmatrix} \frac{14}{3} \\ -\frac{10}{3} \\ \frac{14}{3} \end{pmatrix},$$

and for the second solution,

$$\begin{pmatrix} x_1 \\ x_2 \\ x_3 \end{pmatrix} = \begin{pmatrix} 2 \\ 0 \\ 0 \end{pmatrix}.$$

These solutions are quickly verified by back substitution into the original system.

4.14.1 Definition of Linear Independence

A set of column or row vectors of the same dimension, $\bar{P}_1, \bar{P}_2, \ldots, \bar{P}_n$, is said to be linearly independent if and only if,

$$\sum_{i=1}^{n} c_i \bar{P}_i = \bar{0}$$

$$\text{For } c_i = 0, i = 1, 2, \ldots, n, \tag{4.51}$$

where the c_i are scalars. Otherwise, the set is linearly dependent. For example, the set of column vectors

$$\begin{pmatrix} 3 \\ 0 \\ 0 \end{pmatrix}, \begin{pmatrix} 0 \\ 4 \\ 0 \end{pmatrix}, \begin{pmatrix} 0 \\ 0 \\ 5 \end{pmatrix}$$

is linearly independent because

$$c_1 \begin{pmatrix} 3 \\ 0 \\ 0 \end{pmatrix} + c_2 \begin{pmatrix} 0 \\ 4 \\ 0 \end{pmatrix} + c_3 \begin{pmatrix} 0 \\ 0 \\ 5 \end{pmatrix} = \begin{pmatrix} 0 \\ 0 \\ 0 \end{pmatrix}$$

only for $c_1 = c_2 = c_3 = 0$.

4.14.2 Condition for Singularity of a Square Matrix

A square matrix of size $n \times n$ is singular if any of its rows or columns are linearly dependent on any other rows or columns.

Take, for example, the following matrix

$$\bar{A} = \begin{pmatrix} 3 & 2 & 1 \\ 6 & 4 & 2 \\ 0 & 0 & 1 \end{pmatrix}.$$

The second row is twice the first; therefore, they are linearly dependent on each other. Observe that

$$c_1(3, 2, 1) + c_2(6, 4, 2) + c_3(0, 0, 1) = (0, 0, 0).$$

For $c_1 = -2c_2$ and $c_3 = 0$, thereby not meeting the requirements for linear independence. Reducing the matrix to its canonical form, one finds that its rank is only 2 and that, consequently, the matrix $\bar{A}$, which is a 3×3 matrix, is singular and has no inverse.

$$\begin{pmatrix} 3 & 2 & 1 \\ 6 & 4 & 2 \\ 0 & 0 & 1 \end{pmatrix} \sim \begin{pmatrix} 3 & 2 & 1 \\ 0 & 0 & 0 \\ 0 & 0 & 1 \end{pmatrix} \sim \begin{pmatrix} 3 & 2 & 0 \\ 0 & 0 & 0 \\ 0 & 0 & 1 \end{pmatrix}$$

$$\sim \begin{pmatrix} 1 & \frac{2}{3} & 0 \\ 0 & 0 & 0 \\ 0 & 0 & 1 \end{pmatrix} \sim \begin{pmatrix} 1 & 0 & 0 \\ 0 & 0 & 0 \\ 0 & 0 & 1 \end{pmatrix}; \text{rank} = 2.$$

The fact that $\bar{A}$ has no inverse can be verified with the method of counters,

$$\left(\begin{array}{ccc|ccc} 3 & 2 & 1 & 1 & 0 & 0 \\ 6 & 4 & 2 & 0 & 1 & 0 \\ 0 & 0 & 1 & 0 & 0 & 1 \end{array}\right) \sim \left(\begin{array}{ccc|ccc} 3 & 2 & 1 & 1 & 0 & 0 \\ 0 & 0 & 0 & -2 & 1 & 0 \\ 0 & 0 & 1 & 0 & 0 & 1 \end{array}\right)$$

$$\sim \left(\begin{array}{ccc|ccc} 1 & \frac{2}{3} & \frac{1}{3} & \frac{1}{3} & 0 & 0 \\ 0 & 0 & 0 & -2 & 1 & 0 \\ 0 & 0 & 1 & 0 & 0 & 1 \end{array}\right) \sim \left(\begin{array}{ccc|ccc} 1 & \frac{2}{3} & 0 & \frac{1}{3} & 0 & -\frac{1}{3} \\ 0 & 0 & 0 & -2 & 1 & 0 \\ 0 & 0 & 1 & 0 & 0 & 1 \end{array}\right).$$

One cannot complete the inversion using elementary row transformations alone because the entire second row was reduced to zeros, once again showing that the rank of $\bar{A}$ is only 2, less than its dimensional order.

4.14.3 Existence of Solutions to Systems of Linear Equations

Consider the system of n linear equations in n unknowns,

$$\bar{A}\bar{X} = \bar{B}.$$

If the matrix of coefficients $\bar{A}$ is nonsingular, the system has a unique solution

$$\bar{X} = \bar{A}^{-1}\bar{B}.$$

If $\bar{A}$ is singular, the system of equations can have either an infinite number of solutions or no solutions at all, as will be demonstrated with the following examples.

Examples:

1. Consider the system

$$3x_1 + 2x_2 + x_3 = 5,$$
$$6x_1 + 3x_2 + 2x_3 = -9,$$
$$6x_1 + 4x_2 + 2x_3 = 10.$$

The third equation is twice the first. Therefore, the third equation is redundant and can be eliminated, leaving only two equations in three unknowns. The inverse of the matrix of coefficients does not exist but the system has an infinite number of solutions, as can be shown with the method of counters.

$$\begin{pmatrix} 3 & 2 & 1 & 5 \\ 6 & 3 & 2 & -9 \\ 6 & 4 & 2 & 10 \end{pmatrix} \sim \begin{pmatrix} 1 & \frac{2}{3} & \frac{1}{3} & \frac{5}{3} \\ 0 & -1 & 0 & -19 \\ 0 & 0 & 0 & 0 \end{pmatrix} \sim \begin{pmatrix} 1 & 0 & \frac{1}{3} & -11 \\ 0 & 1 & 0 & 19 \\ 0 & 0 & 0 & 0 \end{pmatrix}.$$

The solution states: From the first row

$$x_1 + \tfrac{1}{3}x_3 = -11,$$

and from the second,

$$x_2 = 19.$$

Let x_3 take on the arbitrary value a; then the solution vector is

$$\begin{pmatrix} x_1 \\ x_2 \\ x_3 \end{pmatrix} = \begin{pmatrix} -11 & -\tfrac{1}{3}a \\ 19 \\ a \end{pmatrix}.$$

Because a is totally arbitrary, this vector represents an infinite number of solutions.

2. Now, consider the system

$$3x_1 + 2x_2 + x_3 = 5,$$
$$6x_1 + 3x_2 + 2x_3 = -9,$$
$$6x_1 + 4x_2 + 2x_3 = 11.$$

This system differs from the first only in that the constant for the third equation is 11 rather than 10. Therefore, the third equation is twice the first except for the constant and, consequently, the linear dependence does not carry into the vector of stipulations. This is an inconsistency and the set has no solutions, which can be demonstrated as follows:

$$\begin{pmatrix} 3 & 2 & 1 & | & 5 \\ 6 & 3 & 2 & | & -9 \\ 6 & 4 & 2 & | & 11 \end{pmatrix} \sim \begin{pmatrix} 1 & \frac{2}{3} & \frac{1}{3} & | & \frac{5}{3} \\ 0 & -1 & 0 & | & -19 \\ 0 & 0 & 0 & | & 1 \end{pmatrix} \sim \begin{pmatrix} 1 & 0 & \frac{1}{3} & | & -11 \\ 0 & 1 & 0 & | & 19 \\ 0 & 0 & 0 & | & 1 \end{pmatrix}.$$

Observe that the third row of the solution matrix states that

$$0x_1 + 0x_2 + 0x_3 = 1.$$

This, of course, is impossible, and there are no solutions to this inconsistent system.

3. As a final example, consider

$$3x_1 + 3x_2 + x_3 = 8,$$
$$5x_1 + 5x_2 + 2x_3 = 4,$$
$$2x_1 + 2x_2 - x_3 = 6.$$

The first and the second columns are equal and, therefore, linearly dependent so that $\bar{A}^{-1}$ does not exist. In this case the system is also inconsistent:

$$\begin{pmatrix} 3 & 3 & 1 & | & 8 \\ 5 & 5 & 2 & | & 4 \\ 2 & 2 & -1 & | & 6 \end{pmatrix} \sim \begin{pmatrix} 0 & 0 & -\frac{1}{2} & | & -1 \\ 0 & 0 & -\frac{1}{2} & | & -11 \\ 1 & 1 & -\frac{1}{2} & | & 3 \end{pmatrix} \sim \begin{pmatrix} 0 & 0 & 1 & | & 2 \\ 0 & 0 & 0 & | & -10 \\ 1 & 1 & 0 & | & 4 \end{pmatrix}.$$

Again, the second row states that

$$0x_1 + 0x_2 + 0x_3 = -10,$$

an impossibility, and the system has no solutions.

In general, then, for

$$\bar{A}\bar{X} = \bar{B},$$

if $\bar{A}^{-1}$ exists, the system possesses a unique solution. If $\bar{A}^{-1}$ does not exist, the system either has an infinite number of solutions, or no solutions at all.

4.15 SETS OF *m* EQUATIONS IN *n* UNKNOWNS, *n* > *m*.

Linear programming models, to be discussed in the following chapters, require the solution of systems of *m* equations in *n* unknowns, with $n > m$. Equations of the type,

$$\bar{A}_{mn}\bar{X}_{n1} = \bar{B}_{m1}. \tag{4.52}$$

Several definitions related to these systems are needed to understand the forthcoming work in linear programming.

1. A *solution* to (4.52) is any *n*-dimensional vector X_{n1} that satisfies the equations.

2. A *basic solution* to (4.52) is one in which $(n - m)$ of the variables are set equal to zero, and that satisfies

$$\bar{A}_{mm} \bar{X}_{m1} = \bar{B}_{m1}. \tag{4.53}$$

Such a solution can have at most m nonzero components.

3. If the basic solution has *exactly* m nonzero components, it is called a *nondegenerate basic solution;* otherwise, it is a *degenerate basic solution*.

4. If *all* the components of a basic solution are non-negative, the solution is called *feasible*. If it has *one or more negative* components, it is called *infeasible*.

The following summary of definitions helps to clarify concepts:

Solution—A vector that satisfies equations (4.52).

Basic Solution—$(n - m)$ variables are set to zero.

Nondegenerate Basic Solution—Exactly $(n - m)$ zero components.

Degenerate Basic Solution—More than $(n - m)$ zero components.

Feasible Basic Solution—m non-negative components.

Infeasible Basic Solution—One or more negative components.

Combinations of these serve to describe all possible types of solutions. For example, a nondegenerate basic feasible solution is a vector with m positive and $(n - m)$ zero components that satisfies equation (4.52). For simplicity, in the following chapters, nondegeneracy will be implied except when stated otherwise; thus, a nondegenerate basic feasible solution will simply be called a basic feasible solution.

This concludes the exposition of matrix analysis topics deemed necessary for understanding the concepts to be presented in the remainder of the book.

A great many texts on this subject are available. A few are listed in this chapter's references.

EXERCISES

4–1. Find $(2\bar{A} + 3\bar{B})$ and $(4\bar{A} - 2\bar{B})$, where

$$\bar{A} = \begin{pmatrix} 2 & 1 & 1 \\ 0 & -2 & 1 \end{pmatrix} \text{ and } \bar{B} = \begin{pmatrix} 2 & -1 & 3 \\ 0 & 0 & 4 \end{pmatrix}.$$

4–2. If $\bar{A} + \bar{B} = \begin{pmatrix} 5 & -1 \\ 0 & 3 \end{pmatrix}$ and $(3\bar{A} - \bar{B}) = \begin{pmatrix} 2 & 1 \\ 1 & 0 \end{pmatrix}$, find $\bar{A}$ and $\bar{B}$.

4–3. Given that $A = \begin{pmatrix} 1 & 2 \\ 3 & 0 \end{pmatrix}$, find $\bar{X} = \begin{pmatrix} x_{11} & x_{12} \\ x_{21} & x_{22} \end{pmatrix}$ such that $\bar{A}\bar{X} = \bar{I}$.

NOTE: The matrix $\bar{X}$ is the inverse of $\bar{A}$.

4-4. Given that

$$\bar{A} = \begin{pmatrix} 3 & 1 \\ 1 & 0 \\ 5 & 4 \end{pmatrix},$$

$$\bar{B} = \begin{pmatrix} 2 \\ 0 \end{pmatrix},$$

and $\bar{C} = (1, 4, 2)$,

compute

a) $(\bar{A}\bar{B})\bar{C}$,

b) $\bar{A}(\bar{B}\bar{C})$.

That is, multiplication is associative.

4-5. Given that

$$x_1 = 4y_1 - y_2 + y_3,$$
$$x_2 = y_1 + 2y_2 - 2y_3,$$
$$y_1 = z_1 - 2z_2,$$
$$y_2 = z_1 + 3z_2,$$
$$y_3 = -2z_1 + 2z_2,$$

express x_1 and x_2 in terms of z_1 and z_2.

4-6. Partition $\bar{A}$ and $\bar{B}$ as indicated and find $\bar{A}\bar{B}$.

a.
$$\bar{A} = \begin{pmatrix} 3 & 4 & \vdots & 5 \\ 0 & -1 & \vdots & 3 \end{pmatrix}; \quad \bar{B} = \begin{pmatrix} 3 & 2 \\ 1 & 1 \\ \cdots & \cdots \\ 0 & 2 \end{pmatrix}.$$

b.
$$\bar{A} = \begin{pmatrix} 1 & \vdots & 1 & \vdots & 0 \\ 1 & \vdots & 1 & \vdots & 0 \\ 0 & \vdots & 2 & \vdots & 1 \end{pmatrix}; \quad \bar{B} = \begin{pmatrix} 3 & 4 \\ \cdots & \cdots \\ 5 & 1 \\ \cdots & \cdots \\ 1 & 0 \end{pmatrix}.$$

4-7. Compute the length of the vector

$$\bar{X} = \begin{pmatrix} 1 \\ 2 \\ 3 \\ 4 \end{pmatrix}.$$

NOTE: The length of $\bar{X}$ is obtained with the formula

$$|\bar{X}| = \sqrt{\bar{X}^t \cdot \bar{X}}.$$

4–8. Given that

$$\bar{A} = \begin{pmatrix} 1 & 3 \\ 4 & 2 \\ 1 & 0 \end{pmatrix}$$

and

$$\bar{B} = \begin{pmatrix} 2 & 1 & 1 \\ 1 & 0 & 2 \end{pmatrix},$$

show that $(\bar{A}\bar{B})^t = \bar{B}^t\bar{A}^t$.

4–9. Determine the rank of $\bar{A}$ by reducing the matrix to its canonical form.

a.
$$\bar{A} = \begin{pmatrix} 2 & 4 & 1 \\ 4 & 8 & 3 \\ 3 & 0 & 1 \end{pmatrix}.$$

b.
$$\bar{A} = \begin{pmatrix} 1 & 2 & 3 \\ 2 & 4 & 5 \\ 1 & 1 & 0 \end{pmatrix}.$$

c.
$$\bar{A} = \begin{pmatrix} 1 & 2 & 4 \\ 2 & 4 & 8 \\ 1 & 1 & 3 \end{pmatrix}.$$

4–10. Compute the inverse of $\bar{A}$ if it exists.

a.
$$\bar{A} = \begin{pmatrix} 2 & 4 & 1 \\ 4 & 8 & 3 \\ 3 & 0 & 1 \end{pmatrix}.$$

b.
$$\bar{A} = \begin{pmatrix} 1 & 2 & 3 \\ 2 & 4 & 5 \\ 1 & 1 & 0 \end{pmatrix}.$$

c.
$$\bar{A} = \begin{pmatrix} 1 & 2 & 4 \\ 2 & 4 & 8 \\ 1 & 1 & 3 \end{pmatrix}.$$

4–11. Solve the following systems of equations:

a.
$$2x_1 + 4x_2 + x_3 = 5,$$
$$4x_1 + 8x_2 + 3x_3 = -2,$$
$$3x_1 \qquad + x_3 = 3.$$

b.
$$-x_1 + 2x_2 \qquad = -5,$$
$$3x_2 + x_3 = 4,$$
$$x_1 \qquad - 2x_3 = -2.$$

c.
$$x_1 + 2x_2 + 4x_3 = 5,$$
$$2x_1 + 4x_2 + 8x_3 = 4,$$
$$x_1 + x_2 + 3x_3 = -3.$$

4-12. Given

$$3x_1 - x_2 + x_3 = 8,$$
$$-3x_1 \qquad + 4x_3 = 2,$$
$$-x_1 + 2x_2 \qquad = 4.$$

   a. Write it in the form $\bar{A}\bar{X} = \bar{B}$.
   b. Find $\bar{A}^{-1}$ by the method of counters.
   c. Compute $\bar{X} = \bar{A}^{-1}\bar{B}$.

4-13. Solve $\bar{A}\bar{X} = \bar{B}$,
   Given that

$$\bar{A} = \begin{pmatrix} 3 & 2 & 1 \\ 2 & 1 & 1 \\ 0 & 1 & 2 \end{pmatrix}; \quad \bar{B} = \begin{pmatrix} 1 \\ 2 \\ 3 \end{pmatrix};$$

and

$$\bar{X} = \begin{pmatrix} x_1 \\ x_2 \\ x_3 \end{pmatrix}.$$

4-14. a. Why does the system

$$3x_1 + 2x_2 + x_3 = 4,$$
$$2x_1 + 3x_2 - x_3 = -1,$$
$$-x_1 + x_2 - 2x_3 = -5,$$

have an infinite number of solutions?
   b. Why does the system

$$3x_1 + 2x_2 + x_3 = 4,$$
$$2x_1 + 3x_2 - x_3 = 2,$$
$$-x_1 + x_2 - 2x_3 = -5,$$

have no solutions?

c. Why does the system

$$x_1 + 2x_2 + 3x_3 = 5,$$
$$2x_1 + 4x_2 + 6x_3 = 4,$$
$$x_1 + x_2 + 2x_3 = -3,$$

have no solutions?

REFERENCES

1. BIRKHOFF, G. and S. MACLANE, *Survey of Modern Algebra.* New York: The Macmillan Company, 1941.
2. BROWNE, E. T., *Introduction to the Theory of Determinants and Matrices.* Chapel Hill: The University of North Carolina Press, 1958.
3. HADLEY, G., *Linear Algebra.* Reading, Mass.: Addison-Wesley, 1961.
4. PERLIS, S., *Theory of Matrices.* Reading, Mass.: Addison-Wesley, 1952.
5. THRALL, R. M. and L. TORNHEIM, *Vector Spaces and Matrices.* New York: John Wiley, & Sons, Inc., 1957.

PART II

Deterministic Systems

CHAPTER 5

Linear Programming: Optimization of Allocation Problems

5.1 INTRODUCTION

Many real world problems in architecture, engineering, construction, and urban and regional planning can be modeled with linear objective functions subject to sets of linear constraints. The constraints consist of systems of linear equations, inequalities, or a combination of these. For most physical systems, the objective is either to minimize cost for a fixed level of performance, or to maximize performance (or profit) at a fixed level of cost. This idea is, in fact, of fundamental importance in defining the cost-benefit method of analysis which has become so popular and which is now standard procedure in many governmental and private organizations for the evaluation of alternative courses of action.

Linear programming deals with the optimization of these types of problems and, because they occur frequently in practice, it merits detailed and careful study. Of all the mathematical optimization techniques, linear programming is perhaps the most used and best understood by the business and industrial community.

5.2 STANDARD FORM OF LINEAR PROGRAMMING PROBLEMS

Linear programming problems are usually expressed in a general form of m linear equations or inequalities involving n variables x_i; $i = 1, \ldots, n$, which are restricted to be non-negative.

The program's goal is to optimize the objective function z which is also a linear function of the variables x_i. Thus, the problem can be stated as follows:

$$a_{11}x_1 + a_{12}x_2 + \cdots + a_{1n}x_n \leq b_1,$$
$$a_{21}x_1 + a_{22}x_2 + \cdots + a_{2n}x_n \geq b_2,$$
$$\vdots \qquad\qquad\qquad \vdots \qquad\qquad (5.1)$$
$$a_{m1}x_1 + a_{m2}x_2 + \cdots + a_{mn}x_n = b_m,$$
$$x_j \geq 0 \qquad (j = 1, 2, \ldots, n), \qquad (5.2)$$

and

$$\text{Opt } z = c_1 x_1 + c_2 x_2 + \cdots + c_n x_n \qquad (5.3)$$

in which the constants c_j $(j = 1, 2, \ldots, n)$ specifying criteria values are called *cost coefficients*; the constant b_i $(i = 1, 2, \ldots, m)$ defining the constraint requirements are called *stipulations*, and the constants a_{ij} $(i = 1, 2, \ldots, m,$ and $j = 1, 2, \ldots, n)$ are called *structural coefficients*.

The objective function in a linear program may be either maximized or minimized. If the objective is to be maximized,

$$\text{Max } v = c_1 x_1 + c_2 x_2 + \cdots + c_n x_n,$$

it may also be maximized by multiplying it through by -1, and by restating it as a minimization problem:

$$\text{Min } (-v) = -c_1 x_1 - c_2 x_2 \cdots - c_n x_n.$$

Letting $z = -v$, the function becomes

$$\text{Min } z = -c_1 x_1 - c_2 x_2 \cdots - c_n x_n$$

because the maximization of a positive quantity can be replaced by the minimization of a negative quantity with the same absolute value.

The actual constraint conditions in (5.1) generally include inequalities of both greater than or equal to and less than or equal to types as well as equations. It is useful to replace inequality constraint conditions by equations with the introduction of slack variables. If the ith condition involving q variables has a less than or equal to sign (type 1 inequality),

$$a_{i1}x_1 + a_{i2}x_2 + \cdots + a_{iq}x_q \leq b_i$$

it can be replaced by

$$a_{i1}x_1 + a_{i2}x_2 + \cdots + a_{iq}x_q + x_{q+i} = b_i$$

where x_{q+i} is a non-negative slack variable. On the other hand, if it has a greater than or equal to sign (type 2 inequality), it can be replaced by

$$a_{i1}x_1 + a_{i2}x_2 + \cdots + a_{iq}x_q - x_{q+i} = b_i$$

where x_{q+i} again is also a non-negative slack variable. Suppose that some of the variables in (5.1) are not physically constrained to be nonnegative; they can always be treated as the difference of two non-negative variables. For instance, if the variable x_j is not restricted in sign, it may be expressed

$$x_j = x_j^+ - x_j^-$$

where

$$x_j^+ \geq 0 \quad \text{and} \quad x_j^- \geq 0.$$

Since x_j^+ may be greater or smaller than x_j^-, the sign of x_j can be positive or negative. Consequently, the problem is again expressible with non-negative variables, inequality (5.2), by decomposing each nonrestricted variable into two non-negative variables.

Example:

Reduce the following problem

$$\text{Min } v = 2x_1 + 4x_2$$

subject to

$$3x_1 - 2x_2 \geq 5,$$
$$x_1 + 4x_2 \geq 7,$$
$$5x_1 + x_2 \geq 4,$$
$$x_1 \geq 0, \ x_2 \text{ unrestricted,}$$

to a maximization problem with equality constraints and non-negative variables. The problem can be expressed as follows:

$$\text{Max } z = -v = -2x_1 - 4x_2$$

subject to

$$x_2 = x_2^+ - x_2^-,$$
$$3x_1 - 2x_2^+ + 2x_2^- - x_3 = 5,$$
$$x_1 + 4x_2^+ - 4x_2^- - x_4 = 7,$$
$$5x_1 + x_2^+ - x_2^- - x_5 = 4,$$

$$x_1 \geq 0,$$
$$x_2^+ \geq 0,$$
$$x_2^- \geq 0,$$
$$x_3 \geq 0,$$
$$x_4 \geq 0,$$

and

$$x_5 \geq 0.$$

5.3 GRAPHICAL SOLUTION OF TWO-DIMENSIONAL PROBLEMS

A two-dimensional (two-variable) problem will help illustrate the graphical technique. An industrial manufacturer produces two types of units, 1 and 2. They are sold at a profit of $40 and $50, respectively. The number of units x_1 and x_2 to be produced daily are the decision variables. If the manufacturer produces x_1 type 1 units and x_2 type 2 units, his daily profit in dollars is

$$40x_1 + 50x_2.$$

This objective function, his total profit, is what the manufacturer wishes to maximize. (Obviously, x_1 and x_2 cannot be negative.) Although one could conclude that the industrialist should produce only the more profitable product type, capacity constraints make this impossible.

Assume that the production is turned out on two production lines—one for unit type 1 and the other for unit type 2, and that the capacity of the lines is limited. A maximum of 10 type 1 units can be produced daily; therefore,

$$x_1 \leq 10.$$

The second line has a production capacity of 7 type 2 units per day; hence,

$$x_2 \leq 7.$$

Also, limitations on labor supply impose an additional constraint. Were this not the case, the manufacturer would produce up to capacity—that is, 10 type 1 and 7 type 2 units each day. The plant employs only 24 people, however, and the available labor amounts to only 24 man-days. Production of a type 1 unit requires 2 man-days and a type 2 unit requires 3 man-days. If x_1 plus x_2 units were assembled each day, they would require a total of $2x_1 + 3x_2$ man-days. Thus, the additional side constraint results

$$2x_1 + 3x_2 \leq 24.$$

This constraint keeps the manufacturer from producing up to capacity, that is, up to $x_1 = 10$ and $x_2 = 7$, because this program would require $2 \times 10 + 3 \times 7 = 41$ man-days, 17 more than available.

This is an example of a linear programming problem with two decision variables x_1 and x_2 which the manufacturer can manipulate at will, within the limits determined by the non-negativity conditions ($x_1 \geq 0$ and $x_2 \geq 0$), by the capacity of the production lines ($x_1 \leq 10$, $x_2 \leq 7$) and by the limited supply of manpower ($2x_1 + 3x_2 \leq 24$). These three constraints are linear in x_1 and x_2. The objective function $40x_1 + 50x_2$, which is to be maximized, is also linear in x_1 and x_2.

With only two decision variables, the problem can be resolved graphically. Consider the Cartesian coordinate system x_1, x_2 of Figure 5.1. Every point in

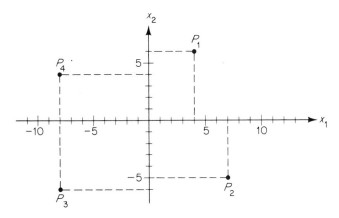

FIGURE 5.1.

the plane represents a production program and vice versa. However, some points represent unusual production schemes. Point P_2, for example, represents the production of a negative number of type 2 units, for this point has coordinates $x_1 = 7$, $x_2 = -5$. This is both physically impossible and mathematically excluded by the non-negativity restriction. A point such as P_2 is said to be infeasible. All points to the left of the vertical axis and below the horizontal axis are also infeasible. The conditions $x_1 \geq 0$ and $x_2 \geq 0$ require that all feasible points be to the right of the vertical axis and above the horizontal axis.

This area is restricted further by the capacity constraints on the production lines; $x_1 \leq 10$ and $x_2 \leq 7$. The point P_1 of Figure 5.1, with $x_1 = 4$, $x_2 = 6$, satisfies these inequalities. Had a point more than 10 to the right of the vertical axis or more than 7 above the horizontal axis been chosen, the capacity restriction would have been violated, because the associated production program could not be implemented.

These considerations generate Figure 5.2 which limits production schedules to the shaded rectangle. The same considerations shall be used to draw the constraint which limits the number of man-days—that is, the constraint

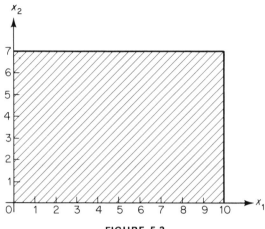

FIGURE 5.2

$2x_1 + 3x_2 \leq 24$. Line Q_1, Q_2 in Figure 5.3 is a plot of the linear equation $2x_1 + 3x_2 = 24$. This line subdivides the plane into two sections. All points in one of the sections satisfy $2x_1 + 3x_2 \leq 24$; this is the side of interest in this problem. All points in the other section satisfy $2x_1 + 3x_2 \geq 24$; this is the infeasible portion. To determine which section is the admissible one, consider whether or not the origin is in the admissible area. If it lies in the admissible half, then the section containing the origin is the portion of interest. The case here is this because $2 \times 0 + 3 \times 0 < 24$. Thus, the admissible region lies under the line Q_1, Q_2 in Figure 5.3.

Because of the non-negativity conditions and of the limited capacities of the production lines, one cannot choose a point outside the shaded rectangle

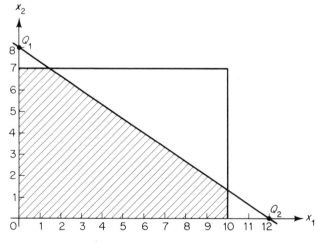

FIGURE 5.3

of Figure 5.2. Also, the point chosen must lie on or below the line Q_1, Q_2 to satisfy the limited labor supply restriction. Figure 5.3 was obtained by combining these conditions.

The admissible or feasible region has been reduced to a pentagon. All points not lying within the pentagon are infeasible.

These considerations take care of the non-negativity conditions and of the side constraints. The problem now is to select a point or points in the feasible region which maximize the profit. To accomplish this, consider, for example, a profit of $150 per day; that is, determine the combinations of values x_1 and x_2 which result in $40x_1 + 50x_2 = 150$.

Because this is a linear equation, its representation is the straight line through R_1 $(x_1 = 0; x_2 = 3)$ and R_2 $(x_1 = 3.75; x_2 = 0)$ in Figure 5.4. Each point on the line R_1, R_2 represents a total profit of $150 per day. Part of the line R_1, R_2 lies inside and part outside the feasible region. The corresponding production programs in the latter case cannot be realized.

Next, consider a $300 per day profit. The profit is $300 daily for all points on the line through S_1 $(x_1 = 0; x_2 = 6)$ and S_2 $(x_1 = 7.5; x_2 = 0)$. The line

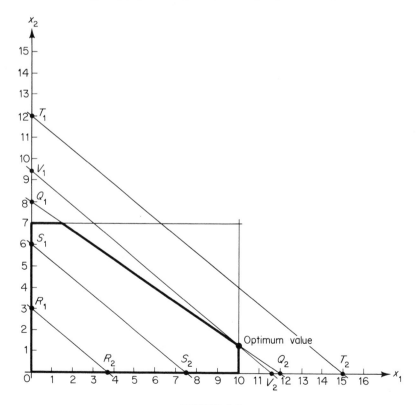

FIGURE 5.4

S_1, S_2 is parallel to R_1, R_2, but further away from the origin of coordinates. Geometrically, this is equivalent to stating that the profit has increased. (The profit is zero at the origin.) The obvious conclusion is that the profit line should lie as far away from the origin as possible, but this has to be done with care. Take the line T_1, T_2, for example; all points on T_1, T_2 represent a profit of $600, but this line does not have a point within the feasible region. Therefore, there does not exist a feasible production schedule yielding a profit of $600 per day. What is needed is to find the profit line which, lying as far away as possible from the origin, still has at least one point within the feasible region. This line is V_1, V_2 and the maximal point, denoted optimum value in Figure 5.4, represents the solution to the linear programming problem. The profit at the optimum is $466.67 per day and the production schedule is $x_1 = 10$, $x_2 = \frac{4}{3}$ (4 type 2 units every 3 days). The production schedule was computed by solving simultaneously the equation of the line Q_1, Q_2 ($2x_1 + 3x_2 = 24$) with the equation of the vertical line through $x_1 = 10$. Then the maximum profit is computed, $40(10) + 50(\frac{4}{3}) = 466.67$.

The graphical solution of the problem has been completed. However, not only has this particular problem been solved, but also an entire class of problems generated simply by considering different values for the profits for type 1 and type 2 units. If the profit for type 1 units remains $40, for example, but the profit for type 2 units increases to $80, a new objective function; $40x_1 + 80x_2$, has to be considered.

Figure 5.5 shows the optimum condition for this new situation. Note that now the optimum value point lies at the intersection of line Q_1, Q_2 with the horizontal line through $x_2 = 7$. The new production schedule is $x_1 = \frac{3}{2}$ (3 type 1 units every 2 days) and $x_2 = 7$, and the new profit is $620 per day.

More type 2 units are manufactured in the new solution, at the expense of the number of type 1 units produced. This is reasonable because the profitability of type 2 units has doubled.

A remarkably interesting situation occurs when the profit margin of type 2 units is $60. The objective function to be maximized is now $40x_1 + 60x_2$, and the profit lines are parallel to Q_1, Q_2. Shifting the profit line away from the origin, one encounters a value for which it coincides with the line Q_1, Q_2 (see Figure 5.5). Further motion away from the origin yields profit lines which do not have any points within the feasible region. Therefore, the line Q_1, Q_2 is as far away as feasible from the origin. Each one of its points, including the two corner points of the region, yields an equal profit of $480 per day. Consequently, if the profit for type 1 units is $40 and for type 2 units it is $60, the optimum production program is *no longer unique*. An infinite number of solutions lead to the same maximum profit.

To conclude, for all the cases investigated, the solution lay on the boundary of the feasible region, not inside. This is invariably the case in linear programming. Furthermore, the solution usually occurs at a corner point, and it may, accidentally, consist of an entire line segment but, even in this case,

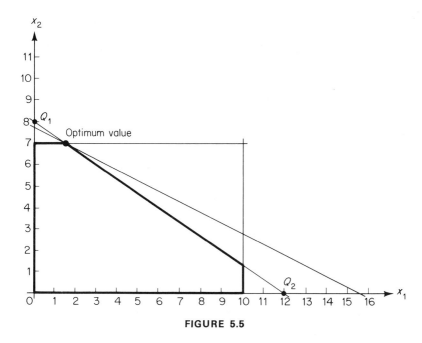

FIGURE 5.5

there exist two corner point solutions. In conclusion, whatever the unit profits are, a corner point of the feasible region is always included in the solution.

It is apparent that for problems involving more than two decision variables the graphical approach is unwieldy, at best, and for problems of four and higher dimensions an analytical solution is indispensable. In the sections that follow, the simplex method of solution will be developed in detail and example problems will be studied.

5.4 ANALYTICAL SOLUTION: THE SIMPLEX ALGORITHM

An algorithm is nothing more than a set of computations which can be performed over and over again following a cyclic pattern, and that because of its repetitive nature can be readily programmed for processing, either manually by an uninformed operator, or automatically in an electronic computer. The simplex algorithm provides an efficient method of solution to linear programming problems.

To continue the study of linear programming, it must be recalled that all problems of this type possess the following structure:

1. An objective function,

$$\text{Max } z = \bar{C}^t \bar{X} \tag{5.4}$$

(Maximization is assumed without loss in generality; see section 5.2.)

2. Constraints, which after the introduction of slack variables become

$$\bar{A}\bar{X} = \bar{B},$$ (5.5)

3. with the non-negativity conditions

$$\bar{X} \geq 0.$$ (5.6)

NOTE: The non-negativity conditions will be assumed implicit to every problem unless stated otherwise.

4. It will also be required that

$$\bar{B} \geq \bar{0}.$$ (5.7)

If necessary, this can be readily accomplished with multiplication by (-1) and with changes of direction in inequality signs.

5. The matrices, vectors, and scalars are

$$\text{structural coefficients, } \bar{A} = \begin{pmatrix} a_{11} & a_{12} & \cdots & a_{1n} \\ a_{21} & a_{22} & \cdots & a_{2n} \\ \cdots\cdots\cdots\cdots\cdots \\ a_{m1} & a_{m2} & \cdots & a_{mn} \end{pmatrix};$$ (5.8)

$$\text{stipulations, } \bar{B} = \begin{pmatrix} b_1 \\ b_2 \\ \cdot \\ \cdot \\ \cdot \\ b_m \end{pmatrix};$$ (5.9)

$$\text{cost coefficients, } \bar{C} = \begin{pmatrix} c_1 \\ c_2 \\ \cdot \\ \cdot \\ \cdot \\ c_n \end{pmatrix};$$ (5.10)

and

$$\text{structural and slack variables, } \bar{X} = \begin{pmatrix} x_1 \\ x_2 \\ \cdot \\ \cdot \\ \cdot \\ x_n \end{pmatrix}.$$ (5.11)

The slack variables used to transform the inequality constraints into equations are preceded by a positive sign for less than or equal to (type 1) inequalities and by a negative sign for more than or equal to (type 2) inequalities.

6. Finally, the student should review at this time the definitions given in chapter 4 for the following concepts:
   a. solution
   b. basic solution
   c. degenerate basic solution
   d. basic feasible solution
   e. degenerate basic feasible solution

Definition e was not explicitly given but follows directly from the previous ones.

5.4.1 Basic Theorems

The understanding of the simplex method of solution to linear programming problems hinges upon the complete understanding of four fundamental theorems, each of which will be discussed in detail and proved in the sections that follow.[1]

5.4.1.1 Theorem No. 1

Consider a linear programming problem with no degenerate basic feasible solutions, and assume that a basic feasible solution has been formed in terms of the first m variables. A new basic feasible solution can be formed by introducing the nonsolution variable x_k, $k > m$, if and only if at least one element of the kth column of the reduced matrix is positive.

Proof:

The restriction that degenerate basic feasible solutions do not exist causes no loss of generality for, as will be seen later, degeneracy can be avoided with a simple perturbation technique. Also, a basic feasible solution can always be formed with the introduction of slack and artificial slack variables. This technique will also be discussed later. Finally, any basic feasible solution can be formed in terms of the first m variables by renumbering the variables as needed, so that this restriction, as well, causes no loss of generality. To construct a basic feasible solution, $n - m$ of the variables must be equal to zero and one that has, in terms of the first m variables,

$$\bar{D}\bar{X}_m = \bar{B} \tag{5.12}$$

from

$$\bar{A}\bar{X} = \bar{B}, \tag{5.13}$$

---

[1] This four-theorem approach was adapted from *Linear Programming* by Robert W. Llewellyn. Copyright (c) 1964 by Holt, Rinehart & Winston, Inc. Adapted and used by permission of Holt, Rinehart & Winston, Inc. (See References.)

where

$$\bar{D} = \begin{pmatrix} a_{11} & a_{12} & \cdots & a_{1m} \\ a_{21} & a_{22} & \cdots & a_{2m} \\ \cdots\cdots\cdots\cdots\cdots\cdots \\ a_{m1} & a_{m2} & \cdots & a_{mm} \end{pmatrix}; \tag{5.14}$$

and

$$\bar{X}_m = \begin{pmatrix} x_1 \\ x_2 \\ \cdot \\ \cdot \\ \cdot \\ x_m \end{pmatrix}. \tag{5.15}$$

The solution to equation (5.12) is

$$\bar{X} = \bar{D}^{-1}\bar{B} \tag{5.16}$$

which can be obtained, thus,

$$(\bar{D}\,|\,\bar{B}) \sim (\bar{D}^{-1}\bar{D}\,|\,\bar{D}^{-1}\bar{B}) = (\bar{I}\,|\,\bar{B}') = \begin{pmatrix} 1 & 0 & \cdots & 0 & b_1' \\ 0 & 1 & \cdots & 0 & b_2' \\ \cdots\cdots\cdots\cdots\cdots & \cdots \\ 0 & 0 & \cdots & 1 & b_m' \end{pmatrix}, \tag{5.17}$$

from which the solution can be written

$$\bar{X}_m = \begin{pmatrix} x_1 \\ x_2 \\ \cdot \\ \cdot \\ \cdot \\ x_m \end{pmatrix} = \begin{pmatrix} b_1' \\ b_2' \\ \cdot \\ \cdot \\ \cdot \\ b_m' \end{pmatrix}. \tag{5.18}$$

Assume that one had augmented the starting matrix $(\bar{D}\,|\,\bar{B})$ with the kth column of $\bar{A}$, purely as a matrix operation. Calling this column

$$\bar{P}_k = \begin{pmatrix} a_{1k} \\ a_{2k} \\ \cdot \\ \cdot \\ \cdot \\ a_{mk} \end{pmatrix}, \tag{5.19}$$

obtain

$$(\bar{D}\,|\,\bar{P}_k\,|\,\bar{B}) \sim (\bar{D}^{-1}\bar{D}\,|\,\bar{D}^{-1}\bar{P}_k\,|\,\bar{D}^{-1}\bar{B}) = (\bar{I}\,|\,\bar{P}_k'\,|\,\bar{B}'), \tag{5.20}$$

where

$$(\bar{I} \mid \bar{P}_k \mid \bar{B}') = \begin{pmatrix} 1 & 0 & \cdots & 0 & a'_{1k} & b'_1 \\ 0 & 1 & \cdots & 0 & a'_{2k} & b'_2 \\ & & \cdots & & & \\ 0 & 0 & \cdots & 1 & a'_{mk} & b'_m \end{pmatrix}. \tag{5.21}$$

Setting $x_k = \Theta$, equation (5.21) is the set

$$\left.\begin{aligned} x_1 + a'_{1k}\Theta &= b'_1, \\ x_2 + a'_{2k}\Theta &= b'_2, \\ \vdots \\ x_m + a'_{mk}\Theta &= b'_m, \end{aligned}\right\} \tag{5.22}$$

with the condition $x_k = \Theta$. $\qquad$ (5.23)

After transposing the $a'_{ik} \Theta$ to the right members of the equations, obtain,

$$\bar{X}_{m+1} = \begin{pmatrix} x_1 \\ x_2 \\ \vdots \\ x_m \\ x_k \end{pmatrix} = \begin{pmatrix} b'_1 - a'_{1k}\Theta \\ b'_2 - a'_{2k}\Theta \\ \vdots \\ b'_m - a'_{mk}\Theta \\ \Theta \end{pmatrix}. \tag{5.24}$$

The last component of $\bar{X}_{m+1}$ reflects the fact that x_k is not zero. The new solution, $\bar{X}_{m+1}$, will be part of a basic feasible solution if Θ meets the following conditions:

$$\Theta = \min \frac{b'_i}{a'_{ik}}, \qquad a'_{ik} > 0. \tag{5.25}$$

Conditions (5.25) are known as the Θ rule.
Say that $i = 2$ meets this condition, then,

$$x_2 = b'_2 - a'_{2k}\Theta = b'_2 - a'_{2k}\frac{b'_2}{a'_{2k}} = 0, \tag{5.26}$$

Because

$$a'_{2k} > 0, \qquad \Theta = \frac{b'_2}{a'_{2k}} = x_k > 0 \tag{5.27}$$

which is necessary for feasibility.
Notice that x_2 leaves the solution, equation (5.26), and that x_k enters it with a positive value, equation (5.27).

Also, since

$$\Theta = \min \frac{b'_i}{a'_{ik}} = \frac{b'_2}{a'_{2k}},$$ (5.28)

then, for

$$j \neq 2, \qquad x_j = b'_j - a'_{jk}\Theta = b'_j - a'_{jk}\frac{b'_2}{a'_{2k}} > 0,$$ (5.29)

because

$$b'_j - a'_{jk}\frac{b'_j}{a'_{jk}} = 0$$

and

$$\frac{b'_j}{a'_{jk}} > \frac{b'_2}{a'_{2k}}.$$ (5.30)

Thus, if at least one element of the kth column of the reduced matrix, $\bar{P}_k$, is positive, a new basic feasible solution can be formed from the existing one. But if all components of $\bar{P}_k$ are negative, no positive Θ exists, and it is impossible to form a new basic feasible solution that includes x_k. This concludes the proof of Theorem No. 1.

5.4.1.2 Theorem No. 2

Consider a linear programming problem with no degenerate basic feasible solutions, with a unique optimal solution, and in which additional basic feasible solutions can be formed, then its maximal solution must be a basic feasible solution.

Proof:

Assume that all existing basic feasible solutions for this problem have been found and evaluated and that the one which maximizes the objective function was formed from the first m variables. This last assumption causes no loss in generality since the variables can be renumbered and rearranged to form the maximizing solution from the first m variables. Consider a second solution, one in which x_k has been introduced at a small positive value Θ. Because it was assumed that additional basic feasible solutions can be formed, x_k is a variable not in the first solution for which $\bar{P}_k$ satisfies Theorem No. 1. Assume that the following are the two solutions discussed, the first being the maximal basic feasible solution.

$$\bar{X}_m = \begin{pmatrix} x_1 \\ x_2 \\ \cdot \\ \cdot \\ \cdot \\ x_m \end{pmatrix} = \begin{pmatrix} b'_1 \\ b'_2 \\ \cdot \\ \cdot \\ \cdot \\ b'_m \end{pmatrix};$$ (5.31)

and

$$
\bar{X}_{m+1} = \begin{pmatrix} x_1 \\ x_2 \\ \cdot \\ \cdot \\ \cdot \\ x_m \\ x_k \end{pmatrix} = \begin{pmatrix} b'_1 - a'_{1k}\Theta \\ b'_2 - a'_{2k}\Theta \\ \cdot \\ \cdot \\ \cdot \\ b'_m - a'_{mk}\Theta \\ \Theta \end{pmatrix}. \tag{5.32}
$$

To prove the theorem by contradiction, assume that the *value* of $\bar{X}_{m+1}$ is *greater* than the value of $\bar{X}_m$.
The value of $\bar{X}_m$ is

$$
z = c_1 b'_1 + c_2 b'_2 + \cdots + c_m b'_m + 0 + 0 + \cdots + 0, \tag{5.33}
$$

and the value of $\bar{X}_{m+1}$ is

$$
z' = c_1(b'_1 - a'_{1k}\Theta) + c_2(b'_2 - a'_{2k}\Theta) + \cdots
$$
$$
+ c_m(b'_m - a'_{mk}\Theta) + 0 + 0 + \cdots + c_k\Theta + \cdots + 0. \tag{5.34}
$$

Let

$$
z_k = c_1 a'_{1k} + c_2 a'_{2k} + \cdots + c_m a'_{mk}, \tag{5.35}
$$

then

$$
z' = z + (c_k - z_k)\Theta. \tag{5.36}
$$

The assumption that

$$
z' > z \text{ implies that } (c_k - z_k) \text{ must be } > 0.
$$

Suppose Θ is increased until the conditions of Theorem 1 are fulfilled. This has been assumed possible for x_k. If this procedure yields the solution $\bar{X}'_m$, its value must be greater than that of $\bar{X}_m$, as z' must increase with Θ because $(c_k - z_k) > 0$. But $\bar{X}'_m$ must be a basic feasible solution and must have a value less than $\bar{X}_m$, because $\bar{X}_m$ is the maximum basic feasible solution. Hence, a contradiction has developed, and the assumption $z' > z$ must be false. The argument can be similarly extended to any number of additional variables, and the theorem has been proven.

5.4.1.3 Theorem No. 3

Assume that a linear programming problem was formulated and a basic feasible solution found. If a variable x_i, for which the Θ calculation can be performed under the Θ rule, leaves the solution, another solution can be generated which changes the value of the objective function by an amount that can be predetermined.

Proof:

It was previously defined that

$$z_k = c_1 a'_{1k} + c_2 a'_{2k} + \cdots + c_m a'_{mk}. \tag{5.37}$$

Equation (5.37) can be written

$$z_k = \sum_{i=1}^{m} c_i a'_{ik}. \tag{5.38}$$

From Theorem No. 2,

$$z' = z + (c_k - z_k)\Theta,$$

and from Theorem No. 1,

$$\Theta = \min \frac{b'_i}{a'_{ik}}, \qquad a'_{ik} > 0.$$

Therefore, one concludes that z', which is the value of the new feasible program, can in fact be determined before the new solution has been actually obtained.

Note that since the maximum solution is sought, it is logical to require that $z' > z$. Because z' depends upon which variable is chosen to be x_k, k could be selected so as to make the change in solution value, $z' - z$, as large as possible. This is not normally done, however, because it requires calculation of all possible Θ's. The most efficient procedure is to select the variable to go into the next solution according to the rule,

$$c_k - z_k = \max_j (c_j - z_j). \tag{5.39}$$

Equation (5.39) is called the rule of *steepest ascent*, and the variable x_k that satisfies it is brought into the next solution. Notice that Θ will be positive according to Theorem No. 1; therefore, the program will be improved as long as equation (5.39) is also positive. If, on the other hand, equation (5.39) is nonpositive, there is no variable that can be introduced into the solution to increase the program value. In this case, the *maximal* solution has already been found.

5.4.1.4 Theorem No. 4

Consider a linear programming problem with a basic feasible solution that for some x_k, not in solution, has a positive $c_k - z_k$ for which all elements of $\bar{P}'_k$ are nonpositive. Then there is *no* upper bound to the value of its objective function.

Proof:

Consider a basic feasible solution $\bar{X}_m$ and another feasible solution $\bar{X}_{m+1}$ which has been formed by introducing $x_k = \Theta$. Then,

$$\bar{X}_{m+1} = \begin{pmatrix} x_1 \\ x_2 \\ \cdot \\ \cdot \\ \cdot \\ x_m \\ x_k \end{pmatrix} = \begin{pmatrix} b'_1 - a'_{1k}\Theta \\ b'_2 - a'_{2k}\Theta \\ \cdot \\ \cdot \\ \cdot \\ b'_m - a'_{mk}\Theta \\ \Theta \end{pmatrix}. \tag{5.40}$$

If the value of $\bar{X}_m$ is z, the value of $\bar{X}_{m+1}$ is

$$z' = z + (c_k - z_k)\Theta. \tag{5.41}$$

With all the $a'_{ik} \leq 0$, all components of $\bar{X}_{m+1}$ remain positive as Θ is increased without limit so that $\bar{X}_{m+1}$ remains feasible for *all* $\Theta > 0$. Since $c_k - z_k$ is positive in equation (5.41), z' increases without limit as Θ is increased.

In real problems unboundness does not occur. Therefore, its presence usually reveals that a mistake has been made in formulation or in computation.

5.4.2 Computational Aspects

The computational aspects of the simplex method and the simplex tableau will be illustrated with the following example:

A developer has the alternative of building two-, three-, and four-bedroom houses. He wishes to establish the number of each, if any, that will maximize his profit, subject to the following constraints:

1. The total budget for the project cannot exceed \$9,000,000.
2. The total number of units must be at least 350 for the venture to be economically feasible.
3. The maximum percentages of each type, based on an analysis of the market, are:
   2-bedroom units, 20% of total
   3-bedroom units, 60% of total
   4-bedroom units, 40% of total

Note that the sum of the percentages exceeds 100, and this is as it should be for they represent upper bounds to the consumer preferences. The information that follows is also required in formulating the mathematical model.

4. Building costs—including land, architectural and engineering fees, landscaping, and so on—are
   2-bedroom unit, $20,000
   3-bedroom unit, $25,000
   4-bedroom unit, $30,000
5. Net profits after interest, taxes, and so on, are
   2-bedroom unit, $2,000
   3-bedroom unit, $3,000
   4-bedroom unit, $4,000

Solution Method:

Let x_1, x_2, x_3 be, respectively, the number of two-, three-, and four-bedroom houses. The profit derived from selling them is, in thousands of dollars,

$$z = 2x_1 + 3x_2 + 4x_3.$$

This constitutes the objective function that must be maximized. If there were no constraints imposed on the project, the answer, to optimize the profit, would be trivial; construct an infinite number of units of each type. However, a set of constraints is always present in this type of problem and for the particular illustration given here it is the following:

1. The total budget cannot exceed $9,000 (in thousands of dollars). This means, using the building costs given, that

   $$20x_1 + 25x_2 + 30x_3 \leq 9,000.$$

2. The total number of units must be at least 350,

   $$x_1 + x_2 + x_3 \geq 350.$$

3. Finally, the market preferences data given can be translated into the following set of constraint inequalities,

   $$x_1 \leq .2 \text{ (Total Number of Units)}$$
   $$x_2 \leq .6 \text{ (T.N.U.)}$$
   $$x_3 \leq .4 \text{ (T.N.U.)}$$

These, then, are the inequalities that define the problem under consideration. Introducing slack variables x_4 to x_8, the expressions become:

1. General budget:

   $$20x_1 + 25x_2 + 30x_3 + x_4 = 9,000. \tag{5.42}$$

2. Number of units:

$$x_1 + x_2 + x_3 - x_5 = 350. \tag{5.43}$$

Note that the total number of units (T.N.U.) is $350 + x_5$.

3. Market preferences data:

$$x_1 - .2x_5 + x_6 = 70, \tag{5.44}$$
$$x_2 - .6x_5 + x_7 = 210, \tag{5.45}$$
$$x_3 - .4x_5 + x_8 = 140. \tag{5.46}$$

A total of 5 equations in 8 unknowns must be investigated to select the configuration that maximizes the profit. This corresponds to a search of all possible combinations of 8 parameters taken 5 at a time, which can be shown to represent, $8!/5!(3!) = 56$ different possibilities. (The symbol (!) indicates the factorial operation.)

An exhaustive search of all cases would prove extremely inefficient and expensive since each trial involves the solution of a system of 5 equations in 5 unknowns.

5.4.2.1 Obtaining an Initial Basic Feasible Solution

Theorems 1, 2, and 3 explained how to optimize a linear programming problem once a basic feasible solution has been obtained. Therefore, it is absolutely essential to form the first basic feasible solution in order to initialize the algorithm. Recall that a basic feasible solution has the following characteristics:

1. In a system of m equations in n unknowns, $n > m$, $n - m$ of the unknowns (called the nonbasic variables) are set equal to zero.
2. The m variables left in solution (also called the basic variables) must each have a positive value for feasibility and nondegeneracy.

Because of this definition, the following rules are given to form the starting basic feasible solution:

RULE NO. 1:

If all the constraints are of the less than or equal to (type 1) inequalities, m slack variables must be introduced preceded by positive signs. Consequently, if all the structural variables are set equal to zero, the slack variables will form a basic feasible solution because of the required non-negativity of the vector of stipulations.

For example:

$$\text{Max } z = 5x_1 + 3x_2 + 2x_3,$$

subject to

$$3x_1 + 3x_2 + 2x_3 \leq 10,$$
$$x_1 - 3x_2 + x_3 \leq 5,$$
$$x_1 \qquad - 3x_3 \leq 12.$$

All inequalities are type 1. Therefore, after introduction of slack variables x_4, x_5, and x_6, the constraint set becomes

$$3x_1 + 3x_2 + 2x_3 + x_4 \qquad\qquad = 10,$$
$$x_1 - 3x_2 - x_3 \qquad + x_5 \qquad = 5,$$
$$x_1 \qquad - 3x_3 \qquad\qquad + x_6 = 12.$$

Setting $x_1 = x_2 = x_3 = 0$, the first basic feasible solution is

$$x_4 = 10,$$
$$x_5 = 5,$$
$$x_6 = 12.$$

Since the cost coefficients c_4, c_5, and c_6 are all equal to zero, the value of this solution is $z = 0$. Better solutions are obtained by applying the theory given in Theorems No. 1 through No. 4. This will be done in the following sections.

RULE NO. 2:

If some of the constraints are equations and/or greater than or equal to (type 2) inequalities, no slack variables are introduced for the equations (there is no slack in them), and the slack variables for the type 2 inequalities are preceded by a negative sign. Therefore, less than m positive slack variables are available. This does not permit the formation of the first basic feasible solution in terms of slack variables alone. To remedy this situation, introduce additional positive slack variables into the equations and type 2 inequalities. These are called *artificial* slack variables because they do not possess any physical meaning, as the other slack variables do. For this reason, one must be assured that all the artificial slack variables will be out of the final optimal solution. This is accomplished by introducing the artificial slack variables into the objective function with large, negative (in the case of maximization) cost coefficients. The negativity of the cost coefficients assures that each artificial slack must take on the value zero to maximize the objective function.

As an example, take

$$\text{Max } z = 5x_1 + 3x_2 + 2x_3,$$

subject to

$$3x_1 + 3x_2 + 2x_3 \leq 10,$$
$$x_1 - 3x_2 + x_3 = 5,$$
$$x_1 \qquad - 3x_3 \geq 12.$$

Introducing slack variables, obtain

$$3x_1 + 3x_2 + 2x_3 + x_4 \quad\quad = 10,$$
$$x_1 - 3x_2 + x_3 \quad\quad\quad = 5,$$
$$x_1 \quad\quad - 3x_3 \quad - x_5 = 12.$$

A basic feasible solution cannot be formed in terms of x_4 and x_5. Therefore, introduce the artificial slack variables x_6 and x_7 into the second and third equations, respectively. Thus,

$$3x_1 + 3x_2 + 2x_3 + x_4 \quad\quad\quad\quad = 10,$$
$$x_1 - 3x_2 + x_3 \quad\quad\quad + x_6 \quad = 5,$$
$$x_1 \quad\quad - 3x_3 \quad - x_5 \quad + x_7 = 12.$$

Now, setting $x_1 = x_2 = x_3 = x_5 = 0$, obtain the basic feasible solution

$$x_4 = 10,$$
$$x_6 = 5,$$
$$x_7 = 12.$$

To assure that x_6 and x_7 will not be in the final solution and because the objective function is to be maximized, form the new objective

$$\text{Max } z = 5x_1 + 3x_2 + 2x_3 - 100x_6 - 100x_7.$$

If for some unforeseen reason, either x_6 or x_7 or both remained in the final solution, the negative cost coefficients would have to be made larger, as large as required to drive both artificial slacks out of the optimal solution. For a minimization problem, the artificial slacks would take on large *positive* coefficients in the objective function. Returning now to the housing problem, observe that equation (5.43) was developed from a type 2 inequality; therefore, one artificial slack variable, x_9, must be introduced, and the problem finally becomes,

Large negative cost coefficient ⌐

$$\text{Max } z = 2x_1 + 3x_2 + 4x_3 - 1,000x_9, \quad\quad (5.47)$$

subject to

$$20x_1 + 25x_2 + 30x_3 + x_4 \quad\quad\quad\quad\quad\quad = 9,000, \quad (5.48)$$
$$x_1 + x_2 + x_3 \quad\quad - x_5 \quad\quad\quad + x_9 = 350, \quad (5.49)$$
$$x_1 \quad\quad\quad - .2x_5 + x_6 \quad\quad\quad = 70, \quad (5.50)$$
$$x_2 \quad\quad - .6x_5 \quad + x_7 \quad\quad = 210, \quad (5.51)$$
$$x_3 \quad - .4x_5 \quad\quad\quad x_8 \quad = 140. \quad (5.52)$$

By setting $x_1 = x_2 = x_3 = x_5 = 0$, the initial basic feasible solution is obtained.

$$x_4 = 9{,}000,$$
$$x_6 = 70,$$
$$x_7 = 210,$$
$$x_8 = 140,$$

and

$$x_9 = 350.$$

The iteration can now be started for an initial basic feasible solution has been found.

5.4.2.2 Organization of the Tableaux

Before proceeding to explain the format of Tableaux No. 1 through No. 5, it is convenient to recall the two fundamental rules of the simplex algorithm (equations [5.25] and [5.39]):

The Θ rule:

$$\Theta_i = \min \frac{b'_i}{a'_{ik}}, \qquad a'_{ik} > 0,$$

and the rule of steepest ascent:

$$c_k - z_k = \max_j (c_j - z_j),$$

where

$$z_j = c_1 a'_{1j} + c_2 a'_{2j} + \cdots + c_m a'_{mj}.$$

Tableau No. 1 in Table 5.1 is set up for the first iteration. Several remarks about its structure follow:

1. The first column, labeled c_i, is for the cost coefficients in the first basic feasible solution.
2. The column labeled "Sol." stands for solution and under it are listed the variables in the current solution. Each variable is listed on the row in which its "1" appears in its unit column vector which is part of the identity matrix.
3. The columns in the next set have $\bar{P}_i$ headings to represent the column vectors which correspond to the variables x_i. Under each $\bar{P}_i$ appear the structural coefficients a_{ij} associated with variables x_i. $\bar{P}_1$, $\bar{P}_2$, and $\bar{P}_3$ are structural vectors; $\bar{P}_4$ through $\bar{P}_8$ are slack variable vectors; and $\bar{P}_9$ is the aritificial slack vector.
4. The next column, $\bar{B}$, is the vector of stipulations.
5. The column labeled "Check" is used to verify the calculations in the other columns. The entries in this column are the totals of the other columns in the tableau, row for row. According to matrix theory, if the same row operations are used in this column as in other columns;

TABLE 5-1. Tableau No. 1.

C_j / C_i	Sol.	2 $\bar{P}_1$	3 $\bar{P}_2$	4 $\bar{P}_3$	0 $\bar{P}_4$	0 $\bar{P}_5$	0 $\bar{P}_6$	0 $\bar{P}_7$	0 $\bar{P}_8$	-1,000 $\bar{P}_9$	0 B	-991 Check	Θ_i
0	x_4	20	25	30	1	0	0	0	0	0	9,000	9,076.	300
-1,000	x_9	1	1	1	0	-1	0	0	0	1	350	353	350
0	x_6	1	0	0	0	-0.2	1	0	0	0	70	71.8	
0	x_7	0	1	0	0	-0.6	0	1	0	0	210	211.4	
0	x_8	0	0	(1)	0	-0.4	0	0	1	0	140	141.6	140 ←
	z_j	-1,000	-1,000	-1,000	0	1,000	0	0	0	-1,000	-350,000	-353,000	
	$c_j - z_j$	1,002	1,003	1,004	0	-1,000	0	0	0	0	350,000	352,009	

(↑ arrow under $\bar{P}_3$ column)

TABLE 5-2. Tableau No. 2.

C_j / C_i	Sol.	2 $\bar{P}_1$	3 $\bar{P}_2$	4 $\bar{P}_3$	0 $\bar{P}_4$	0 $\bar{P}_5$	0 $\bar{P}_6$	0 $\bar{P}_7$	0 $\bar{P}_8$	-1,000 $\bar{P}_9$	0 B	-991 Check	Θ_i
0	x_4	20	(25)	0	1	12	0	0	-30	0	4,800	4,828	192
-1,000	x_9	1	1	0	0	-0.6	0	0	-1	1	210	211.4	210
0	x_6	1	0	0	0	-0.2	1	0	0	0	70	71.8	
0	x_7	0	1	0	0	-0.6	0	1	0	0	210	211.4	210
4	x_3	0	0	1	0	-0.4	0	0	1	0	140	141.6	
	z_j	-1,000	-1,000	4	0	598.4	0	0	1,004	-1,000	-209,440	-210,833.6	
	$c_j - z_j$	1,002	1,003	0	0	-598.4	0	0	-1,004	0	209,440	209,842.6	

(↑ arrow under $\bar{P}_2$ column)

TABLE 5-3. Tableau No. 3.

C_j / C_i	Sol	2 / $\bar{P}_1$	3 / $\bar{P}_2$	4 / $\bar{P}_3$	0 / $\bar{P}_4$	0 / $\bar{P}_5$	0 / $\bar{P}_6$	0 / $\bar{P}_7$	0 / $\bar{P}_8$	-1,000 / $\bar{P}_9$	0 / B	-991 / Check	Θ_i
3	x_2	0.8	1	0	0.04	0.48	0	0	-1.2	0	192	193.12	240
-1,000	x_9	0.2	0	0	-0.04	-1.08	0	0	0.2	1	18	18.28	80
0	x_6	(1)	0	0	0	-0.2	1	0	0	0	70	71.80	70 ←
0	x_7	-0.8	0	0	-0.04	-1.08	0	1	1.2	0	18	18.28	
4	x_3	0	0	1	0	-0.4	0	0	1	0	140	141.60	
	z_j	-197.6	3	4	40.12	1,079.84	0	0	-199.6	-1,000	-16,864	-17,134.34	
	$c_j - z_j$	199.6	0	0	-40.12	-1,079.84	0	0	199.6	0	16,864	16,143.24	

←

TABLE 5-4. Tableau No. 4.

C_j / C_i	Sol	2 / $\bar{P}_1$	3 / $\bar{P}_2$	4 / $\bar{P}_3$	0 / $\bar{P}_4$	0 / $\bar{P}_5$	0 / $\bar{P}_6$	0 / $\bar{P}_7$	0 / $\bar{P}_8$	-1,000 / $\bar{P}_9$	0 / B	-991 / Check	Θ_i
3	x_2	0	1	0	0.04	0.64	-0.8	0	-1.2	0	136	135.68	
-1,000	x_9	0	0	0	-0.04	-1.04	-0.2	0	(0.2)	1	4	3.92	20 ←
2	x_1	1	0	0	0	-0.2	1	0	0	0	70	71.80	
0	x_7	0	0	0	-0.04	-1.24	0.8	1	1.2	0	74	75.72	61.66
4	x_3	0	0	1	0	-0.4	0	0	1	0	140	141.60	140
	z_j	2	3	4	40.12	1,039.92	199.6	0	-199.6	-1,000	-2,892	-2,802.96	
	$c_j - z_j$	0	0	0	-40.12	-1,039.92	-199.6	0	199.6	0	2,892	1,811.96	

←

TABLE 5-5. Tableau No. 5, Optimal Solution.

C_j / C_i	Sol.	2 $\bar{P}_1$	3 $\bar{P}_2$	4 $\bar{P}_3$	0 $\bar{P}_4$	0 $\bar{P}_5$	0 $\bar{P}_6$	0 $\bar{P}_7$	0 $\bar{P}_8$	$-1,000$ $\bar{P}_9$	0 B	-991 Check	Θ_i
3	x_2	0	1	0	-0.2	-5.6	-2.0	0	0	6	160	159.2	
0	x_8	0	0	0	-0.2	-5.2	-1	0	1	5	20	19.6	
2	x_1	1	0	0	0	-0.2	1	1	0	0	70	71.8	
0	x_7	0	0	1	0.2	5.0	2.0	1	0	-6	50	52.2	
4	x_3	0	3	4	0.2	4.8	1	0	0	-5	120	122.0	
	z_j	2	3	4	0.2	2.0	0	0	0	-2	1,100	1,109.2	
	$c_j - z_j$	0	0	0	-0.2	-2.0	⬚0	0	0	-998	-1,100	-2,100.2	

the elements of the check column will equal the row sums at each
iteration.
6. The last column serves to record the Θ_i values when that part of the
iteration is reached.
7. The last two rows are appended for convenience in recording the z_j
and the $c_j - z_j$ computations. Note that the check column is applied
to both the z_j and the $c_j - z_j$ rows.
8. The top row simply lists the cost coefficients for each vector. The
check column is also applied to this row.

The calculations for the housing problem are shown in detail in Tables 5–1
through 5–5. They should be followed closely by the reader.

5.4.2.3   Comments on the Tableaux

The simplex algorithm assumes that a basic feasible solution has been
formed in terms of the first m variables in order to initialize the computations.
This was done with the introduction of slack and artificial slack variables as
described in section 5.4.2.1. The following steps were taken in filling in
Tableau No. 1.

1. The five constraint equations (5.48) through (5.52) are entered in the
tableau with the coefficients shown in the proper columns. For example,
equation (5.48) states:

$$20x_1 + 25x_2 + 30x_3 + x_4 = 9,000.$$

Therefore, the number 20 is entered under the $\bar{P}_1$ column, 25 under $\bar{P}_2$, 30
under $\bar{P}_3$, 1 under $\bar{P}_4$, and 0's under $\bar{P}_5$ through $\bar{P}_9$ because the equation has
no terms in these variables. Finally, 9,000 is entered in $\bar{B}$, the column of
stipulations. The same procedure is repeated for the other four equations.

2. Next, the check column is filled in. For example, the first entry, 9,076,
is computed, thus

$$9,076 = 20 + 25 + 30 + 1 + 0 + 0 + 0 + 0 + 0 + 9,000.$$

Each check column element is always equal to the sum of the corresponding
row elements in columns $\bar{P}_1$ through $\bar{B}$.

3. Note that a canonical form (identity matrix) exists in terms of vectors
$\bar{P}_4, \bar{P}_6, \bar{P}_7, \bar{P}_8$, and $\bar{P}_9$. This is the initial basic feasible solution. As stated
previously, the solution column is filled with the variables in the current solu-
tion, that is, x_4, x_6, x_7, x_8, and x_9. Each variable appears on the row in which
its 1 is located in its unit column vector. Thus, since the 1 in $\bar{P}_4$ appears in
the first row, that is where x_4 is listed. The 1 in $\bar{P}_9$ appears in the second row,
and so on.

4. The cost coefficient associated with each solution variable is then listed in order in the column labeled c_j. This arrangement of the solution variables is tantamount to a renumbering on the subscripts, so that the basic feasible solution is formed in terms of the first five variables as required by the theory.

5. Since $z_j = c_1 a'_{2j} + c_2 a'_{2j} + \cdots + c_m a'_{mj}$, the z_j row is calculated simply by taking each element in the c_i column, multiplying it by the corresponding element in the $\bar{P}_j$ column and adding together all the products thus formed. Consequently, z_1 is computed,

$$-1{,}000 = 0 \times 20 + (-1{,}000) \times 1 + 0 \times 1 + 0 \times 0 + 0 \times 0,$$

and so on for the other elements in this row. The z_j value corresponding to column $\bar{B}$ is, in fact, the value of the objective function for this tableau because the elements of $\bar{B}$ are the values of the x_i variables in the present solution.

6. The $c_j - z_j$ row is calculated simply by taking the c_j value in the top row and subtracting from it the corresponding z_j. For example, $c_1 - z_1$ is computed thus:

From the top row, $c_1 = 2$.

From the z_j row, $z_1 = -1{,}000$.

Consequently, $c_1 - z_1 = 2 - (-1{,}000) = 1{,}002$.

Similarly for all the other elements in this row.

7. According to the rule of steepest ascent, the variable entering the next solution should have the maximum $c_j - z_j$ value, because from equation (5.39) in Theorem No. 3,

$$c_k - z_k = \max_j (c_j - z_j).$$

Therefore, x_3 will be next to come into solution because $c_3 - z_3 = 1{,}004$ is the maximum in the row. This is indicated with an arrow and the box framing column $\bar{P}_3$.

8. Now that it is known that x_3 will be the incoming variable, the outgoing variable must be selected according to the θ rule which, from equation (5.25), Theorem No. 1, states

$$\theta_i = \min \frac{b'_i}{a'_{ik}}, \qquad a'_{ik} > 0.$$

These values are entered in the θ_i column as follows:

Take each element in the $\bar{B}$ column and divide it by the corresponding element in the $\bar{P}_3$ column (the column for x_3, the entering variable). Since the Θ rule requires that $a'_{ik} > 0$, this computation is defined only if the corresponding value in the $\bar{P}_3$ column is positive. Thus, the first entry in θ_i is computed,

$$300 = \frac{9,000}{30}.$$

The second,

$$350 = \frac{350}{1}.$$

The next two positions are empty because the corresponding elements in the $\bar{P}_3$ column are zeros.

Finally,

$$140 = \frac{140}{1}.$$

9. The θ rule requires that the minimum value in the θ_i column be selected for the outgoing variable. Therefore, x_8 is the outgoing variable. The element lying at the intersection of the x_8 row with the $\bar{P}_3$ column is circled because this is the pivotal element that must be brought into solution.

10. The exchanging of variables is accomplished by performing row transformations such that a unit vector develops in column $\bar{P}_3$ with the 1 in the circled position. Because this position happens to be a number 1 already in this instance, the row transformations are easily performed, resulting in Tableau No. 2 of Table 5.2.

11. These computations are repeated until all $c_j - z_j \leq 0$. For, in this case, the objective cannot be increased any further and the maximum has been reached. This is the situation encountered in Tableau No. 5 of Table 5.5. The computations described constitute the simplex algorithm which can be formally stated to consist of the following steps:

5.4.2.4 The Simplex Algorithm

STEP I.

Calculate $c_j - z_j$ for each variable not in the present solution, according to the procedure given in Theorem No. 3.

1. If, for at least one j, $c_j - z_j > 0$ and at least one $a'_{ij} > 0$ for that j, a better feasible solution is possible.
2. If, for at least one j, $c_j - z_j > 0$ but all $a'_{ij} \leq 0$ for that j, the objective function is unbounded (see Theorem No. 4).
3. If $c_j - z_j \leq 0$ for all j, the maximum solution has been found.

STEP II:

If I(1) is the case, identify the variable with the largest $c_j - z_j$ as x_k. Determine θ by equation (5.25)

$$\Theta = \min \frac{b'_i}{a'_{ik}}, \qquad a'_{ik} > 0.$$

Call the variable that will be reduced to zero by the θ rule x_r. The element a'_{rk} is called the pivotal element.

STEP III.

Divide the rth row through by a'_{rk} to reduce a'_{rk}, the corresponding element in the next tableau, to 1. Then perform the row operations that will reduce all the a'_{ik} to zero.

STEP IV:

Repeat steps I, II, and III until, at some iteration, condition I(3) is reached. Then the maximum solution has been found.

This is the simplex algorithm. The theory assures that the solution at each tableau is of greater or, at worst, equal value to the one before. The algorithm converges to the optimum solution in a finite number of iterations.

5.4.2.5   The Optimal Solution

The optimal solution can be read directly from Table 5.5, Tableau No. 5:

$$x_1 = 70,$$
$$x_2 = 160,$$
$$x_3 = 120,$$

and

$$z = 1,100.$$

The value of the objective function is found on the z_j row in the B column of each tableau. Therefore, the solution states: Construct 70 two-bedroom units, 160 three-bedroom units, and 120 four-bedroom units, for a total maximum profit of $1,100,000. A check of the constraints reveals:

a) *Budget* $= 1,400,000 + 4,000,000 + 3,600,000 = 9,000,000$, O.K.
b) *No. of units* $= 70 + 160 + 120 = 350$, O.K.
c) *Market Preferences:*

$$x_1 = \tfrac{70}{350} \times 100 = 20\%, \text{ O.K.};$$
$$x_2 = \tfrac{160}{350} \times 100 = 45.7\%, \text{ O.K.};$$
$$x_3 = \tfrac{120}{350} \times 100 = 34.4\%, \text{ O.K.}$$

The computations that have just been presented were carried out in a desk calculator. They can, however, be readily programmed for solution in a digital computer. Many programs exist and are available at computation centers for immediate implementation. The user should be completely aware of the capabilities of the program he intends to use and of its limitations for, otherwise, he may be making himself vulnerable to all sorts of trouble and even to catastrophic numerical results.

5.4.3 Alternate Optima

According to the theory developed in Theorem No. 2, equation (5.36), when x_k is brought into solution the change in the value of the objective function is given by

$$z' - z = (c_k - z_k)\theta. \qquad (5.53)$$

Suppose that in the final tableau variable x_q is not in the optimal solution, and that for this variable

$$c_q - z_q = 0. \qquad (5.54)$$

Then the change in the objective function would be zero if x_q were brought into the solution. Since for the final table all $c_j - z_j \leq 0$, the solution has been maximized, and since bringing x_q into solution would not change the value of the objective function, the problem is said to have alternate optima. This is the n-dimensional analog of the two-dimensional case where the objective function is parallel to one of the constraints. More specifically, here the objective function hyperplane is parallel to one of the constraint hyperplanes.

Observation of Tableau No. 5 in Table 5.5 reveals that although x_6 is not in the optimal solution,

$$c_6 - z_6 = 0.$$

Therefore, alternate optima exist for this problem—that is, for the same maximum profit of $\$1,100(10)^3$, there exist many (in general, an infinite number) of policies that yield the same profit. From the physical standpoint, this is a very desirable situation for it provides great flexibility to the decision maker.

To determine all alternate optimal policies, proceed as follows:

1. Using equation (5.32) with the information provided in column $\bar{P}_6$ of the final tableau (Tableau No.5), form the vector

$$\bar{X}_6 = \begin{pmatrix} x_2 \\ x_8 \\ x_1 \\ x_7 \\ x_3 \\ x_6 \end{pmatrix} = \begin{pmatrix} 160 + 2.0\theta_6 \\ 20 + 1.0\theta_6 \\ 70 - 1.0\theta_6 \\ 50 - 2.0\theta_6 \\ 120 - 1.0\theta_6 \\ \theta_6 \end{pmatrix}. \qquad (5.55)$$

2. According to the θ rule, θ_6 can vary in the following range,

$$0 \leq \theta_6 \leq 25, \qquad (5.56)$$

where $25 = 50/2.0$ is the minimum b_i'/a_{iq}' as required. This means that

$$160 \leq x_2 \leq 210,$$
$$20 \leq x_8 \leq 45,$$
$$45 \leq x_1 \leq 70,$$
$$0 \leq x_7 \leq 50, \quad \text{this is the outgoing variable,}$$

and

$$95 \leq x_3 \leq 120.$$

3. Since θ_6 can take on an infinite number of values between 0 and 25, there are, in general, an infinite number of alternate optima. However, there are only 26 integer values for θ_6, that is,

$$\theta_6 = 0, 1, 2, 3, \ldots, 25. \tag{5.57}$$

Therefore, if only integer values are admissible, there are only 26 alternate optimal policies. In chapter 7, special cases where only discrete values of the variables are admissible will be studied under the heading of integer programming.

Finally, it must be restated that all of these alternate optima yield, of course, the same maximum value of the objective function: a profit of $\$1,100(10)^3$.

5.5 MINIMIZATION

A parallel development of the theory would result in a similar algorithm for minimization. However, rather than having two separate algorithms, one for maximization and another for minimization, it is preferable to maximize the negative of the objective function in a minimization problem. This makes the theory developed adequate to handle all linear programming problems.

5.5.1 Example:

Consider,

$$\text{Min } y = 3x_1 + 4x_2 + 7x_3,$$

subject to

$$x_1 - 3x_2 + x_3 \geq 7,$$
$$2x_1 + x_2 - x_3 \geq 9,$$

and

$$x_1 \geq 0, \; x_2 \geq 0, \; x_3 \geq 0.$$

This problem is equivalent to the following:

$$\text{Max } z = -y = -3x_1 - 4x_2 - 7x_3,$$

subject to

$$x_1 - 3x_2 + x_3 \geq 7,$$
$$2x_1 + x_2 - x_3 \geq 9,$$

and

$$x_1 \geq 0,\ x_2 \geq 0,\ x_3 \geq 0.$$

One must be careful to note that the cost coefficients must appear with a negative sign in the simplex tableaux.

5.6 DEGENERACY

Degeneracy is caused by having one or more zeros in solution at any time or, equivalently, by having less than m positive variables in a basic feasible solution involving m constraints. The only concern with degenerate solutions is that they can develop cycling loops (visiting the same solutions indefinitely without ever arriving at the optimum). However, in real life problems, cycling due to degeneracy has never been observed and, consequently, it will not be paid a great deal of attention in this text.

If it is anticipated that degeneracy can become a problem, a small perturbation or change in one or more of the stipulations usually avoids it. This perturbation method for avoiding degenerate solutions will be presented in some detail in connection with the transportation problem of chapter 8.

5.7 TWO-PHASE SIMPLEX ALGORITHM

Section 5.4.2.1 discussed one method of handling artificial slack variables by introducing them with large, negative cost coefficients into the objective function of a maximization problem.

The following is an alternative method for obtaining a basic feasible solution when artificial slack variables are required. Consider the linear programming problem,

$$\text{Max } z = \sum_{i=1}^{n} c_i x_i,$$

subject to

$$x_i \geq 0 \text{ for } i = 1, \ldots, n,$$

$$\sum_{i=1}^{n} a_{ki} x_i \leq b_k \text{ for } k = 1, \ldots, f,$$

$$\sum_{i=1}^{n} a_{ki} x_i \geq b_k \text{ for } k = f + 1, \ldots, g,$$

and

$$\sum_{i=1}^{n} a_{ki} x_i = b_k \text{ for } k = g + 1, \ldots, m.$$

(5.58)

In this problem there are f type 1 inequalities, $(g - f)$ type 2 inequalities, and $(m - g)$ equations constraining the objective function. Therefore, one must introduce f positively signed slack variables $(+x_{n+k})$, $k = 1, \ldots, f$; $(g - f)$ negatively signed slack variables $(-x_{n+k})$, $k = f + 1, \ldots, g$; and $(m - f)$ positively signed artificial slack variables $[+\mu_{(k-f)}]$, $k = f + 1, \ldots, m$.

Having done this, the problem becomes

$$\text{Max } z = \sum_{i=1}^{n} c_i x_i,$$

subject to $x_i \geq 0$ for all i,

$$\mu_{(k-f)} \geq 0 \text{ for } k = f + 1, \ldots, m,$$

$$\sum_{i=1}^{n} a_{ki} x_i + x_{n+k} = b_k \text{ for } k = 1, \ldots, f,$$

$$\sum_{i=1}^{n} a_{ki} x_i - x_{n+k} + \mu_{(k-f)} = b_k \text{ for } k = f + 1, \ldots, g,$$

and

$$\sum_{i=1}^{n} a_{ki} x_i + \mu_{(k-f)} = b_k \text{ for } k = g + 1, \ldots, m.$$

Rather than introducing into the objective function artificial slack variables multiplied by large, negative cost coefficients, use the following two-phase simplex algorithm:

PHASE I:

Solve the linear programming problem,

$$\text{Min } y = \sum_{k=f+1}^{m} \mu_{(k-f)},$$

subject to

$$\mu_{(k-f)} \geq 0 \text{ for } k = f + 1, \ldots, m,$$

$$x_i \geq 0 \text{ for all } i,$$

$$\sum_{i=1}^{n} a_{ki} x_i + x_{n+k} = b_k \text{ for } k = 1, \ldots, f, \tag{5.59}$$

$$\sum_{i=1}^{n} a_{ki} x_i - x_{n+k} + \mu_{(k-f)} = b_k \text{ for } k = f + 1, \ldots, g,$$

and

$$\sum_{i=1}^{n} a_{ki} x_i + \mu_{(k-f)} = b_k \text{ for } k = g + 1, \ldots, m.$$

Because $\mu_{(k-f)} \geq 0$, Min $y = 0$. If such a minimum exists, a first basic feasible solution to the original problem (5.58) also exists.

PHASE II:

After a first basic feasible solution has been found, drop the artificial slack variables $\mu_{(k-f)}$ and optimize the original objective function in (5.58) using the simplex algorithm as usual.

Example:

$$\text{Min } z = 2x_1 + 3x_2,$$

subject to

$$x_1 \geq 0,$$
$$x_2 \geq 0,$$
$$x_1 + x_2 \geq 3,$$

and

$$x_1 + 2x_2 \geq 4.$$

Solution:

Proceed as follows with the two-phase algorithm:

PHASE I:

$$\text{Min } y = \mu_1 + \mu_2,$$

subject to

$$\mu_1 \geq 0,$$
$$\mu_2 \geq 0,$$
$$x_1 \geq 0,$$
$$x_2 \geq 0,$$
$$x_3 \geq 0,$$
$$x_4 \geq 0,$$
$$x_1 + x_2 - x_3 + \mu_1 = 3,$$
$$x_1 + 2x_2 - x_4 + \mu_2 = 4.$$

The optimization is performed in Table 5–6 with the transformed objective

$$\text{Max } -y = -\mu_1 - \mu_2.$$

This transformation was required to use the maximization algorithm. The optimum was reached after two iterations.

PHASE II:

Since μ_1 and μ_2 are no longer in solution, they can be dropped and the original objective function

$$\text{Min } z = 2x_1 + 3x_2,$$

or

$$\text{Max } -z = -2x_1 - 3x_2,$$

optimized as usual. This is shown in Table 5.7, where it is apparent that with the new cost coefficients,

$$c_j - z_j \leq 0 \text{ for } j = 1, \ldots, 4.$$

TABLE 5–6. Phase I.

c_i	Sol.	c_j 0 $\bar{P}_1$	0 $\bar{P}_2$	0 $\bar{P}_3$	0 $\bar{P}_4$	-1 $\bar{\mu}_1$	-1 $\bar{\mu}_2$	0 B	-2 Check	Θ_i
-1	μ_1	1	$\boxed{1}$	-1	0	1	0	3	5	3
-1	μ_2	1	$\circled{2}$	0	-1	0	1	4	7	2 ←
	z_j	-2	-3	1	1	-1	-1	-7	-12	
	$c_j - z_j$	2	3 ↑	-1	-1	0	0	7	10	
-1	μ_1	$\circled{\frac{1}{2}}$	0	-1	$\frac{1}{2}$	1	$-\frac{1}{2}$	1	$1\frac{1}{2}$	2 ←
0	x_2	$\frac{1}{2}$	1	0	$-\frac{1}{2}$	0	$\frac{1}{2}$	2	$3\frac{1}{2}$	4
	z_j	$-\frac{1}{2}$	0	1	$-\frac{1}{2}$	-1	$\frac{1}{2}$	-1	$-1\frac{1}{2}$	
	$c_j - z_j$	$\frac{1}{2}$ ↑	0	-1	$\frac{1}{2}$	0	$-1\frac{1}{2}$	+1	$-\frac{1}{2}$	
0	x_1	1	0	-2	1	2	-1	2	3	
0	x_2	0	1	1	-1	-1	1	1	2	
	z_j	0	0	0	0	0	0	0	0	
	$c_j - z_j$	0	0	0	0	-1	-1	0	-2	

TABLE 5–7. Phase II.

c_i	Sol.	c_j -2 $\bar{P}_1$	-3 $\bar{P}_2$	0 $\bar{P}_3$	0 $\bar{P}_4$	0 B	-5 Check	Θ_i
-2	x_1	1	0	-2	1	2	2	
-3	x_2	0	1	1	-1	1	2	
	z_j	-2	-3	1	1	-7	-10	
	$c_j - z_j$	0	0	-1	-1	7	5	

Therefore, the optimization has been accomplished with the following results:

$$\text{Min } z = 7,$$
$$x_1 = 2, \qquad (5.60)$$

and

$$x_2 = 1.$$

The advantage of the two-phase method over the first one proposed is that, if Phase I cannot be successfully completed, this indicates at the outset that a basic feasible solution does not exist for the problem. Such warning is not encountered when using the method of large coefficients. Chapter 6 discusses in detail two examples of application of the linear programming methodology to planning situations.

EXERCISES

5–1. The automobile travel time through existing arterial roads between two towns A and B is at least two hours. To reduce total travel time, an expressway will be constructed linking A to C, an intermediate point between A and B. The time to travel on the new expressway from A to C is to be one-third of the original time on the arterial roads between A and B. Due to limitations in construction funds, the length AC along the expressway must be restricted in such a manner that the new travel time on the expressway should be less than or equal to one half of the travel time on the arterial roads between C and B. Determine graphically the minimum possible travel time between A and B if the expressway is constructed under the given constraints. (See figure below.)

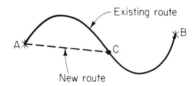

5–2. A chemical company has to produce two million pounds of a mixture of two raw materials. Material costs per pound are the following:
Raw material A . . . $6/pound,
Raw material B . . . $5/pound.
The client requires the mixture to meet the following specifications:
a. The mixture must contain at least 25% of material A.
b. The mixture cannot contain more than 65% of material A.
c. The mixture must contain at least 30% of material B.
    i. Determine graphically the least-cost mixture for the batch of 2,000,000 pounds which will satisfy the customer's specifications.
    ii. Solve with the simplex algorithm.

5–3. A contractor has the finances to build 50 two-bedroom, 100 three-bedroom, and 80 four-bedroom homes. He owns land in Baton Rouge and New Orleans, Louisiana. His Baton Rouge property is large enough for 180 houses and his New Orleans land can accom-

Type	Baton Rouge	New Orleans
2-bedroom	$2,000	$2,500
3-bedroom	$3,000	$3,000
4-bedroom	$4,500	$4,000

modate 120. Set up an initial basic feasible solution for the simplex tableau to determine the number of units of each kind that the contractor should build in each city to maximize his profit according to the profit table on page 132.

5-4. A care food program is to supply at least half of the minimum adult daily requirements of thiamine (1.0 mg.) and of niacin (10 mg.) in breakfast cereal.
Two breakfast foods are competitive:

Cereal	Thiamine per oz.	Niacin per oz.	Cost per oz.
A	0.08 mg.	1.00 mg.	2.5 cents
B	0.12 mg.	0.60 mg.	3.0 cents

Determine the minimum cost mix of cereals that will provide the nutritional value specified above.
a. Solve graphically.
b. Use simplex algorithm.

5-5. In order to assure adequate stability under load repetition, a soil mixture for base and sub-base courses in the construction of a certain highway must have a Liquid Limit, $21 \leq$ L.L. ≤ 28, and a Plasticity Index, $4 \leq$ P.I. ≤ 6.
Two materials, A and B, are available as follows:

Properties	A	B
L.L.	35	20
P.I.	8	3.5
Cost ($/cu. yd.)	$.35	$.65

Assume that the L.L. and the P.I. are linear functions of the combinations of the two materials A and B and determine the optimal proportions, X_A and X_B, of these materials in the mix to minimize cost of construction of base and sub-base.
a. Use graphical analysis.
b. Use the simplex algorithm.

5-6. A student has 32 hours available to prepare for final exams in two subjects. In subject A he has to study 20 chapters of the textbook, but only 15 in subject B. He considers he needs one hour of study per chapter on subject A and two hours per chapter on B. The final exam will count as 50% of the final grade in each course and his average going into the finals is 90% in A and 70% in B. Both courses have the same number of credit hours. He assumes that his final grade in each course will equal the percentage of chapters he has been able to study for each subject—that is, if he studies 10 chapters in subject A, he can expect a 50% mark in his final exam in subject A. How many hours should he spend in each subject such that the sum of the final grades is a maximum, and such that no grade is below the 65% passing mark for each subject?

a. Develop the model.
b. Solve graphically.
c. Use the simplex method.

5–7. The Fair County School Board wishes to combat "de facto" segregation (otherwise, they can kiss federal funds goodbye) by busing children from one school district to another, as necessary, to develop a racially balanced enrollment plan for its grammar school system. Table 5–8 gives a distribution of grammer school children according to district and race.

TABLE 5–8. Children's Distribution by District and Race (No. of Children).

	DISTRICT				
Race	1	2	3	Totals by Race	%
Blue	900	100	0	1,000	40
Green	200	600	100	900	36
Purple	100	100	400	600	24
Totals by District	1,200	800	500	2,500 (Total Children)	

The table shows that 40% of the 2,500 children are blue (highly desirable color, a great deal of prestige and money are associated with this skin pigment), 36% are green (not quite as desirable, but socially acceptable), and 24% are purple (poor kids!). Table 5.9 gives capacities at each of the four existing schools which are geographically arranged as shown in the map of Figure 5.6.

TABLE 5–9. School Capacities.

	CAPACITY	
School No.	Max.	Min.
1	500	400
2	800	600
3	700	500
4	700	600

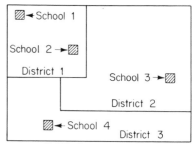

FIGURE 5.6. County Map.

The school board wishes to institute a busing policy which, at a minimum cost, will maintain a racially mixed attendance with 40%

blue, 36 % green, and 24 % purple children at each school. Transportation costs from district i to school j and back, in $ per child per year, are given in Table 5.10.

TABLE 5–10. Transportation Costs in $ Per Child Per Year.

		School j			
		1	2	3	4
District	1	45	48	65	95
i	2	56	48	47	65
	3	100	92	75	52

Develop a mathematical model for busing the kids at a minimum transportation cost to maintain racial balance at each school. In your model, let X_{ij}^k be the number of children of race k, living in district i, attending school j.

5–8. The A.D.H. construction company uses two classes of workers, A and B. Work can be classified into three major types, 1, 2, and 3. Type 1 work can be performed by class A workers alone, or by teams of one class A and two class B workers. Type 2 work can be performed by either class A or class B people working alone. Type 3 work requires teams formed by five class B workers or teams of one class A man supervising three class B workers.

Wages are $10 and $5 per hour for class A and B workers, respectively. Everybody works 40 hours per week, but A and B workers actually yield the weekly equivalent of 40 and 30 productive hours, respectively. The company requires a weekly total of 10,000 productive hours of type 1 work, 20,000 productive hours of type 2 work, and 30,000 productive hours of type 3 work to fulfill its construction schedules. Labor shortages restrict hiring to no more than 400 class A and 800 class B workers. Compute the number of workers of each class that should be employed to minimize the total labor cost per week.

a. Develop the model.

b. Set up the simplex tableau and perform the first iteration *only*.

NOTE: Data on teams is summarized in the table given below:

Team	No. of Class A Workers	No. of Class B Workers	Work Type
1	1	0	I
2	1	2	I
3	1	0	II
4	0	1	II
5	0	5	III
6	1	3	III

5–9. Three types of items must be loaded into a space capsule. The table below gives their weights and values.

Item	Weight	Value
1	4 lbs.	7
2	2 lbs.	4
3	1 lbs.	1

The capsule's payload available for all of these items is 12 pounds, and no more than 2 items of each kind can be used. Determine the loading plan that will maximize the total value. Use the simplex algorithm.

5–10. A manufacturing company produces two types of automobiles. The table given below shows pertinent production data.

Automobile Type	Profit per Unit	Production Belt Capacity	Manufacturing Hours
1	$600	80 units/month	30 hrs./unit
2	$800	60 units/month	40 hrs./unit

The available number of production hours is 3,600/month.
   a. Determine graphically the production mix (or mixes) that maximizes monthly profit and compute the maximum profit.
   b. Use the simplex method.
   c. List all the integer production mixes that maximize the profit. Is there more than one such mix?

5–11. The owner of a new car dealership wishes to invest no more than $20,000 in the purchase of three new models:

   Model A costs $2,000 and sells for $2,250.
   Model B costs $4,000 and sells for $4,900.
   Model C costs $3,000 and sells for $3,600.

The owner wishes to have at least one and no more than three of each model. Determine the buying strategy to maximize profit.

REFERENCES

1. AU, T. and T. E. STELSON, *Introduction to Systems Engineering, Deterministic Models.* Reading, Mass.: Addison-Wesley, 1969.

2. GASS, S. I., *Linear Programming: Methods and Applications.* New York: McGraw-Hill Book Company, 1958.

3. HADLEY, G., *Linear Programming*. Reading, Mass.: Addison-Wesley, 1962.

4. LLEWELLYN, R. W., *Linear Programming*. New York: Holt, Rinehart & Winston, Inc., 1964.

5. REINFELD, N. V. and W. R. VOGEL, *Mathematical Programming*. Englewood Cliffs, N.J.: Prentice-Hall, Inc., 1958.

6. SASIENE, M., A. YASPAN, and L. FRIEDMAN, *Operations Research—Methods and Problems*. New York: John Wiley & Sons, Inc., 1958.

7. WILDE, D. J. and C. S. BEIGHTLER, *Foundations of Optimization*. Englewood Cliffs, N.J.: Prentice-Hall, Inc., 1967.

CHAPTER **6**

Linear Programming
Applications
to Planning

Two examples of the application of linear programming models to architecture and planning are given in this chapter. These examples come from papers originally published by the author in collaboration with Messrs. James E. Hand and Raul S. Gonzalez. The complete references appear at the end of the chapter.

6.1 A GENERALIZED LINEAR MODEL FOR OPTIMIZATION OF ARCHITECTURAL PLANNING

6.1.1 Introduction

In the realm of architectural planning, there exists a type of problem which designers frequently confront when financial return is the most appropriate measure of the system's effectiveness. In this category can be included all rental and speculative housing (single and multiple family dwellings), office buildings, warehouses, stores, many industrial facilities, and so on, and building complexes which combine some or all of these to provide comprehensive services to the tenant.

Traditionally, the problem of planning for capital investment has been handled in a semiempirical way, in which the data serving as basis for decision making may be more or less reliable and up to date, but with the processing of these data (to arrive at the optimum allocation of space for maximum expectation of financial return) entirely intuitive and consequently unreliable. Once the pertinent data are collected, the most profitable design configuration can be ascertained through the optimization of a linear model which incorporates the most salient features of the real life, planning situation.

The model developed in the following section takes into account multiuse facilities and considers realistic design constraints such as zoning regulations, parking requirements, and so on. An example problem is also presented to illustrate application of the theory.

6.1.2 The Model

6.1.2.1 Definition of Terms

The design of buildings and other architectural facilities for financial return is constrained by many internal and external factors such as budget, market restrictions (construction costs, rentability, individual preferences), parking regulations, lot coverage, and so on, in addition to environmental factors, many of which defy quantification.

The procedure described in this section attempts to consider as many as possible of the quantifiable factors in optimizing architectural planning within the framework of economics: return on invested capital.

With the information extracted from the model (a type of *simulation* model, in many respects), rational decisions can be formulated based upon knowledge of the most recent tax depreciation structure, maintenance costs, alternative investment opportunities, and so on.

Let x_{ij}^k be the *floor area* of facility type i, at level j, of architectural quality k.

The "type i facility" refers to the use made of the space considered; offices, apartments, stores, warehouses, laboratories, industrial facilities, classrooms, and so on.

The "j level" denotes the location of the facility; first, second, . . . , nth floor level.

The "k quality" refers to the degree of architectural refinement of the space: types of finishes, environmental control, and many other considerations, the effects of which are reflected in the cost of construction and in the rentability of the facility.

Similarly, let

c_{ij}^k = *cost* per unit of area of facility i, at level j, of architectural quality k.

r_{ij}^k = *rent* per unit of area, per unit of time, from facility i, at level j, and architectural quality k.

p_{ij}^k = *probability of renting or selling* facility i, at level j, and architectural quality k.

and

q_{ij}^k = *probability of not renting or selling* facility i, at level j, and architectural quality k. Hence, $p_{ij}^k + q_{ij}^k = 1$.

It will be assumed that the market is large, and that for the time period under consideration it is in a steady state condition, such that the introduction of additional facilities for rent or sale will *not* significantly affect the parameters defined above.

6.1.2.2   Mathematical Formulation

The total expected rent per unit of time can be expressed as follows:

$$E(R) = \sum_{k=1}^{L} \sum_{i=1}^{m} \sum_{j=1}^{n} p_{ij}^k r_{ij}^k x_{ij}^k. \tag{6.1}$$

This is the objective function to be maximized, subject to constraints of the types described below.

1. *Market*

a. A survey may reveal that, within a specified geographic region, there exist vacancy rates for some or all of the proposed facilities. These vacancy rates are, in fact, the $q_{ij}^k = 1 - p_{ij}^k$ which were already considered in the formulation of the objective function. Nevertheless, the q_{ij}^k should be compared with the upper limits set on them by banks and other funding institutions. If these limits are exceeded, the project may be very difficult if not impossible to finance. Therefore,

$$q_{ij}^k \leq (q_i^k) \text{ max}, \tag{6.2}$$

for all i and k.

b. *Market preferences* data may indicate that the area of each facility type and quality must not exceed a certain fraction of the total building area. Let f_i^k be upper bounds to the area ratios; then the constraints can be written,

$$\sum_j x_{ij}^k \leq f_i^k(\sum_{k=1}^L \sum_{i=1}^m \sum_{j=1}^n x_{ij}^k) + e_i^k, \qquad i = 1, 2, \ldots, m;$$
$$x = 1, 2, \ldots, L. \qquad (6.3)^*$$

In general,

$$\sum_{k=1}^L \sum_{i=1}^m f_i^k > 1 \qquad (6.4)$$

because of the requirement that the f_i^k be upper bounds to the area ratios.

The e_i^k are small positive constants introduced into the constraint equations and inequalities to avert degeneracy.

2. *Zoning regulations* such as:

a. *Maximum building coverage* of total lot area which can be mathematically expressed as follows,

$$\sum_k \sum_i x_{ij}^k \leq A_j, \qquad j = 1, 2, \ldots, n, \qquad (6.5)$$

where A_j = maximum allowable building area of floor level j.

b. *Height restrictions* which could be overall building restrictions such that $j \leq n$ or restrictions on each individual facility; merchandising space, for example, should be located on lowest floors, and so on.

c. *Off-street parking* regulations, usually given as the number, n_i, of parking stalls per building area, a_i, of facility type i. In addition, the area, a, required for each car for parking, drives, and so on, can be easily computed and the off-street, surface parking restriction expressed as,

$$a\left[\sum_{i=1}^m \frac{n_i}{a_i}\left(\sum_{k=1}^L \sum_{j=1}^n x_{ij}^k\right)\right] \leq A_t - \sum_k \sum_i x_{i1}^k, \qquad (6.6)$$

where A_t = total buildable lot area (total lot area minus area required for landscaping, street rights of way, utility easements, set back restrictions, topographically unsuitable land, etc.).

Note that $(A_t - \sum_k \sum_i x_{i1}^k)$ is the site area available for parking (A_t minus first floor area of building).

3. *Design decisions* (massing studies). These decisions are made by designers for aesthetic and other reasons and are often arbitrary. Nevertheless, their effect upon the economic health of the system can be measured by comparing

---

* The range of the sum on the left-hand side of inequality (6.3) varies for each facility type and quality and for this reason it is not given explicitly. This convention will be used for the rest of this chapter.

all optimal solutions from models with different sets of design constraints to one having *no* design constraint. The model without design constraints yields the *global optimal solution*. The massing associated with the global optimal solution may not be aesthetically and/or structurally acceptable and it is at this point that design constraints must be introduced. The constrained model will, in general, yield a lower value of the objective function. Thus, if $E(R)_0$ is the global optimal rent and $E(R)_c$ is the expected rent when the problem is constrained by design considerations,

$$E(R)_c = E(R)_0 - C_c, \tag{6.7}$$

where C_c is the cost of the design decisions. It should be realized that aesthetic values may affect the four parameters c_{ij}^k, r_{ij}^k, p_{ij}^k, and q_{ij}^k, as well as others. If their effect is known, the parameters should be modified accordingly and a new $E(R)_c$ computed.

The ability to ascertain the effect of design decisions upon the system's economic health is a most valuable characteristic of the model, for it measures indirectly the "cost" of beauty and of other aesthetic factors; it represents an attempt to quantify some of the intangibles of architecture and urban planning. Design constraints may take different mathematical forms, depending upon the massing restrictions. If, for example, the planner decides that the building should take the form of a tower with a spread-out, s story base, b times or more larger than each of the tower's floors, the constraints would be written thus:

$$\sum_k \sum_i x_{i(s+1)}^k \leq \frac{1}{b} (\sum_k \sum_i x_{i1}^k) + e_{s+1}, \tag{6.8}$$

$$\sum_k \sum_i x_{ij}^k = \sum_k \sum_i x_{i1}^k + e_j, \quad j = 2, 3, \ldots, s, \tag{6.9}$$

$$\sum_k \sum_i x_{ij}^k = \sum_k \sum_i x_{i(s+1)}^k + e_j, j = s + 2, s + 3, \ldots, n. \tag{6.10}$$

The $e_j, j = 2, \ldots, n$, are, again, small positive constants introduced to avert degeneracy. Many other types of design decisions can be similarly formulated.

4. *Budget*, generally expressed as a fixed amount of money, B, which must not be exceeded by all fees and construction costs, excluding the cost of land.

Construction costs for each facility type and quality are, in general, functions of the total number of floors. Specifically, as the number of floors in the building increases, the costs per units of area could decrease or increase. This variability is not usually too sensitive and can be conveniently expressed as a cost multiplier, step function of the total number of floors, n. Let $\beta_{ij}^k(n) \geq 0$ be such step function. Graphically, it could be typically mapped as in Figure 6.1.

The figure shows that $\beta_{ij}^k(n)$ decreases for $(\alpha_{ij}^k)_2 \leq n < (\alpha_{ij}^k)_3$. It further decreases for $(\alpha_{ij}^k)_3 \leq n < (\alpha_{ij}^k)_4$. Then it increases in the interval $(\alpha_{ij}^k)_4 \leq n$

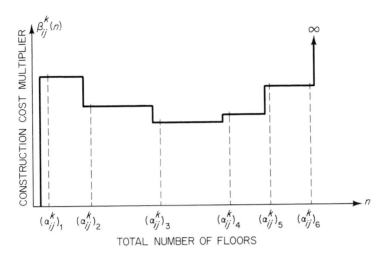

FIGURE 6.1. Total Number of Floors.

$< (\alpha_{ij}^k)_5$, and the increase is even greater for $(\alpha_{ij}^k)_5 \leq n \leq (\alpha_{ij}^k)_6$. These latter increases would reflect the higher costs of foundation and vertical transportation, for example. When $n > (\alpha_{ij}^k)_6$, the cost multiplier becomes infinite, denoting that $(\alpha_{ij}^k)_6$ is the upper bound to the number of floors due to zoning restrictions or other considerations. The variation shown in Figure 6.1 is, of course, only one of an infinite number of possibilities, but all of them will exhibit the same general shape.

In general, there will be $n \times m \times L$ construction cost multipliers with a proportionate number of $(\alpha_{ij}^k)_q$ nodes where changes in the costs occur. Let the nodes be ordered sequentially for all $\beta_{ij}^k(n)$ and renumbered γ_q, $q = 1, 2, 3, \ldots, r$. Then a typical cost multiplier, step function of n, would appear as given in Figure 6.2. This relabeling is necessary to show, for each $\beta_{ij}^k(n)$ multiplier, all points at which construction costs change for any type, level, and quality of facility, thus affording complete control upon the optimization process.

The cost multiplier functions, however, introduce a further complication in the model because the number of floors may not be known a priori. This is, in fact, a nonlinear characteristic of the system that, in this context, exhibits recycle loops. A method will be given later on for handling the nonlinearity in a satisfactory, linear manner. Additional items of cost and cost coefficients must be defined; for example:

a. *Cost of subsurface exploration*, soil investigation, foundation recommendations, and report. Usually given as a lump sum, b_s.

b. *Architectural-engineering fees*, usually expressed as a percentage of total construction costs by a decimal coefficient, c_a.

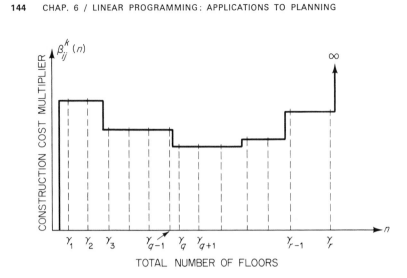

TOTAL NUMBER OF FLOORS

FIGURE 6.2. Total Number of Floors.

c. *Fund for contingencies*, also expressed as a percentage of total construction costs. Let c_c be the decimal equivalent of that percentage.

d. *Cost of movable furniture*, expressed as a percentage of construction cost for each facility type i and quality k. Let c_i^k be the decimal equivalents of those percentages.

e. *Supervision of construction costs*, generally computed as a lump sum, b_c, which is the total salary or salaries of one or more supervisors for the anticipated duration of the construction phase of the building; b_c is assumed constant for a given project.

f. *Cost of parking facilities per unit of area*, c_p. Support space—such as halls, elevators, toilets, and so on— is assumed to be included in the x_{ij}^k and, for this reason, its cost must be apportioned among tenants and/or buyers.

From these considerations, the following constraint inequality can be formulated:

$$(1 + c_a + c_b)\left[\sum_{k=1}^{L}\sum_{i=1}^{m}(1 + c_i^k)\left(\sum_{j=1}^{n}\beta_{ij}^k c_{ij}^k x_{ij}^k\right)\right]$$
$$+ c_p a\left[\sum_{i=1}^{m}\frac{n_i}{a_i}\left(\sum_{k=1}^{L}\sum_{j=1}^{n}x_{ij}^k\right)\right] + b_s + b_c \leq B. \qquad (6.11)$$

The constraint given above presupposes that the total number of floors, n, is known in advance. On this basis, the β_{ij}^k multipliers are obtained without difficulty. In performing an optimization study, however, design constraints would be relaxed at some point, and the total number of floors would become one of the unknown parameters. This, in turn, makes the β_{ij}^k be undetermined and a nonlinear, recycle loop could be generated. The following algorithm is proposed to avoid this complication:

STEP 1:

Assume that the maximum number of floors γ_r will be built (see Figure 6.2). Therefore, $n = \gamma_r$ and the corresponding β_{ij}^k multipliers are used in inequality (6.11) and in the formulation of the mathematical model. A solution is then obtained using a linear programming algorithm.

STEP 2:

Suppose that the solution generated under these conditions yields $n = \gamma_{s1}$, and that $\gamma_r \geq \gamma_{s1} \geq \gamma_{r-1}$. Then the β_{ij}^k used were the correct ones and this is an optimal solution for the constraints imposed.

If, on the other hand, $\gamma_{s1} < \gamma_{r-1}$, the β_{ij}^k used are incorrect and the solution is *not* consistent with its associated construction costs. Proceed to step 3.

STEP 3:

Now, assume that the total number of floors is $n = \gamma_{r-1} - 1$, that is, the maximum value on the next lower interval. Use this value of n to select the β_{ij}^k, formulate the model, and solve, using linear programming. Two possible results follow:

a. The solution yields $n = \gamma_{s2}$ and $\gamma_{s2} < \gamma_{r-2}$. In this case, repeat step 3 for the next lower interval.
b. The solution yields $n = \gamma_{s2}$ where $\gamma_{r-1} > \gamma_{s2} \geq \gamma_{r-2}$. Then the optimal solution for the model has been obtained and $n = \gamma_{s2}$.

5. Many other constraints can be imposed upon the objective function to make the problem conform to reality. In writing the constraints, however, one must keep in mind that they must be linear, if the problem solution is to be obtained with a linear programming algorithm, and one must be careful not to write linearly dependent and/or redundant constraints, for they can be a source of error and inefficiency in carrying out the solution.

6.1.3 Optimization Procedure

The model developed in the previous section is linear: a linear objective function is subjected to linear constraint inequalities and equations. It fits within the framework of problems which can be optimized with "linear programming" methods.

After the introduction of slack and artificial slack variables, the matrix of the structural coefficients must have the same rank as the matrix formed by augmenting the previous matrix with the column vector of stipulations for the system to have basic solutions. This is a good check on the constraint equations and should be performed before embarking upon the task of solving the problem.

6.1.4 Conclusions

The architectural planning model presented in this chapter allows the designer to make rational decisions based on the information provided by the optimal solutions generated. It must be emphasized that this approach reinforces the importance of the role played by the architect-planner in shaping the urban environment. Even with a systematic method such as the one described, he must at all times be deeply involved in the formulation of the model, especially at the "design constraint" level, for the linear model presented is a full-fledged "simulation machine," sensitive to the settings the designer gives it and responsive to the reactions of the system's economic health. A simple but descriptive numerical example problem is given in the next section.

6.1.5 Numerical Example

6.1.5.1 Statement of the Problem

An individual wishes to develop a 250 ft. × 400 ft. commercial piece of property located at the intersection of two major traffic arteries.

6.1.5.1.1. Market studies indicate that, for a development to be rentable, available office space must not exceed 2 floors, stores must not exceed 1 floor, and living units 3 floors for a "walkup" facility.

6.1.5.1.2. The studies further indicate that, within the sector of the city where the property is located, offices have a 20% vacancy rate, stores a 5% vacancy rate, and apartment units a 10% vacancy rate (occupancies must not be less than 70% to obtain financial support from a funding agency).

6.1.5.1.3. Data on market rental preferences reveal that facility areas of individual types and qualities should not exceed the following percentages of total building area:

TABLE 6-1. Rentability.

Type	Quality	% of Total Building Area
1. Offices	1	10
	2	40
	3	25
2. Stores	1	30
3. Apartments	1	40
	2	30

6.1.5.1.4. Zoning regulations for off-street parking are:

a. Offices—1 parking space/ea. 200 sq. ft. of bldg. area.

b. Stores—1 parking space/ea. 1,000 sq. ft. of bldg. area.

c. Apartments—1 parking space/ea. 800 sq. ft. of bldg. area.

Further—building coverage shall not exceed 30% of total lot area. Allow 400 sq. ft./car for parking space, drives, and so on, and a total of 5,000 sq. ft. for landscaping.

6.1.5.1.5.

TABLE 6–2. Rent ($/sq. ft./yr.)

Type	Level	Quality		
		1	2	3
1. Offices	1. Ground floor	4.50	4.00	3.50
	2. Second floor	4.00	3.50	3.00
2. Stores	1. Ground floor	3.00	—	—
3. Apartments	1. Ground floor	3.00	2.50	—
	2. Second floor	3.00	2.50	—
	3. Third floor	2.50	2.00	—

6.1.5.1.6. The $(\beta_{ij}^k c_{ij}^k)$ coefficients are given below in tabular form:

TABLE 6–3. Construction Costs, Including Support Space.

Type	Quality	Total Number of Floors in Building								
		1			2			3		
		Level			Level			Level		
		Ground floor	—	—	Ground floor	Second floor	—	Ground floor	Second floor	Third floor
1. Offices	1	20.00	—	—	20.00	19.00	—	19.00	18.00	—
	2	18.00	—	—	17.00	16.00	—	17.00	15.00	—
	3	15.00	—	—	14.00	14.00	—	13.00	13.00	—
2. Stores	1	13.00	—	—	12.00	—	—	11.00	—	—
3. Apartments	1	20.00	—	—	19.00	18.00	—	18.00	17.00	17.00
	2	17.00	—	—	17.00	16.00	—	16.00	15.00	14.00

Cost of parking area = $0.75 per sq. ft.

6.1.5.1.7. Miscellaneous costs are:

a. Architectural-engineering fees—6% of total construction cost.

b. Contingencies—1.5% of total construction cost.

   c. Movable furniture—
     (1) Offices—2% of total construction cost.
     (2) Stores—1% of total construction cost.
     (3) Apartments—4% of total construction cost.
   d. Soil report—$3,000.
   e. Supervision of construction—$2,000.

   6.1.5.1.8. Budget: Shall not exceed $800,000, excluding cost of land.

Question: What types of units and how many square feet of each shall the investor develop to maximize the rent under present market conditions?

6.1.5.2 The Model

Let

$$i = \begin{cases} 1 \rightarrow \text{Offices} \\ 2 \rightarrow \text{Stores} \\ 3 \rightarrow \text{Apartments} \end{cases}$$

$$j = \begin{cases} 1 \rightarrow \text{Ground floor} \\ 2 \rightarrow \text{Second floor} \\ 3 \rightarrow \text{Third floor} \end{cases}$$

$$k = \begin{cases} 1 \rightarrow \text{Quality 1} \\ 2 \rightarrow \text{Quality 2} \\ 3 \rightarrow \text{Quality 3} \end{cases}$$

Then, from 6.1.5.1.1 and 6.1.5.1.5,

$$x^k_{13} = 0, \quad k = 1, 2, 3;$$
$$x^k_{21} = 0, \quad k = 2, 3;$$
$$x^k_{2j} = 0, \quad j = 2, 3, \quad k = 1, 2, 3;$$
$$x^3_{3j} = 0, \quad j = 1, 2, 3.$$

From 6.1.5.1.2.

$$q^k_{1j} = .20 < .30, \quad j = 1, 2, 3, \quad k = 1, 2, 3. \quad \text{O.K.}$$
$$q^k_{2j} = .50 < .30, \quad j = 1, 2, 3, \quad k = 1, 2, 3. \quad \text{O.K.}$$
$$q^k_{3j} = .10 < .30, \quad j = 1, 2, 3, \quad k = 1, 2, 3. \quad \text{O.K.}$$

From 6.1.5.1.2, 6.1.5.1.5, and equation (6.1),

OBJECTIVE FUNCTION:

$$
\begin{aligned}
\text{Max } E(R) = {} & .80(4.50x^1_{11} + 4.00x^2_{11} + 3.50x^3_{11} \\
& + 4.00x^1_{12} + 3.50x^2_{12} + 3.00x^3_{12}) + .95(3.00x^1_{21}) \\
& + .90(3.00x^1_{31} + 2.50x^2_{31} + 3.00x^1_{32} + 2.50x^2_{32} \\
& + 2.50x^1_{33} + 2.00x^2_{33}).
\end{aligned}
\tag{6.12}
$$

CONSTRAINTS:

1. The lowest *construction cost* is, from 6.1.5.1.6, \$11.00 per sq. ft. Therefore, an upper bound to the total building area is 800,000/11.00 $\cong$ 74,000 sq. ft. Therefore,

$$x_{11}^1 + x_{11}^2 + x_{11}^3 + x_{12}^1 + x_{12}^2 + x_{12}^3 + x_{21}^1 + x_{31}^1$$
$$+ x_{31}^2 + x_{32}^1 + x_{32}^2 + x_{33}^1 + x_{33}^2 \leq 74,000.$$

When slack variable x_1 is introduced, obtain

$$x_{11}^1 + x_{11}^2 + x_{11}^3 + x_{12}^1 + x_{12}^2 + x_{12}^3 + x_{21}^1 + x_{31}^1 + x_{31}^2 + x_{32}^1$$
$$+ x_{32}^2 + x_{33}^1 + x_{33}^2 + x_1 = 74,000. \tag{6.13}$$

Notice that the total building area is $74,000 - x_1$.

2. *Market preferences.* From 6.1.5.1.3, and introducing slack variables x_2 through x_7, obtain

$$x_{11}^1 + x_{12}^1 + .10x_1 + x_2 = 7,400, \tag{6.14}$$

$$x_{11}^2 + x_{12}^2 + .40x_1 + x_3 = 29,600, \tag{6.15}$$

$$x_{11}^3 + x_{12}^3 + .25x_1 + x_4 = 18,500, \tag{6.16}$$

$$x_{21}^1 + .30x_1 + x_5 = 22,200, \tag{6.17}$$

$$x_{31}^1 + x_{32}^1 + x_{33}^1 + .40x_1 + x_6 = 29,600, \tag{6.18}$$

$$x_{31}^2 + x_{32}^2 + x_{33}^2 + .30x_1 + x_7 = 22,200. \tag{6.19}$$

3. *Site building area.* From 6.1.5.1.4, in equation (6.5), and with slack variable x_8,

$$x_{11}^1 + x_{11}^2 + x_{11}^3 + x_{21}^1 + x_{31}^1 + x_{31}^2 + x_8 = 30,000. \tag{6.20}$$

4. *Parking.* From 6.1.5.1.4, in equation (6.6), and introducing slack variable x_9, form

$$3.00(x_{11}^1 + x_{11}^2 + x_{11}^3) + 2.00(x_{12}^1 + x_{12}^2 + x_{12}^3)$$
$$+ 1.40x_{21}^1 + 1.50(x_{31}^1 + x_{31}^2) + .50(x_{32}^1 + x_{32}^2 + x_{33}^1 + x_{33}^2)$$
$$+ x_9 = 95,000. \tag{6.21}$$

5. *Design decisions* (massing study). (All floors must be approximately the same size.) First floor area approximately equal to second floor area:

$$x_{11}^1 + x_{11}^2 + x_{11}^3 + x_{21}^1 + x_{31}^1 + x_{31}^2 = x_{12}^1 + x_{12}^2 + x_{12}^3$$
$$+ x_{32}^1 + x_{32}^2 + e_1.$$

First floor area approximately equal to third floor area:

$$x_{11}^1 + x_{11}^2 + x_{11}^3 + x_{21}^1 + x_{31}^1 + x_{31}^2 = x_{33}^1 + x_{33}^2 + e_2.$$

Let $e_1 = e_2 = 1,000$ to avert degeneracy. This means the second and third floor areas will be within 1,000 sq. ft. of the first floor area.

One can write:

$$x_{11}^1 + x_{11}^2 + x_{11}^3 + x_{21}^1 + x_{31}^1 + x_{31}^2$$
$$- (x_{12}^1 + x_{12}^2 + x_{12}^3 + x_{32}^1 + x_{32}^2) = 1{,}000, \qquad (6.22)$$

and

$$x_{11}^1 + x_{11}^2 + x_{11}^3 + x_{21}^1 + x_{31}^1 + x_{31}^2$$
$$- (x_{33}^1 + x_{33}^2) = 1{,}000. \qquad (6.23)$$

Artificial slack variables must be introduced into equation's (6.22) and (6.23) before proceeding to optimize.

6. *Budget.* Under the assumption that the total number of floors in the building will be three, the budget constraint, from 6.1.5.1.6, 6.1.5.1.7, 6.1.5.1.8, in equation (6.11) and introducing slack variable x_{10}, can be written as follows:

$$1.075[1.02(19x_{11}^1 + 17x_{11}^2 + 13x_{11}^3 + 18x_{12}^1 + 15x_{12}^2 + 13x_{12}^3)$$
$$+ 1.01(11x_{21}^1) + 1.04(18x_{31}^1 + 16x_{31}^2 + 17x_{32}^1 + 15x_{32}^2 + 17x_{33}^1$$
$$+ 14x_{33}^2)] + .75[2.00(x_{11}^1 + x_{11}^2 + x_{11}^3 + x_{12}^1 + x_{12}^2 + x_{12}^3)$$
$$+ .40(x_{21}^1) + .50(x_{31}^1 + x_{31}^2 + x_{32}^1 + x_{32}^2 + x_{33}^1 + x_{33}^2)] + x_{10}$$
$$= 795{,}000. \qquad (6.24)$$

The model is complete and a linear programming maximization of objective function (6.12) subjected to constraint equations (6.13) through (6.24) can now be performed. Because there are only 12 constraint equations, any basic feasible solution, including the optimal one, cannot contain more than 12 nonzero variables. This limitation could be removed by introducing additional constraint conditions.

In accordance with the budget algorithm given in the example, if design decision constraints are removed and the optimal solution shows that

$$x_{33}^1 = x_{33}^2 = 0, \qquad (6.25)$$

the total number of floors could not exceed two. Therefore, set $x_{33}^1 = x_{33}^2 = 0$ in the model and formulate the budget constraint equation as follows:

$$1.075[1.02(20x_{11}^1 + 17x_{11}^2 + 14x_{11}^3 + 19x_{12}^1 + 16x_{12}^2 + 14x_{12}^3)$$
$$+ 1.01(12x_{21}^1) + 1.04(19x_{31}^1 + 17x_{31}^2 + 18x_{32}^1 + 16x_{32}^2)]$$
$$+ .75[2.00(x_{11}^1 + x_{11}^2 + x_{11}^3 + x_{12}^1 + x_{12}^2 + x_{12}^3) + .40(x_{21}^1)$$
$$+ .50(x_{31}^1 + x_{31}^2 + x_{32}^1 + x_{32}^2)] + x_{10} = 795{,}000. \qquad (6.26)$$

In a similar manner, if the new optimal solution shows

$$x_{12}^1 = x_{12}^2 = x_{12}^3 = x_{32}^1 = x_{32}^2 = 0, \qquad (6.27)$$

the building can have only one floor. Hence, set $x_{12}^1 = x_{12}^2 = x_{12}^3 = x_{32}^1$ $= x_{32}^2 = 0$ in the model and the budget constraint becomes

$$1.075[1.02(20x_{11}^1 + 18x_{11}^2 + 15x_{11}^3) + 1.01(13x_{21}^1)$$
$$+ 1.04(20x_{31}^1 + 17x_{31}^2)]$$
$$+ .75[2.00(x_{11}^1 + x_{11}^2 + x_{11}^3) + .40(x_{21}^1)$$
$$+ .50 (x_{31}^1 + x_{31}^2)] + x_{10} = 795,000. \qquad (6.28)$$

The budget algorithm describes the procedure to follow in handling cost coefficient multipliers.

When design constraints are relaxed, the *global optimal solution* is obtained.

6.1.5.3   Solution

The first run, *with no design constraints*, yielded the following data: At a gross expected profit of $140,570 per annum, build

> 12,536 sq. ft. of quality 3, ground floor *offices*
> 16,700 sq. ft. of quality 2, second floor *offices*
> 15,043 sq. ft. of quality 1, ground floor *stores*
> 5,865 sq. ft. of quality 1, second floor *apartments*

This two-level solution was obtained using construction costs for a three-level complex. Therefore, it is necessary to change the cost coefficients in the budget constraint. The third-level variables were assigned large, positive cost coefficients in order to drive them out of solution.

The second run, *with no design constraints*, yielded the following data: At a gross expected profit of $132,640 per annum, build

> 4,348 sq. ft. of quality 2, ground floor *offices*
> 11,660 sq. ft. of quality 3, ground floor *offices*
> 12,711 sq. ft. of quality 2, second floor *offices*
> 13,992 sq. ft. of quality 1, ground floor *stores*
> 3,929 sq. ft. of quality 1, second floor *apartments*

This two-level configuration constitutes the global optimal solution with which all other optima (subject to all design constraints) must be compared.

It is important to note that the global optimal solution established a building area of 30,000 sq. ft. on the ground floor and 16,640 sq. ft. on the second floor.

The third run, for which a design constraint was introduced into the model (that the ground floor area must be approximately equal to that of the second

floor), yielded the following space allocation:
For a gross expected profit of $132,126 per annum, build

> 9,856 sq. ft. of quality 3, ground floor *offices*
> 2,506 sq. ft. of quality 1, second floor *offices*
> 18,712 sq. ft. of quality 2, second floor *offices*
> 1,671 sq. ft. of quality 3, second floor *offices*
> 14,035 sq. ft. of quality 1, ground floor *stores*

Due to *one* decision, the designer is forced to relinquish approximately $600 per annum in profit and to reallocate spaces so that a local optimum may be achieved subject to the constraint imposed upon the building configuration. Notice that the solution is not sensitive to the design constraint imposed, for the annual profit decreases only $514 in $132,640; however, the allocation of space (type and quality as well as floor level) is extremely sensitive, as can be verified by analyzing the last two solutions. Various other design decisions can be incorporated with equal ease.

6.2 OPTIMALITY IN INDUSTRIAL PARK DEVELOPMENT

6.2.1 Introduction

This example shows an application of the systems approach to decision making at the highest level of performance: the planning level. It concentrates on the use of linear programming in the planning of an industrial facility; and, although the problem is specific, it is hoped that the thinking and methodology employed in arriving at the formulation of the mathematical model, leading to the establishment of the optimum strategy, will be sufficiently clear to permit the reader to extend his own thinking into other problems where a linearization of the parameters is equally valid.

Institutional factors have purposely been left out for clarity's sake and no attempt has been made to show the economic venture analysis that must accompany such a study for, once the optimum allocation policy is known, a rate of return or other similarly well-known engineering economy analysis must be performed to ascertain the economic feasibility of the project. This evaluation has to be carried out for the before and after tax situations, where such factors as depreciation and retirement and replacement policies are given careful consideration.

Once the project has been declared profitable, the optimal location of the facility within the site must be investigated. Pattern search is the procedure recommended for this final step in what may be termed the "Optimal Site Development."

The list of references at the end of the chapter gives a publication by the author that describes the application of pattern search for the optimum location of plant facilities (see also chapter 10).

6.2.2 Statement of the Problem

The government of a small city proposes to increase its revenue by developing a 300-acre tract of land for an industrial park. Tentative agreements have been signed with five industries and the city has made commitments to finance, build, and lease on a long-term basis the plant buildings and facilities required by the industries contacted.

The city government now owns all utilities and recently completed an expansion of these facilities to allow for limited increase in demand. The maximum amount of financing available is $6,000,000.

A study of the manpower situation in the area reveals that the available local transient and out-of-town labor force will not exceed the figures given in Table 6–4.

TABLE 6–4. Labor Force.

Type	Max. Number	Annualized Average Income
1. Common Labor	400	$ 5,000
2. Semiskilled Labor	180	$ 7,000
3. Skilled Labor	100	$12,000
4. Clerical Workers	150	$ 5,000
5. Technical Personnel	120	$14,000
6. Managers	33	$30,000

The study also revealed the pattern of local spending of net disposable income, after state and federal taxes. These data are condensed in Table 6–5

TABLE 6–5. Labor Force Spending Pattern.

Type	Average State and Federal Taxes, % of Gross Income	Annualized Net Income	Average % of Net Income Spent in City Area	Cash Flow into City Area
1. Common Labor	10%	$ 4,500	90%	$ 4,050
2. Semiskilled Labor	12%	$ 6,160	85%	$ 5,240
3. Skilled Labor	16%	$10,100	70%	$ 7,070
4. Clerical Workers	10%	$ 4,500	85%	$ 3,830
5. Technical Personnel	18%	$11,500	65%	$ 7,460
6. Managers	24%	$22,800	55%	$12,520

and their computation was based on the average family size for each income level.

It has also been estimated that city revenue from taxes on sales and property and from utilities, fuel, sanitation, and so on will average about 2.5% of the flow of disposable personal income spent in the city area. The revenue that the city will accrue from the employment of each type of worker is summarized in Table 6–6.

TABLE 6–6.   City Revenue from Labor Force.

Type	Annualized Average City Revenue per Worker
1. Common Labor	$101
2. Semiskilled Labor	$131
3. Skilled Labor	$177
4. Clerical Workers	$ 96
5. Technical Personnel	$187
6. Managers	$313

The city-owned utilities, sewerage, and waste disposal systems can satisfactorily handle the following additional loads:

1. Electricity, 14.5×10^6 kw. hrs./yr.
2. Water, 75.0 million gals./yr.
3. Gas, 40.0×10^6 cu. ft./yr.
4. Sewerage, 40.0 million gals./yr.
5. Waste Disposal, 20.0×10^3 tons/yr.

The city's unit profits from the sale of utilities and other services to industrial customers are:

1. Electricity, $0.0007 per kw. hr.
2. Water, $7.00 per million gals.
3. Gas, $0.004 per cu. ft.
4. Sewerage, $1.00 per million gals.
5. Waste, $0.15 per ton.

The five potential customer industries with which the city government has signed tentative agreements will use most of the area's raw materials. Even though this and many other industry-derived benefits will upgrade the community's income level, they cannot be quantified accurately and for this reason will be disregarded in the optimization strategy. Nevertheless, they must be recognized as a plus factor in the decision to implement the plan.

Table 6–7 gives information on the maximum and minimum plant building area requirements for each industry.

Tables 6–8 through 6–10 summarize resource requirements and production levels for each 1,000 square feet of plant building area, for each of the five industries considered.

The city government will impose a 1.5% tax on the gross annual production value of each industry.

The site costs to the city in the location selected for the industrial park will be $2,000 per acre, and the average building construction costs for the various plants are summarized below:

Industry 1 (Electronics)—$15/sq. ft.
Industry 2 (Pressure Vessels)—$17/sq. ft.
Industry 3 (Metal Castings)—$19/sq. ft.
Industry 4 (Plastic Laminates)—$22/sq. ft.
Industry 5 (Recapped Tires)—$16/sq. ft.

The city has agreed to lease the building facilities (including grounds) at an annual cost to the industries of $1.75/sq. ft. of building area. Maintenance costs have been estimated at $0.05/sq. ft. of building area per year.

The city government desires to establish the optimum allocation of space to each industry—that is, the allocation that, consistent with the constraints, will produce the highest annual revenue to the local government.

6.2.3 Mathematical Model

6.2.3.1 The Objective Function

Let x_i, $i = 1, \ldots, 5$, be the plant building areas, in 1,000 sq. ft., to be leased to industries 1 through 5, respectively.

The total annual income the city government can expect to receive as a result of its investment in the industrial park can be broken down as follows:

1. Revenue from labor force spending habits: Table 6–6.
2. Income from utility and other service charges.
3. Revenue from tax on industrial production: Table 6–10.
4. Income from plant building leases minus maintenance costs.

TABLE 6–7. Plant Building Area Requirements.

Industry	Principal Product	Max. Bldg. Area sq. ft.	Min. Bldg. Area sq. ft.
1	Electronics	120,000	40,000
2	Pressure Vessels	80,000	20,000
3	Metal Castings	70,000	26,000
4	Plastic Laminates	100,000	28,000
5	Recapped Tires	150,000	35,000

TABLE 6–8.   Labor Force Requirements in Men/Day (8-hour shifts) per 1,000 sq. ft. of Plant Building Area.

Industry	Principal Product	Common	Semi-skilled	Skilled	Clerical	Technical	Man-agerial
1	Electronics	0.75	0.23	0.38	0.30	0.15	0.05
2	Pressure Vessels	0.90	1.35	0.30	0.60	0.30	0.15
3	Metal Castings	1.30	0.46	0.46	0.58	0.23	0.15
4	Plastic Laminates	0.43	0.64	0.32	0.75	0.32	0.07
5	Recapped Tires	1.37	0.34	0.17	0.17	0.69	0.11

Let $j = 1, \ldots, 6$ be indices representing common, semiskilled, skilled, clerical, technical, and managerial personnel; $r_j, j = 1, \ldots, 6$ be the city revenue from the labor force local spending habits (Table 6–6); $L_{ij}, i = 1, \ldots, 5, j = 1, \ldots, 6$ be the number of workmen employed by industry i of type j per 1,000 sq. ft. of plant building area (Table 6–7); and $p_i, i = 1, \ldots, 5$ be the annual production for industry i in dollars per 1,000 sq. ft. of plant building area (Table 6–10).

Also let $k = 1, \ldots, 5$ be indices representing electric, water, gas, sewerage, and waste disposal services; $C_k, k = 1, \ldots, 5$ be the unit profits from electric, water, gas, sewerage, and waste disposal services; and $u_{ik}, i = 1, \ldots, 5, k = 1, \ldots, 5$ be the requirements of industry i for electric, water, gas, sewerage, and waste disposal services (Table 6–9).

Adhering to these definitions, the objective function can now be written:

$$\text{Max } I = \sum_{i=1}^{5} \left[\sum_{j=1}^{6} L_{ij} r_j + \sum_{k=1}^{5} C_k u_{ik} + 0.015 p_i + 1,700 \right] x_i. \qquad (6.29)$$

6.2.3.2 The Constraints

The region of feasibility is defined by the following set of constraints.

1. Initial capital investment on land and buildings cannot exceed $6,000,000.
2. Maximum available labor force cannot be exceeded: Table 6–4.
3. Maximum utility and service levels cannot be exceeded.
4. Maximum amount of land available is 300 acres: area of industrial park.
5. Maximum and minimum plant building areas must be maintained: Table 6–7.

Let $E_j, j = 1, \ldots, 6$ be the maximum available number of type j workers (Table 6–4); A_i and $a_i, i = 1, \ldots, 5$ be, respectively, the maximum and minimum plant area requirements in 1,000 sq. ft. (Table 6–7); $b_i, i = 1, \ldots, 5$ be the unit building costs in dollars per sq. ft. for plant type i; $s_i, i = 1, \ldots, 5$ be the amount of land required for industry i, in acres per 1,000 sq. ft.

TABLE 6–9. Site and Services Requirements in Units per 1,000 sq. ft. of Plant Building Area.

Industry	Principal Product	Site (acres/ 1,000 ft.²)	Electricity (kw. hrs./yr./ 1,000 ft².)	Water (million gals./ yr./1,000 ft.²)	Gas (cu. ft./yr./ 1,000 ft.²)	Sewerage (million gals./ yr./1,000 ft.²)	Waste (tons/yr./ 1,000 ft.²)
1	Electronics	0.50	40,000	0.050	60,000	0.040	20
2	Pressure Vessels	1.00	32,000	0.100	80,000	0.058	30
3	Metal Castings	1.15	62,000	0.300	120,000	0.200	50
4	Plastic Laminates	1.43	48,000	0.200	75,000	0.150	45
5	Recapped Tires	1.43	58,000	0.100	150,000	0.060	25

TABLE 6–10.   Annualized Industrial Production in $/yr.
per 1,000 sq. ft. of Plant Building Area.

Industry	Principal Product	Annualized Value Added ($/1,000 ft.^{2})
1	Electronics	5,000
2	Pressure Vessels	7,000
3	Metal Castings	12,000
4	Plastic Laminates	15,000
5	Recapped Tires	10,000

of plant building area (Table 6–9); and let $U_k, k = 1, \ldots, 5$ be the maximum available levels of electric, water, gas, sewerage, and waste disposal services.

Then the constraints can be written:

1. Capital Investment

$$1,000 \sum_{i=1}^{5} b_i x_i \leq 5,400,000. \tag{6.30}$$

2. Labor Force

$$\sum_{i=1}^{5} L_{ij} x_i \leq E_j; j = 1, \ldots, 6 \tag{6.31}$$

3. Utility and Service Levels

$$\sum_{i=1}^{5} u_{ik} x_i \leq U_k; k = 1, \ldots, 5 \tag{6.32}$$

4. Available Acreage

$$\sum_{i=1}^{5} s_i x_i \leq 300 \tag{6.33}$$

5. Plant Building Areas

$$a_i \leq x_i \leq A_i; i = 1, \ldots, 5 \tag{6.34}$$

This completes the mathematical description of the model.

The substitution of numerical values into the model and the introduction of slack variables yield the following system.

OBJECTIVE FUNCTION:

Max $I = 2{,}292.03x_1 + 2{,}634.16x_2 + 2{,}831.82x_3 + 2{,}604.56x_4$
$$+ \ 2{,}887.89x_5 \tag{6.35}$$

CONSTRAINTS:

1. Investment

$$15{,}000x_1 + 17{,}000x_2 + 19{,}000x_3 + 22{,}000x_4 + 16{,}000x_5 + x_6$$
$$= 5{,}400{,}000 \tag{6.36}$$

2. Labor Force
   a. Common

$$0.75x_1 + 0.90x_2 + 1.30x_3 + 0.43x_4 + 1.36x_5 + x_7 = 400 \qquad (6.37)$$

   b. Semiskilled

$$0.23x_1 + 1.35x_2 + 0.46x_3 + 0.64x_4 + 0.34x_5 + x_8$$
$$= 180 \qquad (6.38)$$

   c. Skilled

$$0.38x_1 + 0.30x_2 + 0.46x_3 + 0.32x_4 + 0.17x_5 + x_9$$
$$= 100 \qquad (6.39)$$

   d. Clerical

$$0.30x_1 + 0.60x_2 + 0.58x_3 + 0.75x_4 + 0.17x_5 + x_{10}$$
$$= 150 \qquad (6.40)$$

   e. Technical

$$0.15x_1 + 0.30x_2 + 0.23x_3 + 0.32x_4 + 0.69x_5 + x_{11}$$
$$= 120 \qquad (6.41)$$

   f. Managerial

$$0.05x_1 + 0.15x_2 + 0.15x_3 + 0.07x_4 + 0.11x_5 + x_{12}$$
$$= 33 \qquad (6.42)$$

3. Utility and Service Levels
   a. Electricity

$$40{,}000x_1 + 32{,}000x_2 + 62{,}000x_3 + 48{,}000x_4 + 58{,}000x_5$$
$$+ x_{13} = 14.5x10^6 \qquad (6.43)$$

   b. Water

$$0.050x_1 + 0.100x_2 + 0.300x_3 + 0.200x_4 + 0.100x_5$$
$$+ x_{14} = 75.0 \qquad (6.44)$$

   c. Gas

$$60{,}000x_1 + 80{,}000x_2 + 120{,}000x_3 + 75{,}000x_4 +$$
$$150{,}000x_5 + x_{15} = 40.0x10^6 \qquad (6.45)$$

   d. Sewerage

$$0.040x_1 + 0.058x_2 + 0.200x_3 + 0.150x_4 + 0.60x_5$$
$$+ x_{16} = 40.0 \qquad (6.46)$$

   e. Waste

$$20x_1 + 30x_2 + 50x_3 + 45x_4 + 25x_5 + x_{17} = 20.0x10^3 \qquad (6.47)$$

4. Acreage

$$0.50x_1 + 1.00x_2 + 1.15x_3 + 1.43x_4 + 1.43x_5 + x_{18} = 300 \qquad (6.48)$$

5. Plant Building Areas
   a. Maximum Areas

$$x_1 + x_{19} = 120 \qquad (6.49)$$

$$x_2 + x_{20} = 80 \qquad (6.50)$$

$$x_3 + x_{21} = 70 \qquad (6.51)$$

$$x_4 + x_{22} = 100 \qquad (6.52)$$

$$x_5 + x_{23} \doteq 150 \qquad (6.53)$$

   b. Minimum Areas

$$x_1 - x_{23} = 40 \qquad (6.54)$$

$$x_2 - x_{24} = 20 \qquad (6.55)$$

$$x_3 - x_{25} = 26 \qquad (6.56)$$

$$x_4 - x_{26} = 28 \qquad (6.57)$$

$$x_5 - x_{27} = 35 \qquad (6.58)$$

6.2.4 Solution

The solution, obtained with a standard linear programming package, yielded the following results:

1. *Maximum Annual Income* = $810,046
2. *Initial Investment* = $5,238,314
3. *Optimum Plant Building Areas*
   Industry 1 (Electronics)—120,000 sq. ft.
   Industry 2 (Pressure Vessels)—73,228 sq. ft.
   Industry 3 (Metal Castings)—26,000 sq. ft.
   Industry 4 (Plastic Laminates)—28,000 sq. ft.
   Industry 5 (Recapped Tires)—67,714 sq. ft.
4. *Optimum Land Area* = 300 acres
5. *Labor Utilization*
   Common—294
   Semiskilled—179
   Skilled—100
   Clerical—127
   Technical—102
   Managerial—30
6. *Utility and Service Levels*
   Electricity—14,026,743 kw. hr./yr.

Water—33.49 million gals./yr.
Gas—28,435,428 cu. ft./yr.
Sewerage—22.51 million gals./yr.
Waste—8,850 tons/yr.

All-solution values are feasible; they do not violate any of the limitations imposed upon the resources.

6.2.5 Conclusions and Observations

In building mathematical models for real life problems, one must be careful to verify that the available number of resources is sufficient to fulfill their minimum levels of utilization, for otherwise there will *not* exist any feasible alternatives to implement the plan.

It must also be pointed out that the industries used in the example as well as the data presented, although realistic, are strictly hypothetical. The principal concern here has been to guide the reader through a modeling situation and to stress the need to use the systems approach to planning in optimizing real problems even when they exhibit only a moderate level of complexity.

EXERCISES

6–1. The southern California area is faced with the problem of ultimately supplying water to a future population of 40 million people. Assuming that the future water demand will be 190 gal./day/capita, the water required by the area will be 7,600 m.g.d. Local surface water and ground water can supply only 990 m.g.d. Other possible sources of water are as follows:

   1. An aqueduct system to the Colorado River and to the Feather River located at a distance of some 200 miles, with a total capacity of 4,000 m.g.d.
   2. Reclaimed waste water from domestic sewerage treatment with a maximum of 3,200 m.g.d.
   3. Desalinized water from the ocean, which can be supplied in unlimited quantities by the electrolytic ion-exchange method.

   The quality of the water to be supplied must contain no more than 700 p.p.m total dissolved solids; no more than 250 p.p.m sulfates; and no more than 110 p.p.m chlorides.

Data on the possible sources are given below:

	Colorado River	Feather River	Reclaimed Waste	Desalinized Water	Local Source
Cost ($/1,000 gal.)	0.21	0.70	0.50	0.80	0.15
T.D.S. (p.p.m.)	805.	720.	634.	500.	650.
Sulfates (p.p.m.)	335.	132.	366.	30.	150.
Chlorides (p.p.m.)	118.	137.	20.	30.	60.

The problem is to adequately supply the water required at a minimum cost.

6–2. Develop planning problems utilizing the models described in this chapter. Introduce variations and try to expand the model formulations to fit other physical situations.

6–3. A lamp manufacturer is investigating the possibility of expanding his plant. For expansion to be feasible, the additional facility must (1) have a minimum capacity to produce 300 lamps per month; and (2) produce a minimum profit of $1,500 per month from a maximum budget of $5,000 per month.

The lamp manufacturer produces four types of lamps: (1) ceramic lamps, (2) pine lamps, (3) oak lamps, and (4) mahogany lamps. The unit cost to produce the lamps is $3, $5, $10, and $15, respectively. The manufacturer's unit profit is $1, $2, $3, and $5, respectively. The resources available constrain the monthly production to no more than 130 ceramic lamps, 400 pine lamps, 240 oak lamps, and 130 mahogany lamps. The availability of skilled lathe operators limits the production of wooden lamps to 630 per month.

Based on a market survey, the maximum demand for additional lamps is 750 per month with the following breakdown: 20% ceramic, 40% oak, 25% pine, and 15% mahogany lamps, respectively. Determine whether or not it is feasible to expand the plant.

REFERENCES

1. AGUILAR, R. J., "Decision Making in Building Planning," *Computers in Engineering Design Education*, III, 26–33, College of Engineering, University of Michigan. Ann Arbor, April 1, 1966.

2. ———, *Optimum Location of Plant Facilities*, Division of Engineering Research, Louisiana State University, Bulletin No. 98. Baton Rouge, 1969.

3. ———, *The Mathematical Formulation and Optimization of Architectural and Planning Functions*, Division of Engineering Research, Louisiana State University, Bulletin No. 93. Baton Rouge, 1967.

4. ———, GONZALEZ, R. S. and J. E. HAND, "Optimality in Industrial Park Development; A Systems Approach," *Proceedings, Computer Applications to Environmental Design*, School of Architecture, University of Kentucky. Lexington, April 1970.

5. ———, and J. E. HAND, "A Generalized Linear Model for Optimization of Architectural Planning," *Proceedings, 1968 Spring Joint Computer Conference* (Atlantic City), XXXII, 81–88.

CHAPTER 7

Integer Programming

7.1 INTRODUCTION

The usual linear programming problem specifies the variables to be non-negative real numbers. This may, at times, not be a sufficiently stringent requirement because it is also necessary in many situations that the variables take on non-negative integer values.

The need for integer quantities exists, for example, in an allocation problem involving a small number of whole units; if one tries to allocate seven trucks to two jobs, it would be physically meaningless to divide the trucks into fractions, and the problem may be sensitive enough that rounding off could be too gross an approximation to be valuable. Nevertheless, linear programming models used for allocation assignment purposes, as all linear programming types of problems solved with the simplex algorithm, usually generate noninteger solution vectors.

The integer solution system developed here was originated by R. A. Gomory and its basis is simple number theory.

7.2 THE STRUCTURE OF INTEGER PROGRAMMING

The material in this chapter concerns itself with the class of problems defined to have the following structure:

Obtain the vector $\bar{X}$ that satisfies

$$\bar{A}\bar{X} \leq \bar{B}, \tag{7.1.}$$

that maximizes

$$z = \bar{C}^t\bar{X},$$

and that also satisfies the integer constraints

$$x_j = 0 \text{ or } 1 \text{ or } 2 \text{ or } \ldots \text{ for all } j. \tag{7.2}$$

The constraint set represented by equation (7.2) requires all variables to take on integer values at the optimum. The problem has to be defined in terms of inequalities because even when only *one* equality constraint exists, it is extremely rare for it to have a solution. Expression (7.1) is not meant to imply that all constraints have to be type 1 inequalities. Instead, the initial $\bar{B}$ vector will be assumed unrestricted as to sign, which permits any of the constraints to be of type 2. The definition also admits minimizing as well as maximizing problems because the cost coefficients can be multiplied by minus one. In general, linear programming problems require the variables to be non-negative real numbers. Integer programming specifies the additional requirement that they be *non-negative integers*.

7.2.1 Basis for Development of Algorithm

The most familiar number system is based on the number 10. However, special sets of numbers, much more general than the decimal system, can be developed. Define a number, a, to be *congruent* to another number, b, if their difference is a multiple of another number, c, called a *modulus*. For example,

$$26 \equiv 20 \text{ (mod. 6)},$$

$$8 \equiv -10 \text{ (mod. 6)},$$

$$\tfrac{24}{3} \equiv \tfrac{96}{3} \text{ (mod. 6)}.$$

The triple bar sign is used to signify congruence. Integer programming concerns itself with the system whose modulus is 1. The following examples belong to this system:

$$\tfrac{1}{9} \equiv \tfrac{10}{9} \text{ (mod. 1)},$$

$$-\tfrac{1}{5} \equiv \tfrac{9}{5} \text{ (mod. 1)},$$

and many others. More precisely, $b \equiv a$ (mod. 1) if and only if $b - a$ is an integer. The following notation will also be used:

$$f(b) = a, \qquad 0 \leq a < 1. \tag{7.3}$$

This implies that $b - a$ is an integer; that is, every number b is congruent, modulus 1, with a number between 0 and .99999. . . . A few examples using this notation are

$$f(\tfrac{1}{5}) = \tfrac{1}{5},$$

$$f(\tfrac{9}{5}) = \tfrac{4}{5},$$

$$f(-\tfrac{9}{5}) = \tfrac{1}{5},$$

$$f(-\tfrac{6}{7}) = f(\tfrac{1}{7}) = f(\tfrac{43}{7}) = \tfrac{1}{7}.$$

A special case of importance is

$$f(1) = f(3) = f(510) = f(0) = 0.$$

In (7.3), *a is said to be the congruent equivalent to b*, modulus 1; a is always non-negative.

To proceed, assume that the optimal solution vector of a linear programming problem is nonintegral in violation of the integral condition requirement. The variables can be forced to take an integral form, as follows: Take any row of the final tableau in which the current b_i' is nonintegral:

$$b_i' = a_{ip}' x_p + a_{iq}' x_q + \ldots + a_{ir}' x_r + x_s. \tag{7.4}$$

Here the primes are meant to indicate that the constraint came out of the final tableau, x_s is the solution variable on the ith row (this is indicated by its coefficient of 1), and $x_p, x_q, \ldots, x_r$ are the variables which are not in the final, optimal tableau.

Equation (7.4) can be arranged as follows:

$$b_i' - a_{ip}' x_p - a_{iq}' x_q - \ldots - a_{ir}' x_r = x_s. \tag{7.5}$$

x_s, which now equals the noninteger b_i, must be made to take on an integer value. This can be accomplished by forcing the left side of equation (7.5) to be congruent to zero, that is,

$$b_i' - a_{ip}' x_p - a_{iq}' x_q - \ldots - a_{ir}' x_r \equiv 0. \tag{7.6}$$

This expression can be rewritten as

$$a_{ip}' x_p + a_{iq}' x_q + \ldots + a_{ir}' x_r \equiv b_i'. \tag{7.7}$$

Equation (7.7) can be written as a congruent equivalent expression as follows:

$$f(a_{ip}') x_p + f(a_{iq}') x_q + \ldots + f(a_{ir}') x_r = f(b_i'). \tag{7.8}$$

However, equation (7.8) holds only if $x_p, x_q, \ldots, x_r$ are *integers*, not all equal to zero. It does not hold at first since the variables not in solution are all equal to zero. Because (7.8) is violated, it will act as a constraint that has not been satisfied when introduced into the linear programming tableau. Some of the variables $x_p, x_q, \ldots, x_r$ will be forced into the solution in the following iterations. Thus they become nonzero. Because of this new constraint and others similar to it, the variables are forced to integral levels and eventually satisfy equation (7.8) unless the constraint becomes redundant. The explanation of this last remark will be made later, first, the task of describing exactly how equation (7.8) is introduced into the tableau must be finished.

The coefficients of $x_p, x_q, \ldots, x_r$ in equation (7.8) are all positive by the definition of the congruence functions. Therefore, if $f(b_i')$ equals ϕ, the non-negativity condition, in addition to the integral condition on the variables not in solution, implies that the right side can be equal to ϕ, or to $\phi + 1$, or to $\phi + 2$, or to $\phi + 3$, or . . . , and not violate the constraint. Therefore, when the numerical values of the congruent equivalents are substituted into equation (7.8), one must change the equation into a greater than or equal to inequality, that is, the constraint must be transformed into a type 2 inequality,

As stated previously, this new constraint is not satisfied by the present solution, so one has to develop a new optimal feasible solution through additional iterations. The new optimal solution may not be integral, in which case the process is repeated. It may become necessary to add many constraints of this type before the integral condition is satisfied. The new constraint is usually formed from the b_i' that has the *largest fractional part* in the present solution. Assume that a constraint is formed for a row in which x_s is located in some iteration and that it is necessary to construct another constraint for the same variable in a later iteration. Then the first one has become redundant and may be dropped from the tableau and the new one added in its place. This observation is useful to keep the tableau from becoming excessively large.

These new constraints, called *cutting-plane constraints*, "cut out" some of the feasible region that lies between the portion containing no integer points and the portion containing the integer points. Their function is to force the integral condition. Because the simplest integer problem is four-dimensional, cutting-plane constraints cannot be shown graphically.

When sufficient cutting-plane constraints are provided, only solutions in which all of the variables are non-negative integers are possible. The only variables with nonzero cost coefficients are the ones in the original objective function, therefore, they control the optimal integral solution.

7.2.2 Example Problem

To clarify the concepts developed, solve the following problem: Maximize

$$z = x_1 + 2x_2,$$

subject to

$$x_1 + 3x_2 \leq 9,$$
$$x_1 + x_2 \leq 8, \tag{7.9}$$
$$x_1 = 0, 1, 2, \ldots,$$

and

$$x_2 = 0, 1, 2, \ldots.$$

7.2.2.1 Graphical Solution

Because this is a two-dimensional problem the graphical solution, given in Figure 7.1, is easy to obtain. From the figure it is seen that the maximum noninteger solution occurs at the intersection of the lines defined by the equations

$$x_1 + 3x_2 = 9,$$

and

$$x_1 + x_2 = 8.$$

Solving simultaneously, find

$$x_1 = 7\tfrac{1}{2}$$

and

$$x_2 = \tfrac{1}{2},$$

for which

$$\max z = 8\tfrac{1}{2}.$$

One maximum integer solution is seen to lie at the intersection of the line defined by the equation $x_1 + x_2 = 8$ with the x_1 coordinate axis. The coordinates of this point are, obviously,

$$x_1 = 8$$

and

$$x_2 = 0,$$

for which

$$\max z = 8.$$

Note that this integer optimal solution, being more constrained than the noninteger one, yields a lower value of the objective function. Also note that there is another integer optimum at $x_1 = 6, x_2 = 1$.

7.2.2.2 Analytical Solution by Integer Programming

The final tableau for the optimal noninteger solution is given in Table 7.1. It shows that

$$\text{Max } z = \tfrac{17}{2} = 8\tfrac{1}{2},$$

with

$$x_1 = \tfrac{15}{2} = 7\tfrac{1}{2},$$

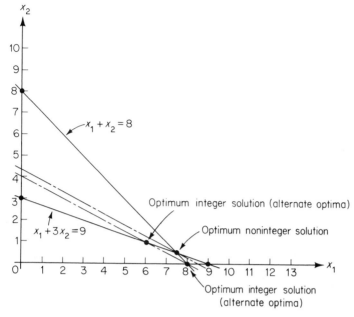

FIGURE 7.1

and

$$x_2 = \tfrac{1}{2}.$$

The x_1 row of this tableau has the largest fractional part and it reads,

$$x_1 - \tfrac{1}{2}x_3 + \tfrac{3}{2}x_4 = \tfrac{15}{2}. \tag{7.10}$$

In accordance with the theory given previously, a new constraint must be formed from equation (7.10) so as to force x_1 to take on an integer value. The congruence relation is obtained as follows:

$$\tfrac{15}{2} + \tfrac{1}{2}x_3 - \tfrac{3}{2}x_4 = x_1, \tag{7.11}$$

and for an integer solution to develop,

$$\tfrac{15}{2} + \tfrac{1}{2}x_3 - \tfrac{3}{2}x_4 \equiv 0. \tag{7.12}$$

TABLE 7–1

c_j		1	2	0	0	0	3	
c_i	Sol.	$\bar{P}_1$	$\bar{P}_2$	$\bar{P}_3$	$\bar{P}_4$	B	Check	θ_i
1	x_1	1	0	$-\tfrac{1}{2}$	$\tfrac{3}{2}$	$\tfrac{15}{2}$	$\tfrac{19}{2}$	
2	x_2	0	1	$\tfrac{1}{2}$	$-\tfrac{1}{2}$	$\tfrac{1}{2}$	$\tfrac{3}{2}$	
	z_j	1	2	$\tfrac{1}{2}$	$\tfrac{1}{2}$	$\tfrac{17}{2}$	$\tfrac{25}{2}$	
	$c_j - z_j$	0	0	$-\tfrac{1}{2}$	$-\tfrac{1}{2}$	$-\tfrac{17}{2}$	$-\tfrac{19}{2}$	

Therefore,

$$\tfrac{1}{2}x_3 - \tfrac{3}{2}x_4 \equiv -\tfrac{15}{2}, \tag{7.13}$$

which can be rewritten

$$f(\tfrac{1}{2})x_3 + f(-\tfrac{3}{2})x_4 = f(-\tfrac{15}{2}). \tag{7.14}$$

From this last expression, obtain

$$\tfrac{1}{2}x_3 + \tfrac{1}{2}x_4 \geq \tfrac{1}{2}. \tag{7.15}$$

Because (7.15) is a type 2 inequality, a slack variable as well as an artificial slack variable must be introduced for solution. Thus,

$$\tfrac{1}{2}x_3 + \tfrac{1}{2}x_4 - x_5 + x_6 = \tfrac{1}{2}. \tag{7.16}$$

Constraint (7.16) must be introduced into the tableau of Table 7–1 to force the noninteger optimal solution to take on an integer value, and solution proceeds following the standard simplex algorithm. For convenience, in this case a large negative cost coefficient for the artificial slack variable will be introduced into the objective function, rather than proceeding with the two-phase simplex algorithm described in chapter 5.

The optimization is accomplished in Table 7–2 with the following results:

TABLE 7–2

c_i	Sol.	1 $\bar{P}_1$	2 $\bar{P}_2$	0 $\bar{P}_3$	0 $\bar{P}_4$	0 $\bar{P}_5$	-10 $\bar{P}_6$	0 B	-7 Check	θ_i
1	x_1	1	0	$-\tfrac{1}{2}$	$\tfrac{3}{2}$	0	0	$\tfrac{15}{2}$	$\tfrac{19}{2}$	—
2	x_2	0	1	$\tfrac{1}{2}$	$-\tfrac{1}{2}$	0	0	$\tfrac{1}{2}$	$\tfrac{3}{2}$	1
-10	x_6	0	0	$\textcircled{\tfrac{1}{2}}$	$\tfrac{1}{2}$	-1	1	$\tfrac{1}{2}$	$\tfrac{3}{2}$	1 $\leftarrow$
	z_j	1	2	$-\tfrac{9}{2}$	$-\tfrac{9}{2}$	10	-10	$\tfrac{7}{2}$	$-\tfrac{5}{2}$	
	$c_j - z_j$	0	0	$\tfrac{9}{2}$	$\tfrac{9}{2}$	-10	0	$-\tfrac{7}{2}$	$-\tfrac{9}{2}$	
				$\uparrow$						
1	x_1	1	0	0	2	-1	1	8	11	
2	x_2	0	1	0	-1	1	-1	0	0	
0	x_3	0	0	1	1	-2	2	1	3	
	z_j	1	2	0	0	1	-1	8	11	
	$c_j - z_j$	0	0	0	0	-1	-9	-8	-18	

$$\text{Max } z = 8;$$

$$x_1 = 8, \text{ an integer value};$$

and

$$x_2 = 0, \text{ an integer value}.$$

These results, of course, confirm those obtained in the graphical solution.

Note that the integer solution yields alternate optima on slack variable x_4 because, although x_4 is not in solution, $c_4 - z_4 = 0$, denoting that the introduction of this variable into solution would not change the value of the objective function. Also, the optimal integer solution is degenerate because $b'_2 = 0$. This example was selected for its simplicity, and the reader must be warned that he will seldom be able to find the optimal integer solution with so little effort. (Only one iteration beyond the noninteger optimum was required.) Integer routines have been included into computer codes and integer programming packages are available at most computer centers.

EXERCISES

7-1. A contractor wishes to rent heavy construction equipment for a particularly large job. Two types of equipment are available and the personnel required for each is given below:

	Operators	Helpers
Type 1	1	1
Type 2	1	3

The contractor can employ *no more* than 5 operators and 6 helpers.
After paying all rental and labor costs, he calculates that his profits from the use of the equipment will be \$20/hr. for type 1 and \$30/hr. for type 2. How many units of each type should the contractor rent to maximize his hourly profit?
a. Solve graphically.
b. Solve analytically using integer programming.

7-2. Obtain the *integer* program that *maximizes*.

$$y = 4x_1 + 3x_2$$

subject to $x_1, x_2 = 0, 1, 2, \ldots$,

$$x_1 + 2x_2 \leq 4,$$

and

$$2x_1 + x_2 \leq 6.$$

7-3. Three types of items are available for packing into a survival kit which has a maximum volumetric capacity of *10 cu. ft.* The table below gives pertinent information on the items; U_i and V_i are the per-unit utility and volume of item i, respectively:

i	U_i	V_i
1	\$3	2 cu. ft.
2	\$4	3 cu. ft.
3	\$1	1 cu. ft.

Assuming one *cannot* select more than *three* units of each item, find the selection policy (*S*) which *maximizes* the total kit *utility*. Solve using *integer programming*.

a. Develop the model.

b. Use the simplex method to obtain the noninteger optimum.

c. Use the integer programming algorithm.

d. Obtain all alternate optima, if they exist.

REFERENCES

1. GARVIN, W. W., *Introduction to Linear Programming*. New York: McGraw-Hill Book Company, 1960.

2. GOMORY, R. A., "Outline of an Algorithm for Integer Solutions to Linear Programs," Bulletin American Mathematical Society (September 1958), 64.

3. ———, "An Algorithm for Integer Solutions to Linear Programs," Technical Report No. 1, IBM Mathematical Research Project. Princeton, 1958.

4. ———, and W. J. BAUMOL, "Integer Programming and Pricing," *Econometrica*, XXVIII (1960), 521–50.

5. LLEWELLYN, R. W., *Linear Programming*. New York: Holt, Rinehart & Winston, Inc., 1964.

6. MARKOWITZ, H. M. and A. S. MANNE, "On the Solution of Discrete Programming Problems," *Econometrica*, XXV, No. 1 (1957).

CHAPTER **8**

Transportation, Transshipment, and Assignment Problems

8.1 INTRODUCTION

This chapter will be devoted to linear programming problems which belong to two special classes: the transportation and the assignment categories. The matrix formulations of these problems are such that although they form part of the linear programming group, special algorithms, easier to implement than the simplex method, can be developed and used in their solutions. These algorithms are excellent examples of the simplifications that result when one takes advantage of the special structure of the matrices encountered in the problem formulation.

8.2 THE TRANSPORTATION PROBLEM

Although transportation type models are applicable to the solution of problems having nothing to do with the transportation of products, the transportation problem can best be paraphrased by considering m factories (sources) which produce goods at levels $a_1, a_2, \ldots, a_m$, and n warehouses (sinks) with demands for the same goods in amounts $b_1, b_2, \ldots, b_n$. The cost of transporting one unit of product from factory i to warehouse j is $c_{ij}, i = 1, 2, \ldots, m; j = 1, 2, \ldots, n$. The objective of the solution is to establish the shipping pattern that exhausts all factory supplies and fulfills all warehouse demands at a minimum cost. To accomplish this, it is necessary to assume that the total production of units exactly equals the total demand. The number of units shipped from factory i to warehouse j will be symbolized by $x_{ij}, i = 1, 2, \ldots, m; j = 1, 2, \ldots, n$. Figure 8.1 shows a schematic description of the physical model.

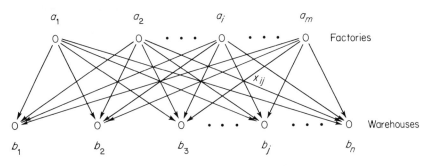

FIGURE 8.1. Model Description.

In formulating the mathematical model one must recognize the existence of two matrices:

1. The *cost matrix*, giving the costs of transporting one unit from each factory to each warehouse.

Warehouses

		1	2	$\cdots$	j	$\cdots$	n
	1	c_{11}	c_{12}	$\cdots$	c_{1j}	$\cdots$	c_{1n}
	2	c_{21}	c_{22}	$\cdots$	c_{2j}	$\cdots$	c_{2n}
	.	$\cdots$	$\cdots$	$\cdots$	$\cdots$	$\cdots$	$\cdots$
	.	$\cdots$	$\cdots$	$\cdots$	$\cdots$	$\cdots$	$\cdots$
Factories	.	$\cdots$	$\cdots$	$\cdots$	$\cdots$	$\cdots$	$\cdots$
	i	c_{i1}	c_{i2}	$\cdots$	c_{ij}	$\cdots$	c_{in}
	.	$\cdots$	$\cdots$	$\cdots$	$\cdots$	$\cdots$	$\cdots$
	.	$\cdots$	$\cdots$	$\cdots$	$\cdots$	$\cdots$	$\cdots$
	.	$\cdots$	$\cdots$	$\cdots$	$\cdots$	$\cdots$	$\cdots$
	m	c_{m1}	c_{m2}	$\cdots$	c_{mj}	$\cdots$	c_{mn}

2. The *distribution matrix* showing the number of units to be shipped from each factory to each warehouse. These quantities are, of course, the unknowns x_{ij}. However, note that the sum of the elements along each row must be equal to the production at the factory

represented by that row, and that the sum of the elements along each column must be equal to the demand at the warehouse represented by that column. These entries, and production and demand levels, are respectively called the row requirements and the column requirements. In a balanced situation the sum of the row requirements (the total production) equals the sum of the column requirements (the total demand). An unbalanced condition can be easily transformed into a balanced one. This topic will be discussed later in the chapter. The distribution matrix has the following form:

Warehouses

	1	2	$\cdots$	j	$\cdots$	n	Row Requirements
1	x_{11}	x_{12}	$\cdots$	x_{1j}	$\cdots$	x_{1n}	a_1
2	x_{21}	x_{22}	$\cdots$	x_{2j}	$\cdots$	x_{2n}	a_2
.	$\cdots$	$\cdots$	$\cdots$	$\cdots$	$\cdots$	$\cdots$	.
.	$\cdots$	$\cdots$	$\cdots$	$\cdots$	$\cdots$	$\cdots$	.
i	x_{i1}	x_{i2}	$\cdots$	x_{ij}	$\cdots$	x_{in}	a_i
.	$\cdots$	$\cdots$	$\cdots$	$\cdots$	$\cdots$	$\cdots$	.
.	$\cdots$	$\cdots$	$\cdots$	$\cdots$	$\cdots$	$\cdots$	.
.	$\cdots$	$\cdots$	$\cdots$	$\cdots$	$\cdots$	$\cdots$	.
m	x_{m1}	x_{m2}	$\cdots$	x_{mj}	$\cdots$	x_{mn}	a_m

Factories (label at left, rows i)

| b_1 | b_2 | $\cdots$ | b_j | $\cdots$ | b_n |

Column Requirements

The mathematical model can now be stated as follows: Minimize

$$z = \sum_{i=1}^{m} \sum_{j=1}^{n} c_{ij}x_{ij}, \qquad (8.1)$$

subject to

$$x_{ij} \geq 0 \quad \text{for all } i, j, \qquad (8.2)$$

and

$$\sum_{j=1}^{n} x_{ij} = a_i, \qquad i = 1, 2, \ldots, m, \qquad (8.3)$$

$$\sum_{i=1}^{m} x_{ij} = b_j, \qquad j = 1, 2, \ldots, n. \qquad (8.4)$$

Thus, one has a linear objective function subject to linear constraints.

The objective function (8.1) represents the total cost of transportation. This is the reason for the double summation. The non-negativity condition (8.2) is required because it makes no physical sense to ship negative units. Constraints (8.3) state that the sum of the products leaving each factory must be equal to what the factory produces. Constraints (8.4) state that the sum of the products arriving at each warehouse must be equal to the demand at the warehouse.

The constraint set, equations (8.3) and (8.4), is seen to consist of $m + n$ equations in $m \times n$ unknowns. However, because of the existence of the balanced condition (production equals demand) in the problem, any one of the equations is linearly dependent upon the others. Note that from equations (8.3),

$$\sum_{i=1}^{m} a_i = \sum_{i=1}^{m} \sum_{j=1}^{n} x_{ij}, \qquad (8.5)$$

and from equations (8.4),

$$\sum_{j=1}^{n} b_j = \sum_{j=1}^{n} \sum_{i=1}^{m} x_{ij}. \qquad (8.6)$$

Because the order of the summation does not affect the result, equations (8.5) and (8.6) are equal to each other, which means that

$$\sum_{i=1}^{m} a_i = \sum_{j=1}^{n} b_j. \qquad (8.7)$$

Therefore, the balanced condition is, in fact, a consistency condition which must be satisfied if a solution is to exist. Equation (8.7), in turn, establishes a relationship between equations (8.3) and (8.4), thereby eliminating one degree of freedom from the system and making one of the equations linearly dependent upon the rest. Consequently, the set (8.3) and (8.4) consists of $m + n - 1$ linearly independent equations in $m \times n$ unknowns. A basic feasible (nondegenerate) solution must, therefore, have $m + n - 1$ positive components.

In attempting to solve the transportation problem with the simplex algorithm, one would have to introduce $m + n - 1$ artificial slack variables to form an initial basic feasible (nondegenerate) solution, because the constraint set consists of equations only. However, due to the special structure of the constraint matrix, a simpler, easier-to-apply method of solution can be developed. Note that the constraint system (8.3) and (8.4) possesses two very important features:

1. The coefficients of all the variables in the equations are ones.
2. Any variable x_{ij} appears only once in the first m equations and only once in the last n equations.

These special characteristics of the constraint set permit the development of a special transportation algorithm which will be discussed later in the chapter.

8.2.1 An Example Problem

A company has three plants which manufacture a certain type of product. The first plant has 15 units in inventory, the second has 25, and the third 20. The company supplies two warehouses. Warehouse number one requires 35 units to meet the expected demand, and the second warehouse needs 25. The dollar costs of shipping one unit of product from factory i to warehouse j are given below.

Warehouses

		1	2	Inventories
	1	15	20	15
Factories	2	18	22	25
	3	25	19	20
Demands		35	25	

The plant inventories and the warehouse demands are shown with the cost matrix. The problem is balanced—that is, the total production of 60 units equals the total demand. What shipping plan will meet the warehouse requirements at a minimum cost?

8.2.2 The Mathematical Model

Let x_{ij} be the number of units shipped from factory i to warehouse j, $i = 1, 2, 3$; $j = 1, 2$. Then the model is
Minimize

$$z = 15x_{11} + 20x_{12} + 18x_{21} + 22x_{22} + 25x_{31} + 19x_{32}$$

subject to

$$x_{ij} \geq 0 \qquad \text{for all } i \text{ and } j,$$

and

$$\left. \begin{array}{l} x_{11} + x_{12} = 15, \\ x_{21} + x_{22} = 25, \\ x_{31} + x_{32} = 20, \end{array} \right\} \text{Factory Inventories}$$

$$\left. \begin{array}{l} x_{11} + x_{21} + x_{31} = 35, \\ x_{12} + x_{22} + x_{32} = 25. \end{array} \right\} \text{Warehouse Demands}$$

Because the sum of the row requirements equals the sum of the column requirements, only four of the five equations in the constraint set are linearly independent. Thus one can eliminate any one of the equations as redundant. Say the last equation is eliminated. One is left with the set

$$x_{11} + x_{12} = 15,$$
$$x_{21} + x_{22} = 25,$$
$$x_{31} + x_{32} = 20,$$
$$x_{11} + x_{21} + x_{31} = 35.$$

This set of four linearly independent equations contains six unknowns. A basic feasible (nondegenerate) solution will have four positive components. Therefore, two of the variables must be set equal to zero. Any two, as long as one does not create an inconsistency. For example, if one sets $x_{11} = 0$ and $x_{12} = 0$, one would have for the first equation

$$0 + 0 = 15,$$

an obvious inconsistency. However, one can arbitrarily set, say, $x_{12} = 0$ and $x_{31} = 0$. The system is now reduced to

$$x_{11} = 15,$$
$$x_{21} + x_{22} = 25,$$
$$x_{32} = 20,$$
$$x_{11} + x_{21} = 35.$$

It can be rearranged as follows:

$$x_{11} = 15,$$
$$x_{11} + x_{21} = 35,$$
$$x_{21} + x_{22} = 25,$$
$$x_{32} = 20.$$

With this arrangement, one very important property of the matrix comes to the surface. The matrix is triangular; that is, all elements above the "main" diagonal are zeros. Depending upon which variables are set to zero and how the equations are arranged, one could generate different types of triangular matrices. But the triangularity would always be present. What this means, of course, is that one can solve the system quickly and easily. Observe: From the first equation, $x_{11} = 15$. Because of this, the second equation immediately yields $x_{21} = 35 - 15 = 20$. Then, from the third, $x_{22} = 25 - 20 = 5$. Finally, from the fourth equation, $x_{32} = 20$.

The same solution can be obtained more systematically by a simple procedure which yields the so-called *northwest corner solution*.

8.2.3 The Northwest Corner Solution

The solution obtained before can be put in the following format:

$$\bar{X} = \begin{array}{|c|c|} \hline 15 & \\ \hline 20 & 5 \\ \hline & 20 \\ \hline \end{array}$$

This is an example of the northwest corner solution which can also be formed by following an independent process that accomplishes the same objective:

to solve a triangular set of linear equations. The process is implemented as follows:

1. Lay out a blank matrix with the problem's row and column requirements.

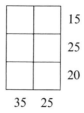

35   25

2. Let the northwest corner element x_{11} be equal to the smaller of the first row requirement and the first column requirement. Having done this, reduce the row and column requirements by the value of x_{11}.

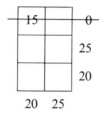

20   25

3. Eliminate the row or column whose requirement has been reduced to zero. In the example problem the first row is eliminated from further consideration.
4. Repeat step 2 using the northwest corner element of the matrix left after elimination of the first row or column. This element is x_{12} in the example because the first row was eliminated in step 3.

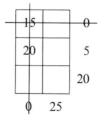

0   25

5. Repeat the process by applying step 2 until the solution is completed.

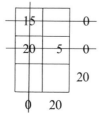

0   20

15		0
20	5	0
	20	0

　　0　　0

This solution is, of course, the same obtained previously from the triangular set. Its value is computed from the cost and distribution matrices, as follows:

$$z = 15(15) + 20(0) + 18(20) + 22(5) + 25(0) + 19(20) = 1,075.$$

The zero entries are $x_{12} = 0$ and $x_{31} = 0$ because they are not in the present solution.

In general, the northwest corner solution is not optimal (minimal). Optimality can be ascertained and an optimal solution developed with a special transportation algorithm which will be discussed in section 8.5.

Note that the system of four linearly independent equations represented by the matrix given above has four positive components in solution and, therefore, it is a basic feasible (nondegenerate) solution. In the following section degeneracy will be discussed in detail.

8.2.4 Degeneracy

It is possible for the northwest corner solution to yield a degenerate solution —that is, a solution with less than $n + m - 1$ positive components. This occurs whenever the sum of some ordered subset of the row requirements is equal to the sum of some ordered subset of the column requirements. For example,

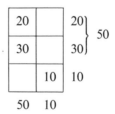

　　50　　10

In the above matrix, $a_1 + a_2 = b_1$, and $a_3 = b_2$. Consequently, those ordered subsets of row and column requirements have equal sums. The result of this condition is that when the northwest corner solution is being formed, a row and a column requirement are fulfilled simultaneously at element x_{21}. Thus, both a row and a column are eliminated and one less variable appears in the solution. This creates one degree of degeneracy. The number of degrees of degeneracy is equal to the number of such ordered subsets of row and column requirements with equal sums.

To avoid degeneracy a perturbation technique can be used. The technique consists of adding a small positive constant, ϵ, to *each* of the row requirements and $m\epsilon$ to *one* of the column requirements. For the example given above, assume $\epsilon = 0.01$. This is a small number when compared to the values of the requirements. The method yields the following nondegenerate solution (four variables in solution):

20.01		$20.01 = 20 + \epsilon$
29.99	0.02	$30.01 = 30 + \epsilon$
	10.01	$10.01 = 10 + \epsilon$

50.00 $10.03 = 10 + m\epsilon$

A nondegenerate solution can also be obtained by initiating the computations at a cell for which the equality of the sums of the ordered subsets of row and column requirements is destroyed. Thus, for the same problem, one can form a nondegenerate solution by choosing a different sequence (order) of steps to form it. Thus, if one starts at x_{22},

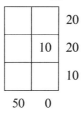

and proceeds from there,

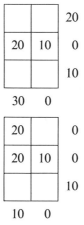

one obtains

20		0
20	10	0
10		0
0	0	

Although this is *not* a northwest corner solution, the procedure yielded a basic feasible (nondegenerate) one. This last approach is, in the author's opinion, preferable to the perturbation technique for forming a starting non-degenerate solution. Basically, the method rearranges the rows and columns and thereby eliminates the cause of degeneracy.

Once an initial basic feasible (nondegenerate) solution has been found, the remaining iterations, to obtain a minimal solution, can be performed with the simplex algorithm. However, the special transportation algorithm, to be discussed in section 8.5, is much simpler and easier to apply.

8.3. THE VOGEL APPROXIMATION METHOD

Frequently, the northwest corner method yields an initial solution which is far from minimal. A special procedure called the Vogel Approximation Method generates an initial basic feasible solution which is very close to being minimal in most cases.

The procedure for forming this solution is the following:

1. Determine the penalty for each row and column that one would have to pay if a variable were not placed in the initial solution inside the cell that has the smallest cost on that row or column. These penalties are defined as the differences between the next smallest and the smallest costs in each row and column. If two elements in a row or a column are tied for the rank of smallest, the penalty is zero.

2. After the penalties have been calculated for all rows and columns, locate the largest, whether a row or a column penalty, and place the variable in the cell that has the smallest cost in the row or column with the largest penalty. The value of the variable is set equal to the smaller of the row and and column requirements corresponding to the variable being brought into solution. The row or column whose requirement is satisfied is deleted from further consideration and the requirement of the other (row or column) is reduced by the value assigned to the variable entering the solution.

3. If a row requirement has been satisfied, the column penalties must be

recomputed because the elements of the cost matrix corresponding to the row deleted are no longer considered in the calculation of column penalties. If a column requirement is satisfied, the row penalties must be recomputed.

4. If a tie develops between two or more row or column penalties, select arbitrarily (using good judgment) which among the tied rows and/or columns will be used to introduce the next variable into solution. As the Vogel method is only an approximation, the author does not consider it necessary to develop a more refined procedure for this case.

5. Iterate steps 2 through 4 until the solution is completed.

Although it cannot be proved mathematically that the Vogel method yields a near minimal solution, extensive computational experience has shown this to be the case. Application of the method will be demonstrated with the problem given in section 8.2.1.

Cycle 1:

STEP 1:

Compute the row and column penalties on the cost matrix.

Row Penalties

15	20	5
18	22	4
25	(19)	6 ←

Column Penalties   3   1

The largest penalty is 6 (marked with the arrow). Therefore, the first variable will be entered in the third row in position x_{32} because 19 is the lowest cost cell in this row.

STEP 2:

Referring to the distribution matrix and assigning a maximum value to x_{32}, the row requirement is satisfied and the column requirement is reduced to 5.

Row Requirements

		15
		25
	20	2̶0̶ 0

Column Requirements   35   2̶5̶
                              5

STEP 3:

Because row 3 has been satisfied, it is eliminated from further consideration and the column penalties must be recomputed.

Cycle 2:

STEP 1:

Recompute the column penalties on the cost matrix.

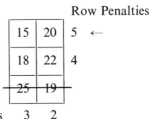

Row Penalties

15	20	5 ←
18	22	4
~~25~~	~~19~~	

Column Penalties   3   2

The largest penalty corresponds to row 1.

STEP 2:

Enter variable x_{11} (c_{11} is the smallest cost in row 1) in the distribution matrix.

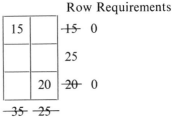

Row Requirements

15		~~15~~ 0
		25
	20	~~20~~ 0

~~35~~ ~~25~~

Column Requirements   20   5

STEP 3:

Row 1 has been satisfied. Recompute the column penalties.

Cycle 3:

STEP 1:

The column penalties are now zero because there is only one row left to consider. From the cost matrix one has,

Row Penalties

~~15~~	~~20~~	
18	22	4 ←
~~25~~	~~19~~	

Column Penalties   0   0

The largest penalty corresponds to row 2.

STEP 2:

Enter variable x_{21} in the distribution matrix.

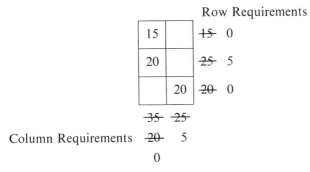

Row Requirements

Column Requirements

STEP 3:

Column 1 has been satisfied. Recompute the row penalties.

Cycle 4:

STEP 1:

All penalties in the cost matrix are now zero.

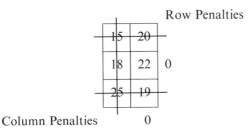

Row Penalties

Column Penalties

STEP 2:

Enter variable x_{22} into solution in the distribution matrix.

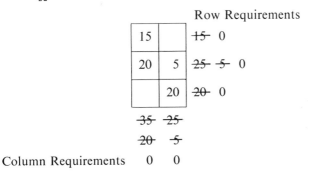

Row Requirements

Column Requirements

STEP 3:

A basic feasible solution has been formed.

Comments:

For this particular problem, the Vogel solution is seen to be identical to the northwest corner solution, indicating that the northwest corner solution was, to begin with, either minimal or near minimal. In most problems, however, especially in larger ones, Vogel yields a much better initial solution than the northwest corner, because this last method depends on the ordering of the rows and columns and, if the ordering happens to be a poor one, the number of iterations required to optimize an initial northwest corner solution may be very large. For most problems, because of the nature of the procedure, Vogel yields either a minimal initial solution or one that is very close to being minimal, such that just a few iterations are needed to achieve optimality.

8.4 UNBALANCED CONDITIONS

When the sum of the row requirements is not equal to the sum of the column requirements the transportation problem is said to be unbalanced. There are two possibilities:

1. *Overproduction.* If the sum of the row requirements is larger than the sum of the column requirements, that is,

$$\sum_{i=1}^{m} a_i > \sum_{j=1}^{n} b_j, \tag{8.8}$$

the production at the factories or sources exceeds the demand at the warehouses or sinks, and a condition of overproduction exists.

The problem can be balanced by creating an artificial warehouse where the excess units are sent. This is equivalent to adding one column to the cost and distribution matrices. This additional column is given a requirement equal to the difference between the sum of the row and column requirements, that is, the demand at the artificial warehouse is given by

$$\sum_{i=1}^{m} a_i - \sum_{j=1}^{n} b_j. \tag{8.9}$$

For the purposes of this book, the cost coefficients for the artificial column will be assumed to be all zeros. However, they could have positive values equal to the cost of storing the excess inventory at each warehouse, for example. Their true values depend on the nature of the problem being studied.

As an example of overproduction, consider a problem with the following cost matrix and row and column requirements:

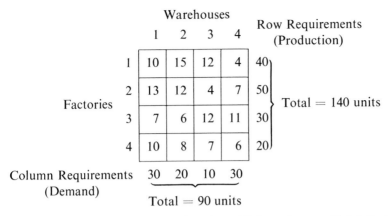

Warehouse

		1	2	3	4	Row Requirements (Production)
Factories	1	10	15	12	4	40
	2	13	12	4	7	50
	3	7	6	12	11	30
	4	10	8	7	6	20

Total = 140 units

Column Requirements (Demand)   30   20   10   30

Total = 90 units

The overproduction is 50 units. With the addition of an artificial factory (column), the balanced problem becomes,

Warehouses

		1	2	3	4	5	Row Requirements (Production)
Factories	1	10	15	12	4	0	40
	2	13	12	4	7	0	50
	3	7	6	12	11	0	30
	4	10	8	7	6	0	20

Total = 140 units

Column Requirements (Demand)   30   20   10   30   50

Total = 140 units

Warehouse 5 is the artificial warehouse. Its demand is 50 units. This balances the problem and standard solution procedures can be applied without any difficulty.

2. *Underproduction.* When the sum of the row requirements is less than the sum of the column requirements, that is,

$$\sum_{i=1}^{m} a_i < \sum_{j=1}^{n} b_j, \qquad (8.10)$$

the demand at the warehouses (sinks) exceeds the production at the factories (sources) and underproduction exists. To balance the problem, add an artificial factory with a scheduled production equal to the unsatisfied demand,

$$\sum_{j=1}^{n} b_j - \sum_{i=1}^{m} a_i. \qquad (8.11)$$

Again, for the purposes of this book, the cost coefficients will be assumed all zeros. Their true values depend upon the situation at hand.

The following is an example of underproduction:

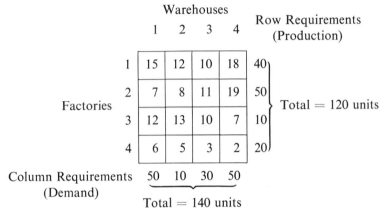

The level of underproduction is 20 units. To balance the problem, add an artificial factory (row) with a scheduled production of 20 units.

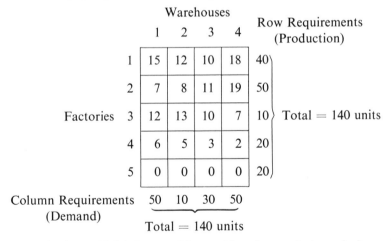

Factory 5 is the artificial factory. The problem is now balanced. A special transportation algorithm for minimizing the balanced transportation problem will be developed in the next section.

8.5 THE TRANSPORTATION ALGORITHM

By taking advantage of the triangularity of the constraint matrix and the fact that all its structural coefficients are ones and zeros, a special algorithm can be formulated for balanced transportation problems. The algorithm is, basically, only a simplification of the simplex method. Proof of the following theorems leads to its development.

8.5.1   Theorem No. 1

Every transportation problem with degenerate basic feasible solutions can be transformed into a problem with only nondegenerate basic feasible solutions.

Proof:

The proof is immediate. By applying one of the methods discussed in Section 8.2.4, perturbation and/or rearrangement of the rows and columns, the cause of degeneracy (equality of partial sums of ordered row and column requirements) is eliminated.

8.5.2   Theorem No. 2

If one has obtained an initial basic feasible solution to a transportation problem with no degenerate solutions, one can construct a second basic feasible solution from the first by introducing a new variable from the solution set. This second basic feasible solution is also unique.

Proof:

The application of Theorem No. 1 eliminates degeneracy. To prove the theorem, one must simply recall that the transportation problem is only a linear programming problem and, consequently, the four basic theorems discussed in chapter 5 also apply to it. The development of the proof of this theorem, however, yields specialized results.

Consider a basic feasible solution to a typical problem:

	1	2	3	4	
1		x_{12}			a_1
2	x_{21}		x_{23}		a_2
3			x_{33}	x_{34}	a_3
4		x_{42}	x_{43}		a_4
	b_1	b_2	b_3	b_4	

The matrix has seven variables in solution as required for nondegeneracy. The value of the solution is given by

$$z = c_{12}x_{12} + c_{21}x_{21} + c_{23}x_{23} + c_{33}x_{33} + c_{34}x_{34} + c_{42}x_{42} + c_{43}x_{43}.$$

Suppose one wished to introduce variable x_{14} into the solution at a small positive value θ_{14}. To keep the row and column requirements satisfied, θ_{14} must be added and subtracted in the following manner:

	1	2	3	4	
1		$x_{12-\theta}$		θ	a_1
2	x_{21}		x_{23}		a_2
3			$x_{33+\theta}$	$x_{34-\theta}$	a_3
4		$x_{42+\theta}$	$x_{43-\theta}$		a_4
	b_1	b_2	b_3	b_4	

Note that the addition and subtraction of θ_{14} traces a path through the matrix. This path is called the plus-minus $(+-)$ path. Because there must be at least one variable in solution in every row and in every column, the $(+-)$ path is always a closed path.

Now assume that θ_{14} is allowed to increase gradually from zero. In order to maintain feasibility, the maximum value that θ_{14} can achieve is equal to the value of the smallest variable in the present solution from which θ_{14} is *subtracted*. Once θ_{14} reaches this value, that variable leaves the solution and one has, again, $n + m - 1 = 7$ positive variables in the solution set. Thus, a new basic feasible (nondegenerate) solution has been formed. To prove that such new basic feasible solution exists always, observe that if θ_{ik} is added to one vacant cell, it must be subtracted from at least two solution variables (one in the ith row and one in the j column). Therefore, no less than two solution variables will be driven toward zero as θ_{ij} increases. The assumption of nondegeneracy guarantees that two of the decreasing variables will not be driven to zero simultaneously. A final word can be said about existence and uniqueness: Because the transportation problem is a linear programming problem, these properties are guaranteed by the four theorems proved in chapter 5.

8.5.3 Theorem No. 3

The minimal solution to a transportation problem with no degenerate basic feasible solutions and with a unique optimum must be a basic feasible solution.

Proof:

It has been shown that a transportation problem is a special type of linear programming problem, with an objective function to be minimized (or the negative of the objective function maximized) subject to $m + n - 1$

linearly independent constraint equations. Therefore, the optimal solution must contain $m + n - 1$ positive variables (it was assumed that degenerate basic feasible solutions are impossible) and, consequently, be a basic feasible solution.

8.5.3.1 Corollary

As for linear programming problems, if there exist two basic feasible solutions of the same minimum cost, then there exist an infinite number of minimal solutions. These solutions are called alternate optima.

8.5.4 Theorem No. 4

Once a basic feasible solution to a transportation problem has been generated, the value of each basic feasible solution which can be developed from the first can be determined a priori.

Proof:

From the work done in Theorem No. 2, one has that the value of the initial basic feasible solution for the example problem

	1	2	3	4	
1		$-$ x_{12}		$+$	a_1
2	x_{21}		x_{23}		a_2
3			$+$ x_{33}	$-$ x_{34}	a_3
4		$+$ x_{42}	$-$ x_{43}		a_4
	b_1	b_2	b_3	b_4	

is given by

$$z = c_{12}x_{12} + c_{21}x_{21} + c_{23}x_{23} + c_{33}x_{33} + c_{34}x_{34} + c_{42}x_{42} + c_{43}x_{43}.$$

By introducing x_{14} into solution at a small positive value θ_{14}, the cost of the new solution is, from the $(+-)$ path,

$$z' = c_{12}(x_{12} - \theta_{14}) + c_{14}\theta_{14} + c_{21}x_{21} + c_{23}x_{23} + c_{33}(x_{33} + \theta_{14})$$
$$+ c_{34}(x_{34} - \theta_{14}) + c_{42}(x_{42} - \theta_{14}) + c_{43}(x_{43} - \theta_{14})$$
$$= c_{12}x_{12} + c_{21}x_{21} + c_{23}x_{23} + c_{34}x_{34} + c_{42}x_{42} + c_{43}x_{43}$$
$$+ [c_{14} - c_{12} + c_{42} - c_{43} + c_{33} - c_{34}]\theta_{14}$$
$$= z + [c_{14} - c_{12} + c_{42} - c_{43} + c_{33} - c_{34}]\theta_{14}.$$

Let

$$c_{14} - c_{12} + c_{42} - c_{43} + c_{33} - c_{34} = D_{14},$$

then,

$$z' = z + D_{14}\theta_{14}.$$

The change in the cost is, consequently, given by $D_{14}\theta_{14}$. In general, if θ_{ij} is introduced into solution, the $(+-)$ path will trace a D_{ij} as follows:

$$D_{ij} = c_{ij} - c_{iq} + c_{tq} - c_{tp} + \cdots - c_{rj}, \qquad (8.12)$$

where D_{ij} is, in fact, the sum of the costs around the $(+-)$ path. The value of the new solution is given by

$$z' = z + D_{ij}\theta_{ij}. \qquad (8.13)$$

Because $\theta_{ij} \geq 0$, the new solution cost, z', will be smaller than the old solution cost, z, if and only if $D_{ij} < 0$.

All of the elements in equation (8.13) can be determined without actually forming any of the new basic feasible solutions. This proves the theorem.

8.5.5 The Rule of Steepest Descent

With the information developed in Theorem No. 4, it can now be concluded that in order to develop the best new basic feasible solution one would only have to compute all $D_{ij}\theta_{ij}$ for each variable not in the present solution. This procedure, however, involves computation of two quantities, D_{ij} and θ_{ij}. Rather than doing this, because $\theta_{ij} \geq 0$, the standard procedure is to compute all possible D_{ij} and then select the smallest $D_{ij} < 0$. This rule is called the *rule of steepest descent* and parallels the rule of steepest ascent developed in chapter 5 for the simplex method. Therefore,

$$D_{ij} = \min D_{kl}, \ D_{kl} < 0. \qquad (8.14)$$

Still, the computation of all D_{kl} would involve the tracing of $(+-)$ paths for all variables not in the present solution. To avoid this lengthy and tedious procedure, a method that utilizes Lagrange multipliers (also called simplex multipliers) will be developed in the following section.

8.5.6 Simplex Multipliers and the Kuhn-Tucker Conditions

Consider the typical transportation model,

$$\text{Min } z = \sum_{i=1}^{m} \sum_{j=1}^{n} c_{ij}x_{ij}$$

subject to

$$x_{ij} \geq 0 \text{ for all } i \text{ and } j,$$

$$\sum_{j=1}^{n} x_{ij} = a_i, \ i = 1, \ldots, m, \text{ (row equations)}$$

and

$$\sum_{i=1}^{m} x_{ij} = b_j, j = 1, \ldots, n. \text{ (column equations)}$$

Let u_i and v_j be the Lagrange (also called simplex) multipliers for the row and column equations, respectively. The Lagrangian function (see chapter 3) for the problem is

$$L = \sum_{i=1}^{m} \sum_{j=1}^{n} c_{ij} x_{ij} - \sum_{i=1}^{m} u_i \left(\sum_{j=1}^{n} x_{ij} - a_i \right) - \sum_{j=1}^{n} v_j \left(\sum_{i=1}^{n} x_{ij} - b_j \right).$$

(8.15)

For the Lagrangian to be at a minimum, its total differential must be non-negative for every possible perturbation of the variable x_{ij}. Thus,

$$dL = \sum_{i=1}^{m} \sum_{j=1}^{n} \frac{\partial L}{\partial x_{ij}} dx_{ij} \geq 0.$$

(8.16)

Two distinct variables x_{ij} must be recognized. Solution variables, $x_{st} > 0$, and nonsolution variables, $x_{pr} = 0$.

Consider all possible perturbations of a solution variable, $x_{st} > 0$, while all other variables are held constant. Because x_{st} is positive in the present solution, its perturbation dx_{st} can be positive or negative and still maintain feasibility. Therefore, in order to guarantee the non-negativity of dL, the corresponding partial derivative must vanish at the minimum.

$$\frac{\partial L}{\partial x_{st}} = 0.$$

(8.17)

It follows that the product of x_{st} and $\partial L / \partial x_{st}$ must also vanish at the minimum.

$$x_{st} \left(\frac{\partial L}{\partial x_{st}} \right) = 0.$$

(8.18)

Now consider all possible perturbations of a nonsolution variable, $x_{pr} = 0$, while all other variables are held constant. Because x_{pr} is zero in the present solution, its perturbation dx_{pr} must be positive to maintain feasibility. To guarantee the non-negativity of dL, it is only necessary that its corresponding partial derivative be non-negative at the minimum. Thus,

$$\frac{\partial L}{\partial x_{pr}} \geq 0.$$

(8.19)

Because $x_{pr} = 0$, it, again, follows that

$$x_{pr} \left(\frac{\partial L}{\partial x_{pr}} \right) = 0$$

(8.20)

at the minimum.

Equations (8.17) through (8.20) are the Kuhn-Tucker conditions for a minimum. These are necessary conditions for minimum points in mathematical programming problems. Note that they can be grouped as follows:

1. Non-negativity conditions [from equations (8.17) and (8.19)],

$$\frac{\partial L}{\partial x_{ij}} \geq 0, \text{ for all } i \text{ and } j, \text{ at the minimum.} \tag{8.21}$$

2. Complementary slackness conditions [from equation (8.18) and (8.20)],

$$x_{ij}\left(\frac{\partial L}{\partial x_{ij}}\right) = 0, \text{ for all } i \text{ and } j \text{ at the minimum.} \tag{8.22}$$

Having developed the Kuhn-Tucker conditions, it is now possible to establish a relationship between the cost coefficients for the solution variables and the simplex multipliers u_i and v_j.

Consider the partial derivative of the Lagrangian function with respect to a solution variable x_{st}. From equation (8.15)

$$\frac{\partial L}{\partial x_{st}} = c_{st} - u_s - v_t = 0 \tag{8.23}$$

at the minimum, because $x_{st} > 0$ [equation (8.17)].

Therefore, for all solution variables,

$$c_{st} = u_s + v_t. \tag{8.24}$$

This, of course, is not true for nonsolution variables because the corresponding partial derivatives for the Lagrangian function do not have to vanish [equation (8.19)]. Now consider the rule of steepest descent, equation (8.14), and the definition of D_{ij}, equation (8.12). The D_{ij} coefficients must be computed for all nonsolution variables; thus, in equation (8.12),

$$D_{ij} = c_{ij} - c_{iq} + c_{tq} - c_{tp} + \cdots - c_{rj},$$

with c_{ij} the cost coefficient that corresponds to the nonsolution variable x_{ij}, whereas all other cost coefficients correspond to solution variables. Substitution of equation (8.24) for each cost coefficient corresponding to a solution variable in the expression for D_{ij} yields,

$$D_{ij} = c_{ij} - (u_i + v_q) + (u_t + v_q) - (u_t + v_p) + \cdots - (u_r + v_j). \tag{8.25}$$

Note that $-v_q$ cancels $+v_q$, $+u_t$ cancels $-u_t$, and so on, telescoping because of the nature of the $(+-)$ path with the result that equation (8.25) reduces to

$$D_{ij} = c_{ij} - (u_i + v_j). \tag{8.26}$$

Consider D_{14} in the example problem of section 8.5.2. It was computed that

$$D_{14} = c_{14} - c_{12} + c_{42} - c_{43} + c_{33} - c_{34},$$

where c_{14} is the cost coefficient that corresponds to the nonsolution variable x_{14}. All other cost coefficients belong to solution variables. Therefore, one can write,

$$D_{14} = c_{14} - (u_1 + v_2) + (u_4 + v_2) - (u_4 + v_3) + (u_3 + v_3) - (u_3 + v_4),$$

or $D_{14} = c_{14} - (u_1 + v_4)$, as predicted by equation (8.26). Consequently, one does not need to trace any $(+ -)$ path to compute all D_{ij}. Equation (8.26) provides a convenient formula for doing it. One more question remains unanswered. How does one compute the u_i and v_i multipliers?

The answer is immediate. There are $m - u_i$'s (one for each row equation) and $n - v_j$'s (one for each column equation). There is one redundant equation, thus only $m + n - 1$ multipliers are linearly independent. Because there must be $m + n - 1$ positive variables in solution, to solve for the u_i and v_j one has a total of $m + n - 1$ equations in $m + n$ unknowns. One of the simplex multipliers is redundant and can be given any value whatsoever. It is common practice to set $v_1 = 0$ and then solve for the other multipliers. In the example problem of section 8.5.2, one has for the variables in solution,

$$c_{12} = u_1 + v_2,$$
$$c_{21} = u_2 + v_1,$$
$$c_{23} = u_2 + v_3,$$
$$c_{33} = u_3 + v_3,$$
$$c_{34} = u_3 + v_4,$$
$$c_{42} = u_4 + v_2,$$

and

$$c_{43} = u_4 + v_3.$$

A total of seven equations in eight unknowns. Setting $v_1 = 0$, obtain the solution,

$$v_1 = 0,$$
$$u_2 = c_{21},$$
$$v_3 = c_{23} - c_{21},$$

and so on. Easy to obtain because the system of equations is triangular. As a final comment, equation (8.26) can be rewritten,

$$D_{ij} = c_{ij} - z_{ij} \qquad (8.27)$$

where

$$z_{ij} = u_i + v_j. \qquad (8.28)$$

This form is reminiscent of the equation developed for the rule of steepest ascent in linear programming (the simplex method).

8.5.7 Application of the Transportation Algorithm

The application of the theory developed will be demonstrated with an example problem. Let the cost matrix and the row and column requirements of the problem be given as follows:

Sinks

		1	2	3	4	Row Requirements
	1	5	7	8	9	25
Sources	2	15	2	3	7	15
	3	5	4	3	8	10
	4	9	6	6	4	40
Column Requirements		35	10	10	35	90 ⟶ Balanced

First, an initial solution must be developed. With the Vogel method, the following basic feasible (nondegenerate) solution was obtained.

	1	2	3	4	Row Requirements
1	25				25
2		10	5		15
3	5		5		10
4	5			35	40
Column Requirements	35	10	10	35	

Now one is ready to initiate the algorithm.

I. Set up three matrices side by side as shown below.

	1	2	3	4		v_j	0	−3	−2	−5						
					u_i											
1	⑤	7	8	9	5	5	2	3	0			0	5	5	9
2	15	②	③	7	5	5	2	3	0	−	=	10	0	0	7
3	⑤	4	③	8	5	5	2	3	0			0	2	0	8
4	⑨	6	6	④	9	9	6	7	4			0	0	−1	0
		c_{ij}				$z_{ij} = u_i + v_j$							$D_{ij} = c_{ij} - z_{ij}$			

1. The first one is the original cost matrix, c_{ij}. Circle in this matrix the cost elements corresponding to the present solution (the initial basic feasible solution for this iteration).

2. The second matrix will be used to compute the $z_{ij} = u_i + v_j$ (simplex multipliers). It was shown that for the solution variables, $c_{ij} = u_i + v_j$. Therefore, write in the corresponding positions of the second matrix the cost coefficients of the variables in the present solution (those circled in the c_{ij} matrix).

3. Now one is ready to compute the simplex multipliers u_i and v_j. This is done along the top and left side of the z_{ij} matrix. Starting with $v_1 = 0$, note that for $c_{11} = 5$ it is necessary that $u_1 = 5$ because $u_1 + v_1 = c_{11} = 5 + 0 = 5$. Similarly, along column 1, $v_1 = 0$ requires $u_3 = 5$ and $u_4 = 9$ because $c_{31} = 5$ and $c_{41} = 9$. Thus,

$$u_3 + v_1 = c_{31} = 5 + 0 = 5,$$

and

$$u_4 + v_1 = c_{41} = 9 + 0 = 9.$$

Now, because

$$u_3 = 5 \text{ and } c_{33} = 3, v_3 = -2.$$

Hence,

$$u_3 + v_3 = c_{33} = 5 - 2 = 3.$$

Similarly, $u_4 = 9$ and $c_{44} = 4$; therefore, $v_4 = -5$. Now, since $c_{23} = 3$ and $v_3 = -2$, $u_2 = 5$. This in turn generates $v_2 = -3$ because $c_{22} = 2$.

All of the u_i and v_j have been computed. In fact, a system of seven equations (for the seven c_{ij} corresponding to the solution variables) in eight unknowns (four row multipliers, u_i, and four column multipliers, v_j) has been solved. v_1 was set arbitrarily equal to zero (the multiplier chosen as redundant), thereby fixing the values of the others.

4. Now the $z_{ij} = u_i + v_j$ for the nonsolution variables can be computed. Thus,

$$z_{13} = u_1 + v_3 = 5 - 2 = 3,$$
$$z_{14} = u_1 + v_4 = 5 - 5 = 0,$$
$$z_{21} = u_2 + v_1 = 5 + 0 = 5,$$
$$z_{24} = u_2 + v_4 = 5 - 5 = 0,$$
$$z_{32} = u_3 + v_2 = 5 - 3 = 2,$$
$$z_{34} = u_3 + v_4 = 5 - 5 = 0,$$
$$z_{42} = u_4 + v_2 = 9 - 3 = 6,$$

and

$$z_{43} = u_4 + v_3 = 9 - 2 = 7.$$

It is well to verify at this point that all $z_{ij} = u_i + v_j$ in the second matrix.

5. The third matrix, $D_{ij} = c_{ij} - z_{ij}$ is, cell by cell, simply the algebraic difference between the corresponding elements of the first and second matrices. Thus,

$$D_{11} = c_{11} - z_{11} = 5 - 5 = 0,$$
$$D_{12} = c_{12} - z_{12} = 7 - 2 = 5,$$

and so on, until all its elements are computed. The numerical values of all D_{ij} are shown in the matrix setup.

6. It was proved that if z is the value of the present solution, the value of a new solution formed by introducing θ_{ij}, a small positive quantity, into cell ij, is given by equation (8.13) as

$$z' = z + D_{ij}\theta_{ij}.$$

Because $\theta_{ij} > 0$, for minimization, a new solution should be formed only if one or more D_{ij} are negative. This is the case because $D_{43} = -1$. Note that all $D_{ij} = 0$ correspond to variables in the present solution, with the exception of $D_{42} = 0$, which indicates that although x_{42} is not in solution, the value of the objective function would not change if it were brought into solution. Thus, alternate solutions of the same value exist. This can lead to alternate optima in the same manner as for linear programming. Because $D_{43} = -1$, introducing x_{43} into solution will yield a lower value of the objective function. In general, if there were several negative D_{ij}, one would select the most negative to bring the corresponding variable into solution. This is merely the application of the rule of steepest descent given by equation (8.14).

II. Returning to the distribution matrix for the problem and tracing the $(+ -)$ path for θ_{43} one finds,

	Sinks 1	2	3	4	Row Requirements
1	25				25
2		10	5		15
3	+5		⁻5		10
4	⁻5		+	35	40
Column Requirements	35	10	10	35	

Sources (label at left of rows)

Therefore, $\theta_{14} = 5$ (the smallest element from which it is subtracted), and the new solution becomes,

Sinks

	1	2	3	4	Row Requirements
1	25				25
2		10	5		15
3	10				10
4			5	35	40

Sources (label at left, rows 1–4)

Column Requirements   35   10   10   35

The new solution is seen to be degenerate (one degree of degeneracy, $m + n - 1 = 7$ and only 6 variables are in solution). Degeneracy causes a slight difficulty in computing the simplex multipliers. A method for overcoming it will be discussed in detail later. ⌐

III. Now step I is repeated to check for optimality. Again, the three matrices are set up as before in order to compute all D_{ij} for the new solution matrix.

c_{ij}

	1	2	3	4
1	⑤	7	8	9
2	15	②	③	7
3	⑤	4	3	8
4	9	6	⑥	④

v_j 0   −1   0   −2

$z_{ij} = u_i + v_j$

u_i				
5	5	4	5	3
3	3	2	3	1
5	5	④	5	3
6	6	5	6	4

$D_{ij} = c_{ij} - z_{ij}$

0	3	3	6
12	0	0	6
0	0	−2	5
3	1	0	0

Note that after setting $v_1 = 0$, one can compute immediately $u_1 = 5$ and $u_3 = 5$. Now, however, one finds it impossible to continue the computation of the simplex multipliers. A gap has developed that precludes one from obtaining more values. This gap is the result of degeneracy. Only six variables are in solution instead of seven. To bridge the gap, it is necessary to bring into solution, at zero value, any variable that will act as an appropriate pivot. Because the variable is brought in at zero value, the solution is not altered, but the computation of the simplex multipliers can continue. This was done with x_{32} because c_{32} is small and it does the required job. Thus, $c_{32} = 4$ is entered in the z_{ij} matrix. Now the rest of the u_i and v_j can be computed as shown in the second matrix. The z_{ij} for the nonsolution variables are computed also and the D_{ij} for the new solution obtained. $D_{33} = -2$ indicates that bringing x_{33} into solution will decrease the value of the objective function. This negative value of D_{33} might have resulted simply because of a poor selection of the bridging variable x_{32}.

IV. Going through another iteration, one computes the $(+\,-)$ path for the new solution, thus,

Sinks

		1	2	3	4	Row Requirements
	1	25				25
	2		10⁺	5⁻		15
Sources	3	10	0⁻	+		10
	4			5	35	40
Column Requirements		35	10	10	35	

The $(+\,-)$ path shows that $\theta_{33} = 0$, because $x_{32} = 0$ is the smallest element from which θ_{33} is subtracted. This verifies that x_{32} was, indeed, a poor choice in bridging the gap to compute the simplex multipliers.

V. The new D_{ij} are then calculated as before but with $x_{33} = 0$ in solution and x_{32} out of solution.

	1	2	3	4	v_j 0	−3	−2	−4				
					u_i							
1	⑤	7	8	9	5 5	2	3	1	0	5	5	8
2	15	②	③	7	5 5	2	3	1	10	0	0	6
3	⑤	4	③	8	5 5	2	3	1	0	2	0	7
4	9	6	⑥	④	8 8	5	6	4	1	1	0	0

c_{ij} $z_{ij} = u_i + v_j$ $D_{ij} = c_{ij} - z_{ij}$

VI. Because all D_{ij} are non-negative, the minimal solution has been found and is

Sinks

		1	2	3	4	Row Requirements
	1	25				25
Sources	2		10	5		15
	3	10		0		10
	4			5	35	40
Column Requirements		35	10	10	35	

The value of the minimal solution is

$$z^* = 25(5) + 10(2) + 5(3) + 10(5) + 0(3) + 5(6) + 35(4),$$

or $z^* = 380$.

This completes the presentation of the transportation algorithm. The computations can be carried out using different formats. The three-matrix format is clear and for this reason is the one adopted in this book. The algorithm can be summarized as follows:

1. Develop an initial basic feasible solution (with the Vogel method, for example).
2. Set up the c_{ij} matrix and circle the cost elements for all solution variables.
3. Enter the cost elements for all solution variables in the $z_{ij} = u_i + v_j$ matrix and compute the simplex multipliers u_i and v_j.
4. Compute the $z_{ij} = u_i + v_j$ for the nonsolution variables and enter their values in the second matrix also, thereby completing all entries in it.
5. Compute the $D_{ij} = c_{ij} - z_{ij}$ matrix. If one or more D_{ij} are negative, select the most negative one (rule of steepest descent) and go to step 6. If all D_{ij} are non-negative, the minimum solution has been found.
6. Trace the $(+-)$ path for the θ_{ij} which corresponds to the negative D_{ij} selected in step 5.
7. Compute a new solution and start another iteration at step 2.

SOME OBSERVATIONS ABOUT THE TRANSPORTATION ALGORITHM:

1. The transportation algorithm involves addition and subtraction only. Therefore, it is computationally free of round-off error.

2. The process is self-correcting because each iteration starts with the constant c_{ij} matrix.
3. Each solution is easy to check against the row and column requirements (constraints) of the problem.
4. Any one route can be eliminated a priori by assigning it an artificially high cost. Thus, if one wished to eliminate branch $i - j$, one would only have to assign a high cost, c_{ij}, to it.
5. Degeneracy causes a gap in the computation of the simplex multipliers. This problem can be overcome by employing a perturbation technique or by introducing a nonsolution variable into solution at zero value to bridge the computational gap. This latest method was demonstrated in the example' problem.

8.6 TRANSPORTATION WITH TRANSSHIPMENT

In a transportation problem, it was assumed that a factory or source acts only as a producer of goods and that a warehouse or sink acts only as a consumer. The model will now be generalized to allow for the transshipment of goods at all points in the network—that is, instead of forcing a direct shipment from source to sink, the goods produced at some source may arrive at their final destination (sink) via some other sources and/or sinks which act as transshipment points.

If a point is a source, the net amount of goods leaving it must equal a_i. If a point is a sink, the net amount arriving must equal b_j. It is unnecessary to classify pure transshipment points in a separate category because a pure transshipment point can be considered a source with zero production, $a_i = 0$, or a sink with zero demand, $b_j = 0$. However, it is now necessary to establish all possible links joining the nodes (sources and sinks) in the network. Each branch or link in the network has to be considered as two separate links because it makes a difference whether goods travel from node i to node j or from node j to node i, even when the cost coefficient $c_{ij} = c_{ji}$.

Figure 8.2 shows graphically the added complexity of the problem when transshipment is allowed.

The transportation problem with transshipment can be formulated as a standard transportation problem, as follows:

Let

$$x_{ij} = \text{quantity shipped from city } i \text{ to city } j,$$

$$t_i = \text{quantity transshipped through city } i,$$

$$c_{ij} = \text{cost of shipping from city } i \text{ to city } j,$$

$$c_i = \text{cost of transshipping through city } i,$$

$$a_i = \text{net supply at city } i,$$

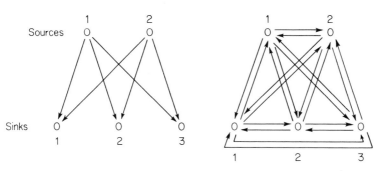

FIGURE 8.2

and

$$b_i = \text{net demand at city } i.$$

Recall that

$$c_{ij} \text{ may or may not equal } c_{ji}.$$

Assume that cities $i = 1, \ldots, m$ are sources and transshipment points and that cities $i = m + 1, \ldots, m + n$ are sinks and transshipment points. Recall, also, that pure transshipment points can be considered sources with $a_i = 0$ or sinks with $b_i = 0$, and that, by definition, a source k has $a_k \geq 0$ and $b_k = 0$, and a sink p has $a_p = 0$ and $b_p \geq 0$.

The matrix of sources and sinks is arranged as shown in Figure 8.3. Note that, as for the standard transportation problem, the sum of the elements along each row represents goods leaving, and the sum of the elements along each column represents goods arriving. Also note that, at least for the time being, the diagonal elements x_{ii} do not appear in the matrix. This is because these elements symbolize the number of goods transshipped and they were already given the notation $t_i, i = 1, \ldots, m + n$.

Now, the following observations can be made:

1. What *leaves* a *source* is equal to what the source produces plus what it transships; therefore,

$$\sum_{j=1}^{m+n}{}' x_{ij} = a_i + t_i, i = 1, \ldots, m. \tag{8.29}$$

The prime alongside the summation sign is meant to indicate that $i = j$ is *not* included.

2. What *leaves* a *sink* is equal to what the sink transships, that is, $a_i = 0$ for $i = m + 1, \ldots, m + n$; therefore,

$$\sum_{j=1}^{m+n}{}' x_{ij} = a_i + t_i, i = m + 1, \ldots, m + n. \tag{8.30}$$

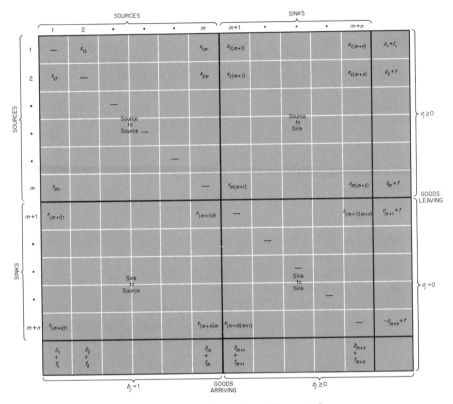

FIGURE 8.3. Transshipment Array.

3. What *arrives* at a *source* is equal to what the source transships, that is, $b_j = 0$ for $j = 1, \ldots, m$; therefore,

$$\sum_{i=1}^{m+n}{}' x_{ij} = b_j + t_j, j = 1, \ldots, m. \tag{8.31}$$

4. What *arrives* at a *sink* is equal to what the sink consumes plus what it transships; therefore,

$$\sum_{i=1}^{m+n}{}' x_{ij} = b_j + t_j, j = m + 1, \ldots, m + n. \tag{8.32}$$

5. The objective function is

Minimize

$$z = \sum_{i=1}^{m+n}{}' \sum_{j=1}^{m+n}{}' c_{ij} x_{ij} + \sum_{i=1}^{m+n} c_i t_i. \tag{8.33}$$

The objective function is constrained by the non-negativity conditions on the variables and by equations (8.29) through (8.33). Note that the primes besides the summation signs indicate again that the diagonal terms x_{ii} are not included. The model equations (8.29) through (8.33) resemble the transportation model with two notable exceptions:

1. The diagonal elements x_{ii} are not included.
2. Variables t_i appear on the right-hand side of equations (8.29) through (8.32).

Both of these difficulties can be overcome as follows:

a. Let t be the upper bound to transshipment, that is, $t \geq t_i$ for $i = 1, \ldots, m + n$. If t is made equal to the total production or to the total demand (which are the same because of the balanced condition of the problem), one can be assured that t will be the upper bound to the transshipment. Thus,

$$t = \sum_{i=1}^{m+n} a_i = \sum_{j=1}^{m+n} b_j. \tag{8.34}$$

b. Introduce slack variables x_{ii} such that

$$t_i + x_{ii} = t, \, i = 1, \ldots, m + n$$

or

$$t_i = t - x_{ii}, \, i = 1, \ldots, m + n. \tag{8.35}$$

c. Substitution of equations (8.34) and (8.35) into (8.29) through (8.33) yields,

1. For equations (8.29) and (8.30) together,

$$\sum_{j=1}^{m+n} x_{ij} = a_i + t \text{ for } i = 1, \ldots, m + n. \tag{8.36}$$

Note that the prime has disappeared from the summation sign because the diagonal elements x_{ii} are now included.

2. For equations (8.31) and (8.32) together,

$$\sum_{i=1}^{m+n} x_{ij} = b_j + t \text{ for } j = 1, \ldots, m + n. \tag{8.37}$$

3. For the objective function (8.33),
   Minimize

$$z = \sum_{i=1}^{m+n} {}' \sum_{j=1}^{m+n} {}' c_{ij} x_{ij} + \sum_{i=1}^{m+n} c_i(t - x_{ii}),$$

or Minimize

$$z = \sum_{i=1}^{m+n} \sum_{j=1}^{m+n} c_{ij} x_{ij} + \sum_{i=1}^{m+n} c_i t, \tag{8.38}$$

with

$$c_{ii} = -c_i. \tag{8.39}$$

In summary, the final model of the transportation problem with transshipment is

Minimize

$$z = \sum_{i=1}^{m+n} \sum_{j=1}^{m+n} c_{ij} x_{ij} + k$$

where

$$k = \text{constant} = \sum_{i=1}^{m+n} c_i t,$$

$$t = \sum_{i=1}^{m+n} a_i = \sum_{j=1}^{m+n} b_j \ (\text{balanced system}),$$

and

$$c_{ii} = -c_i \text{ for } i = 1, \ldots, m+n.$$

The objective function is subject to

$$x_{ij} \geq 0 \text{ for all } i \text{ and } j,$$

and

$$\sum_{j=1}^{m+n} x_{ij} = a_i + t, \ i = 1, \ldots, m+n,$$

$$\sum_{i=1}^{m+n} x_{ij} = b_j + t, \ j = 1, \ldots, m+n.$$

Note that the x_{ii} must always be in solution because

$$x_{ii} = t - t_i$$

and

$$t_i < t$$

not to restrict transshipment artificially at any node; hence, $x_{ii} > 0$ (the principal diagonal is always in solution).

The final forms of the cost and distribution matrices for the transportation problem with transshipment are:

	1	2	$\cdots$	m	$\cdots$	$m+n$
1	$-c_1$	c_{12}	$\cdots$	c_{1m}	$\cdots$	$c_{1(m+n)}$
2	c_{21}	$-c_2$	$\cdots$	c_{2m}	$\cdots$	$c_{2(m+n)}$
.	.	.		.		.
.	.	.		.		.
.	.	.		.		.
m	c_{m1}	c_{m2}	$\cdots$	$-c_m$	$\cdots$	$c_{m(m+n)}$
.	.	.		.		.
.	.	.		.		.
.	.	.		.		.
$m+n$	$c_{(m+n)1}$	$c_{(m+n)2}$	$\cdots$	$c_{(m+n)m}$	$\cdots$	$-c_{(m+n)}$

Cost Matrix

	1	2	$\cdots$	m	$\cdots$	$m+n$	Row Requirements
1	x_{11}	x_{12}	$\cdots$	x_{1m}	$\cdots$	$x_{1(m+n)}$	$a_1 + t$
2	x_{21}	x_{22}	$\cdots$	x_{2m}	$\cdots$	$x_{2(m+n)}$	$a_2 + t$
.	.	.		.		.	.
.	.	.		.		.	.
.	.	.		.		.	.
m	x_{m1}	x_{m2}	$\cdots$	x_{mm}	$\cdots$	$x_{m(m+n)}$	$a_m + t$
.	.	.		.		.	.
.	.	.		.		.	.
.	.	.		.		.	.
$m+n$	$x_{(m+n)1}$	$x_{(m+n)2}$	$\cdots$	$x_{(m+n)m}$	$\cdots$	$x_{(m+n)(m+n)}$	$a_{m+n} + t$
Column Requirements	$b_1 + t$	$b_2 + t$	$\cdots$	$b_m + t$	$\cdots$	$b_{(m+n)} + t$	

Distribution Matrix

It must be pointed out that the diagonal element x_{ii} represents the additional number of goods that could have been transshipped through node i (the slack variable), rather than the number of goods actually transshipped. For example, if $t = 100$ and in the final solution $x_{pp} = 55$, the number of goods transshipped through node p is $100 - 55 = 45$.

8.6.1 Example Problem

A network can be defined as an arrangement of nodes and branches with each node connected to at least one other node by one or more branches.

Consider the network given below with cost coefficients and inputs and outputs of goods as shown.

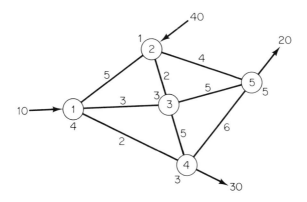

In this problem:

1. The transportation costs given alongside the branches are

	1	2	3	4	5
1	▨	5	3	2	M
2	5	▨	2	M	4
3	3	2	▨	5	5
4	2	M	5	▨	6
5	M	4	5	6	▨

M is a large, positive cost coefficient meant to exclude that branch from solution (for example, it is impossible to go directly from node 1 to node 5).

2. The transshipment costs, given next to the nodes, are

$$c_1 = 4,$$
$$c_2 = 1,$$
$$c_3 = 3,$$
$$c_4 = 3,$$
$$c_5 = 5.$$

3. The source productions shown as input arrows are

$$a_1 = 10,$$
$$a_2 = 40,$$
$$a_3 = 0,$$
$$a_4 = 0,$$
$$a_5 = 0,$$

4. The sink demands shown as output arrows are

$$b_1 = 0,$$
$$b_2 = 0,$$
$$b_3 = 0,$$
$$b_4 = 30,$$
$$b_5 = 20.$$

Therefore, nodes 1 and 2 are sources, node 3 is a pure transshipment point, and nodes 4 and 5 are sinks.

5. The transshipment matrices are set up as follows:

$$t = \sum_{i=1}^{5} a_i = \sum_{j=1}^{5} b_j = 10 + 40 = 30 + 20 = 50. \text{ (balanced)}$$

a. The cost matrix is (recall $c_{ii} = -c_i$),

	1	2	3	4	5
1	-4	5	3	2	M
2	5	-1	2	M	4
3	3	2	-3	5	5
4	2	M	5	-3	6
5	M	4	5	6	-5

b. The distribution matrix is

	1	2	3	4	5	Row Requirements
1	x_{11}	x_{12}	x_{13}	x_{14}	x_{15}	$a_1 + t = 60$
2	x_{21}	x_{22}	x_{23}	x_{24}	x_{25}	$a_2 + t = 90$
3	x_{31}	x_{32}	x_{33}	x_{34}	x_{35}	$a_3 + t = 50$
4	x_{41}	x_{42}	x_{43}	x_{44}	x_{45}	$a_4 + t = 50$
5	x_{51}	x_{52}	x_{53}	x_{54}	x_{55}	$a_5 + t = 50$

Column Requirements

$$b_1 + t \quad b_2 + t \quad b_3 + t \quad b_4 + t \quad b_5 + t$$
$$\| \qquad \| \qquad \| \qquad \| \qquad \|$$
$$50 \qquad 50 \qquad 50 \qquad 80 \qquad 70 \qquad \boxed{300} \text{ (balanced)}.$$

6. Now the solution proceeds as for the standard transportation problem.

8.7 ASSIGNMENT PROBLEMS

Consider the problem of assigning n jobs to n machines with the stipulation that there be a one-to-one correspondence between jobs and machines; that is, every job cannot be performed by more than one machine, and every machine cannot perform more than one job. All problems which can be modeled within the framework outlined above are called assignment problems. The goal of the assignment problem is to find the combination of jobs and machines that minimizes the total cost of assignment, given an $n \times n$ cost of performing the ith job with the jth machine. Instead of talking of jobs and machines, one can generalize the model by stating the problem in terms of requirements (jobs) and methods of satisfying them (machines).

If in the assignment matrix $x_{ij} = 1$, then it will be understood that the ith requirement is being satisfied by the jth method; if $x_{ij} = 0$, then the ith requirement is *not* being satisfied by the jth method.

Because of the one-to-one correspondence stipulated between requirements (jobs) and methods (machines), the mathematical model for the assignment problem can be stated as follows:
Minimize

$$z = \sum_{i=1}^{n} \sum_{j=1}^{n} c_{ij} x_{ij}, \tag{8.40}$$

subject to

$$x_{ij} \geq 0 \text{ for all } i \text{ and } j,$$

$$\sum_{j=1}^{n} x_{ij} = 1, i = 1, 2, \ldots, n, \tag{8.41}$$

and

$$\sum_{i=1}^{n} x_{ij} = 1, j = 1, 2, \ldots, n.$$

The model, equations (8.40) and (8.41), belongs to the transportation class. The constraints (8.41) define a so-called permutation matrix which has the following structure:

$$\begin{pmatrix} 0 & 1 & 0 & \cdots & 0 \\ 0 & 0 & 0 & \cdots & 1 \\ 0 & 0 & 1 & \cdots & 0 \\ & & \cdots & & \\ 1 & 0 & 0 & \cdots & 0 \end{pmatrix} \tag{8.42}$$

Because of the constraints imposed upon the problem, every solution must be a permutation matrix. Matrix (8.42) is called a permutation matrix because when it multiplies on the left a conformable matrix $\bar{B}$, it interchanges the rows of $\bar{B}$; if it multiplies $\bar{B}$ on the right, it interchanges its columns (see section 4.10). There are $n!$ ways of ordering either the rows or the columns of an $n \times n$ matrix; therefore, there are $n!$ permutation matrices of order n.

Returning to the assignment problem model, equations (8.40) and (8.41), one can observe that since any of its solutions must be a permutation matrix, each has only n positive variables and for this reason is highly degenerate (a nondegenerate solution must contain $n + n - 1 = 2n - 1$ positive variables). Because of this high degree of degeneracy, it is not convenient or efficient to use the transportation algorithm for the assignment problem.

Taking advantage of the simplified structure of the assignment matrix, the Hungarian mathematician Konig developed a very efficient, special algorithm for its solution. His algorithm is known as the Hungarian Method.

8.7.1 The Hungarian Method

The Hungarian Method is based on the premise that if a constant is added or subtracted from every element in a row or a column of the cost matrix of an assignment problem, the optimal solution matrix of the problem thus modified must be the same as the optimal solution matrix of the original problem. This statement is correct because if, for example, one subtracts the constant, k, from each element of the rth row, since one and only one element on the rth row appears in any solution matrix and that element must always have the value of 1, the value of *all* solutions has been decreased by the constant k. Therefore, the value of the optimal solution must have decreased exactly as much as the value of all other solutions and must have remained optimal for the modified problem. The same manipulation can be performed simultaneously on several rows and columns with the identical conclusion holding by induction.

The algorithm will be presented with the aid of an example. Consider the cost matrix given below. The row and column requirements are always ones and will not be repeated.

	1	2	3	4	5	Row Requirements
1	-3	4	5	2	-1	1
2	4	7	2	1	4	1
3	5	4	3	2	2	1
4	-5	4	7	2	5	1
5	8	-2	-3	4	5	1

Column Requirements   1   1   1   1   1

1. Subtract the smallest cost element in each row from each element in that row.

	1	2	3	4	5
1	0	7	8	5	2
2	$\cdot 3$	6	1	0	3
3	3	2	1	0	0
4	0	9	12	7	10
5	11	1	0	7	8

A zero has been developed on each row of the cost matrix.

2. Investigate whether or not a solution to the modified problem has been obtained already. This solution exists only if n (in this case, 5) independent zeros appear in the modified cost matrix such that if each zero were replaced by a 1 and all other elements by 0's, a permutation matrix would be formed. The most efficient procedure for establishing whether or not $n(5)$ independent zeros exist is to draw the set of minimum number of lines through the zeros resulting from the initial step. If, to cover all zeros, fewer than $n(5)$ lines are needed, the optimal solution has not as yet been obtained.

	1	2	3	4	5	
1	0	7	8	5	2	
2	3	6	1	0	3	←
3	3	2	1	0	0	←
4	0	9	12	7	10	
5	11	1	0	7	8	←

The arrows indicate the position of the lines. Note that 4 are sufficient to cover all zeros. Therefore, the optimal solution has not yet been found and one must proceed to step 3. (Note that the set of minimum number of lines is not unique, that is, there are many combinations of 4 lines that will cover all zeros; however, the minimum set contains only 4 lines in any case.)

3. Repeat step 1, this time operating on the columns, with the following results:

	1	2	3	4	5
1	0	6	8	5	2
2	3	5	1	0	3
3	3	1	1	0	0
4	0	8	12	7	10
5	11	0	0	7	8

4. Draw the set of minimum number of lines.

	1	2	3	4	5	
1	0	6	8	5	2	
2	3	5	1	0	3	←
3	3	1	1	0	0	←
4	0	8	12	7	10	
5	11	0	0	7	8	←

↑

Again, only 4 lines are needed to cover all zeros. The optimal solution has not as yet been found. Proceed to step 5.

5. Call the set of elements *covered* by the lines in step 4, matrix $\bar{C}$, and the set of elements *not* covered by the lines, matrix $\bar{N}$. Find the smallest element in $\bar{N}$, subtract it from all elements in $\bar{N}$, and add it to the elements in $\bar{C}$ that lie on the intersection of two lines (if any exist). All other elements remain unchanged. The logic behind this procedure will be fully explained later.

	1	2	3	4	5	
1	0	6	8	5	2	
2	3	5	1	0	3	←
3	3	1	1	0	0	←
4	0	8	12	7	10	
5	11	0	0	7	8	←

↑

In the matrix above, the elements constituting matrix $\bar{N}$ are shown with a heavy outline. All other elements constitute matrix $\bar{C}$. The procedure yields,

	1	2	3	4	5
1	0	4	6	3	0
2	5	5	1	0	3
3	5	1	1	0	0
4	0	6	10	5	8
5	13	0	0	7	8

6. Draw a new set of minimum number of lines.

	1	2	3	4	5
1	0	4	6	3	0
2	5	5	1	0	3
3	5	1	1	0	0
4	0	6	10	5	8
5	13	0	0	7	8	←

Again, only 4 lines are needed. Still, the optimal solution has not been formed. Go to step 7.

7. This step and all succeeding steps are simply a repetition of steps 5 and 6, operating on the latest results. It can be said, therefore, that the iterations (and, consequently, the algorithm) really start at step 5. The procedure yields

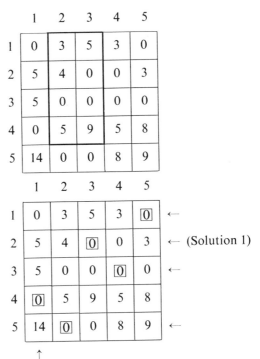

	1	2	3	4	5
1	0	3	5	3	0
2	5	4	0	0	3
3	5	0	0	0	0
4	0	5	9	5	8
5	14	0	0	8	9

	1	2	3	4	5	
1	0	3	5	3	[0]	←
2	5	4	[0]	0	3	← (Solution 1)
3	5	0	0	[0]	0	←
4	[0]	5	9	5	8	
5	14	[0]	0	8	9	←

The set of minimum number of lines contains 5 lines; therefore, the optimal solution has been found ($n = 5$). To locate a set of 5 independent zeros (there may be more than one such set, in which case the problem has alternate optima), consider a row or column containing only one zero; row 4, for example, because of its uniqueness in this row, the zero at position (4, 1) must be in the solution. A box is drawn around this zero. Because of the one-to-one correspondence between requirements (jobs) and methods (machines), the zero in position (1, 1) cannot be used; therefore, the zero at position (1, 5) must be in solution, and so forth. One set of five independent zeros is shown above marked by the boxes. Two other sets are given below.

	1	2	3	4	5
1	0	3	5	3	[0]
2	5	4	0	[0]	3
3	5	0	[0]	0	0
4	[0]	5	9	5	8
5	14	[0]	0	8	0

(Solution 2)

	1	2	3	4	5
1	0	3	5	3	[0]
2	5	4	0	[0]	3
3	5	[0]	0	0	0
4	[0]	5	9	5	8
5	14	0	[0]	8	9

(Solution 3)

The sets have been labeled solutions 1, 2, and 3, respectively. The corresponding assignment matrices are,

	1	2	3	4	5
1	0	0	0	0	1
2	0	0	1	0	0
3	0	0	0	1	0
4	1	0	0	0	0
5	0	1	0	0	0

Solution 1

	1	2	3	4	5
1	0	0	0	0	1
2	0	0	0	1	0
3	0	0	1	0	0
4	1	0	0	0	0
5	0	1	0	0	0

Solution 2

	1	2	3	4	5
1	0	0	0	0	1
2	0	0	0	1	0
3	0	1	0	0	0
4	1	0	0	0	0
5	0	0	1	0	0

Solution 3

The minimum cost of these alternate optima is (refer to the original cost matrix)

$$z^* = -4.$$

To justify the procedure given in step 5, one must show that it entails only addition and subtraction of a constant to all elements of several rows and columns. It has already been proven that this method of addition and subtraction does *not* alter the optimal solution matrix.

Consider the modified cost matrix of step 5. The minimum cost element in $\bar{N}$ (outlined) was found to be 2. Let it be called k to generalize the results. Then,

1. Subtract k from all elements in the matrix.
2. To avoid destroying existing zeros in the $\bar{C}$ matrix,
   a. Add k to all row elements in $\bar{C}$.
   b. Add k to all column elements in $\bar{C}$.

The net result is that k has been subtracted from all elements in $\bar{N}$, added to the elements in $\bar{C}$ lying on the intersection of two lines, and all other elements

	1	2	3	4	5	
1	$-k$ $+k$	$-k$	$-k$	$-k$	$-k$	
2	$-k$ $+k$ $+k$	$-k$ $+k$	$-k$ $+k$	$-k$ $+k$	$-k$ $+k$	←
3	$-k$ $+k$ $+k$	$-k$ $+k$	$-k$ $+k$	$-k$ $+k$	$-k$ $+k$	←
4	$-k$ $+k$	$-k$	$-k$	$-k$	$-k$	
5	$-k$ $+k$ $+k$	$-k$ $+k$	$-k$ $+k$	$-k$ $+k$	$-k$ $+k$	←

↑

have not been altered, exactly as required by the procedure outlined in step 5. All additional matrix manipulations can be justified on the same basis.

As a final comment, the following point must be made: If more requirements than methods of satisfying them, or vice-versa, exist in a problem, artificial columns or rows must be added to the problem matrices to make them square. All cost elements in artificial rows or columns are set equal to zero. Artificial rows simply mean that some requirements cannot be met; artificial columns mean that some methods cannot be used.

8.8 OPTIMAL NETWORK ROUTING

The assignment algorithm (Hungarian Method) can be used to establish the shortest or the longest path from any node i to any other node j through a network. The branches can have time, dollars, distance, or some other unit cost dimensions. In chapter 9, for critical path method applications, the branches will be assumed to have time dimensions.

Consider the network of Figure 8.4. The cost matrix, also called incidence matrix, for this network is assembled as shown in Figure 8.5. Note that:

1. If the path from node i to node j is defined by an arrow directed from i to j, the cost coefficient c_{ij} for that branch has the value specified on

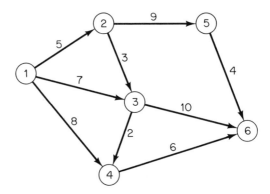

FIGURE 8.4. A Typical Network.

	1	2	3	4	5	6
1	0	5	7	8	M	M
2	M	0	3	M	9	M
3	M	M	0	2	M	10
4	M	M	M	0	M	6
5	M	M	M	M	0	4
6	M	M	M	M	M	0

FIGURE 8.5. Network Incidence Matrix.

the network, that is,

$$c_{12} = 5,$$
$$c_{23} = 3,$$

etc.

2. All cost coefficients $c_{ii} = 0$.
3. If a path from node i to node j is *not* defined,

$$c_{ij} = M,$$

where M is an arbitrarily large, positive cost intended to keep the connection from i to j out of the solution (because it does not exist.)
4. The cost coefficient $c_{ji} = M$ when the arrow is directed from i to j. Flow against the direction of the arrows is *not* allowed.

8.8.1 Shortest Paths Through Networks

The first problem to be considered will be that of establishing shortest paths through networks. Let x_{ij} represent the branch from node i to node j. If

$x_{ij} = 1$, the branch from i to j will be assumed in the solution; if $x_{ij} = 0$, the branch from i to j will be assumed *not* to be in the solution; if $x_{ii} = 1$, node i will be assumed *not* to be part of the shortest path. The interpretation of this last assumption is that $x_{ii} = 1$ represents a loop from node i to node i; this is a blind loop and, consequently, node i falls outside the shortest path. With these assumptions, it is clear that the application of the Hungarian Method to the incidence matrix of Figure 8.5 will yield the minimum cost, shortest route. The question is, from what node to what node? In fact, one can observe that a solution already exists for the network of Figure 8.4 in terms of the diagonal elements. The method appears to say: "Do nothing, that is the minimum cost policy." To resolve this predicament, assume that one wishes to compute the shortest path from node p to node r. To accomplish this, force into the solution an artificial branch from node r to node p by specifying the cost element c_{rp} equal to a large negative coefficient, $-M$. The branch from r to p will not be considered part of the solution; it is used only to force the other solution variables out of the principal diagonal. Mathematically, branch $r \longrightarrow p$ acts as a catalyst; because a branch *from r* exists now in the solution, a branch *to r* is forced into the solution; similarly, because a branch *to p* has been forced into the solution, a branch *from p* must also be formed. The intermediate branches completing the circuit between p and r are also forced into the solution yielding the least cost (shortest) path from node p to node r.

The mathematical model is
Minimize

$$z = \sum_{i=1}^{n} \sum_{j=1}^{n} c_{ij} x_{ij}$$

subject to

$$x_{ij} \geq 0 \text{ for all } i \text{ and } j,$$

$$\sum_{j=1}^{n} x_{ij} = 1, i = 1, 2, \ldots, n;$$

$$\sum_{i=1}^{n} x_{ij} = 1, j = 1, 2, \ldots, n;$$

and

$$x_{rp} = 1, \tag{8.43}$$

when the shortest path desired is that linking p to r. This is the model of the assignment problem with the additional constraint (8.43).

Returning to the example network of Figure 8.4, assume that one wishes to compute the shortest path from node 1 to node 6, the network and its

associated incidence matrix are

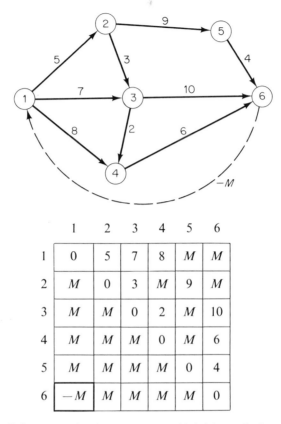

	1	2	3	4	5	6
1	0	5	7	8	M	M
2	M	0	3	M	9	M
3	M	M	0	2	M	10
4	M	M	M	0	M	6
5	M	M	M	M	0	4
6	$-M$	M	M	M	M	0

The cost coefficient $-M$ is given to the artificial branch $6 \longrightarrow 1$ to force a solution from node 1 to node 6. Application of the Hungarian Method to the incidence matrix given would yield the least cost (shortest) path from 1 to 6.

No calculations are performed here since they are of the same type developed before. The reader can verify that the minimal route is $1 \longrightarrow 4 \longrightarrow 6$ with a least cost of 14 units.

8.8.2 Longest Paths Through Networks

In construction projects the path of longest time through the work flow network controls total project duration. Although this topic will be discussed in detail in chapter 9, it is interesting to consider it here as another application of the Hungarian Method.

Because the Hungarian Method is a minimization algorithm, in order to obtain the maximum cost (longest) path through a network, one must minimize the negative of the objective function. Care must be exercised, however, to introduce the negative costs in the appropriate cells. To establish the longest route from node p to node r, the incidence (cost) matrix is set up as follows:

1. If the path from node i to node j is defined by an arrow directed from i to j, the cost coefficient c_{ij} for that branch is given a value equal to the negative of the one specified on the network. For the one of Figure 8.4, for example,

$$c_{12} = -5,$$

$$c_{23} = -3,$$

etc.

2. All cost coefficients $c_{ii} = 0$.
3. If a path from node i to node j is *not* defined,

$$c_{ij} = M,$$

where M is, again, an arbitrarily large, positive cost intended to keep the connection from i to j out of the solution, because it does *not* exist.
4. The cost coefficient $c_{ji} = M$ when the arrow is directed from i to j. As before, flow opposite to the direction of the arrows is not allowed.
5. The cost coefficient $c_{rp} = -M$ because one wishes to establish the longest path from node p to node r.

Statements 3, 4, and 5 are the same as for the shortest route problem because, in both cases, the Hungarian Method minimizes the objective function.

The incidence matrix required to initialize the algorithm for establishing the longest route from node 1 to node 6 in the network of Figure 8.4 is

	1	2	3	4	5	6
1	0	-5	-7	-8	M	M
2	M	0	-3	M	-9	M
3	M	M	0	-2	M	-10
4	M	M	M	0	M	-6
5	M	M	M	M	0	-4
6	$-M$	M	M	M	M	0

Chapter 9 will present a more efficient and informative method for computing longest routes through networks.

EXERCISES

8–1. A contractor has the financial resources to build 50 two-bedroom, 100 three-bedroom, and 80 four-bedroom houses. He owns land in Baton Rouge and New Orleans. His Baton Rouge property is large enough for 180 houses and his New Orleans land can accommodate 120. Determine the number of units of each kind the contractor should build in each city to maximize his profit in accordance with the following *profit table:*

	Baton Rouge	New Orleans
2-bedroom	2,000	2,500
3-bedroom	3,000	3,000
4-bedroom	4,500	4,000

8–2. On a certain day, Olympic Airways has scheduled chartered trips out of three cities:

New Orleans (NO)—5

Los Angeles (LA)—18

New York   (NY)—25

The airline has available (uncommitted) jets at five airports:

Los Angeles (LA)—10

New York   (NY)—15

Chicago   (CH)—15

San Francisco (SF)—12

Air miles between cities are given below (in hundreds of miles):

	NO	LA	NY
LA	19	0	28
NY	13	28	0
CH	9	21	8
SF	23	4	30

Develop a schedule to fulfill chartered flight requirements for planes at minimum (air miles) cost.

a. Use the Vogel Approximation Method.

b. Check optimality with the special transportation algorithm.

8–3. Determine the minimum cost of transportation plus transshipment for the network given below. The inputs are also specified.

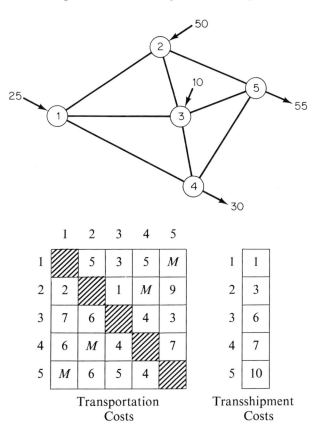

Use the Vogel method to obtain the first basic feasible solution. *M* stands for a large, positive cost coefficient.

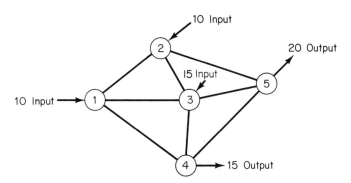

8-4. The transportation and transshipment costs for the network above are given below:

	1	2	3	4	5
1	5	14	16	12	M
2	14	7	12	M	18
3	15	20	4	18	16
4	12	M	15	0	18
5	M	17	16	20	8

Route the inputs at nodes 1, 2, and 3 to nodes 4 and 5 such that the demands be fulfilled at a minimum cost. *M* stands for a large, positive cost coefficient.

8-5. Use the assignment algorithm to establish the shortest route(s) from node 1 to node 5.

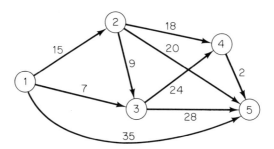

8-6. Write the incidence matrix for the network given below (durations given) and use the Hungarian Method to compute the longest (critical) path from node 1 to node 7.

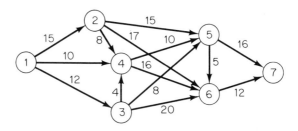

8-7. Use the assignment algorithm to compute the shortest route from node 2 to node 5 in the network given on the next page.

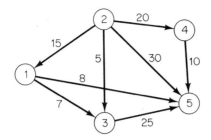

8-8. Use the assignment problem to determine the longest path from node 1 to node 6.

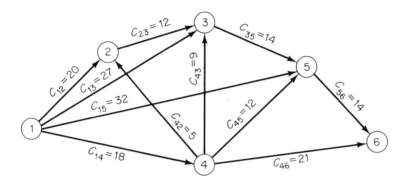

REFERENCES

1. Au, T. and T. E. Stelson, *Introduction to Systems Engineering, Deterministic Models.* Reading, Mass.: Addison-Wesley Publishing Company, 1969.

2. Hadley, G., *Linear Programming.* Reading, Mass.: Addison-Wesley Publishing Company, 1962.

3. Llewellyn, R. W., *Linear Programming.* New York: Holt, Rinehart & Winston, Inc., 1964.

9

Management of Construction Projects

9.1 INTRODUCTION

Planning and scheduling problems can be represented by networks depicting the activities to be performed in their proper order of execution. Especially for large and complex projects, networks can be used very effectively to control completion time as well as cash flow. In addition, a network of the activities involved in a project provides an excellent tool for establishing the work sequence much in advance of actual performance, thereby allowing the planner sufficient time to modify or correct potential problem areas, to allocate resources efficiently, and to redefine work relationships if necessary for expediting the completion of the project objectives.

Two basic methods of control have been developed to date: the Critical Path Method (CPM) and Program Evaluation and Review Technique (PERT). Because these two methods are very similar, the study of one is sufficient to fully understand the other. For this reason, only CPM will be discussed in detail in this chapter.

9.2 DEVELOPMENT OF CPM AND PERT

The Critical Path Method was developed in 1956 by the E. I. du Pont de Nemours' Construction Division in collaboration with Sperry-Rand Corporation. The method was originally called Project Planning and Scheduling System. Since its inception, CPM caught the eye of the construction industry and is now in wide and general use for the management and control of construction activities.

Program Evaluation and Review Technique was developed in 1958 by Booz-Allen and Hamilton, a systems consulting firm, working under the sponsorship of the Special Projects Office, Bureau of Ordnance, U.S. Navy. PERT was originally called Program Evaluation Research Task. The U.S. Navy is, of course, committed to PERT and as an example of its usefulness and effectiveness it is pointed out that successful completion of the Polaris Missile development program, two years ahead of schedule, is largely attributed to its application.

9.3 THE MANAGEMENT FUNCTION

CPM is useful because it permits management by exception; that is, it allows the manager to concentrate his attention on the critical activities of the project, those upon which project completion time depends. Before tools such as CPM became available, management had to be extensive in character; that is, all activities had to be controlled with the same degree of attention. By focusing his skills upon the critical activities only, the administrator has ample time to plan the project more accurately and reliably. The management function involves four interrelated steps, as follows:

1. *Planning.* Answers the question, What should be done?
2. *Programming or scheduling.* Answers the question, When should it be done?
3. *Budgeting.* Answers the question, How much will it cost?
4. *Implementation.* Carries to fruition the first three steps.

CPM has several levels of application in project control, as follows:

1. In planning the project by explicitly defining all dependencies among operations.
2. In programming by scheduling occurrence of all events and performance of all activities.

3. In determining all critical events and activities, those which control target dates.
4. In analyzing, modifying, and refining the work plan.
5. In allocating resources in an optimal manner.
6. In establishing optimal time-cost trade-offs.

9.4 PLANNING WITH CPM

Two types of networks can be used to implement CPM:

1. Activity on node or circle networks.
2. Activity on branch or arrow networks.

A considerable amount of experience with CPM has led the author to conclude that the first type, an activity on node or circle network, is the more flexible, useful, and efficient of the two and also the easiest to implement and computerize. For this reason, this chapter will concentrate on activity on node networks. However, mention will be made of activity on branch or arrow networks primarily because they are also in general use and the reader should understand exactly how they work. As will be seen, understanding one type immediately leads to a complete understanding of the other.

9.5 CPM ACTIVITY ON NODE NETWORKS

A network of the type being discussed here is composed of nodes and branches. The nodes (circles) represent activities and events and the branches (arrows) indicate explicitly the relationships among the activities and events. In the context used here, an activity is an operation with positive duration and an event, one with zero duration.

It is clear that every activity or event is related to every other activity or event as follows:

1. An activity or event must immediately precede another activity or event (except for the last one in the network).
2. An activity or event must immediately follow some other activity or event (except for the first one in the network).
3. An activity or event can be performed or occur simultaneously with another activity or event.

These three are the only possible relationships among the nodes in a network and are the ones that establish its connectivity or topology. Consider the following statements (logic) and their corresponding network diagrams:

Logic	*Network Diagram*
1. D can begin only after C is completed.	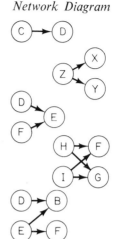
2. Neither X nor Y can start before Z is completed, but X and Y can be performed simultaneously.	
3. E must wait for D and F to be completed.	
4. Neither F nor G can start before H and I are completed; however, F and G can be performed simultaneously.	
5. B cannot start until both D and E are completed, but F can start after *only* E is completed.	

A full network is simply an assembly, in the correct order, of a number of statements such as the ones given above.

Consider the following additional logical statement and its corresponding diagram.

Logic	*Network Diagram*
A and B can be performed simultaneously.	
C cannot start before A is completed.	

Although the network diagram expresses the logic correctly, it is disconnected and therefore can be confusing. This situation occurs at the beginning and at the end of a network that does not have single beginning and ending operations. To avoid this problem, two dummy events (with zero durations) are used. They are a "start" node and an "end" node. Thus, the previous diagram can be connected to appear as follows:

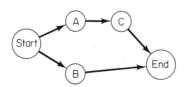

As a matter of course, a "start" and an "end" node will always be used in all completed networks.

9.6 NUMBERS FOR OPERATIONS

To make the use of networks simpler and more convenient, activities and events (nodes) can be given numbers, as follows:

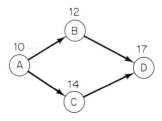

For programming purposes and to simplify checking, it is a good practice to number operations in increasing order (tail operations showing a smaller number than head operations), as shown in the preceding network.

9.7 ASSEMBLING NETWORKS

The typical procedure for assembling a complete network is for the analyst to organize a meeting with the project leader. In a construction job, for example, a meeting is held usually with a corporate officer or company owner, as the case may be, and with the superintendent of construction: the man who knows how the building will be erected. During this meeting, the analyst questions the project leader about the organization of the job. This results in a series of statements and partial network diagrams which have to be carefully assembled into a full net depicting the entire job. This initial network is studied later on, analyzed, refined, and modified to conform to the most accurate representation of efficient field operations. Much of the success of CPM as a management tool depends upon this first effort.

As an example, consider the following network logic.

	Logic	*Network Diagrams*

1. A is the first operation.

2. F cannot start until B and C are completed.

3. J follows I.

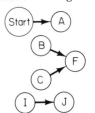

4. E precedes I.

5. Q cannot begin before P and K are completed.

6. R depends on L and Q.

7. T follows R and S.

8. O precedes S.

9. K depends on J and N.

10. M follows F and G.

11. B must follow A.

12. B, C, D, and E can be performed simultaneously.

13. C precedes G.

14. H follows D.

15. L follows K.

16. O follows M.

17. P can start only after N and M are completed.

18. N cannot start until F, G, and H are finished.

19. C, D, and E can be performed simultaneously and must follow A.

20. T is the last operation.

21. B precedes F.

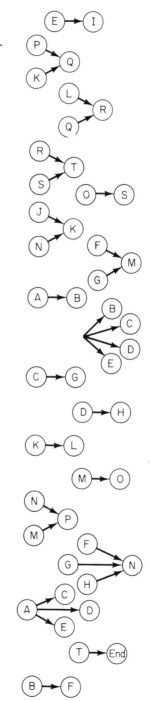

Note that many different statements can be used to express the same logic. However, once the network diagrams are drawn, there is no doubt as to what the relationships are among the operations involved.

To assemble the entire network proceed as follows:

1. Start by laying out the first operation (A) and check out the corresponding statement because, once used, a statement must be excluded from further consideration.

2. Scan all statements and select those that contain the first operation (A). Lay out the operations involved and, again, check out the statements already considered.

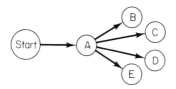

3. Now consider another operation—B, for example—and repeat the procedure until all statements that contain B have been systematically laid out.

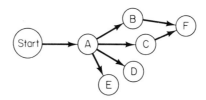

4. Proceed in the same manner to scan all statements in the logic and to lay out the entire network. Note that statements can be repetitive or redundant. This causes no difficulty so long as they do not contradict each other. If they do, there must be an error in the logic. The completed network is depicted in Figure 9.1.

The geometry or appearance of the network is irrelevant since no scale has to be specified to lay it out. Only the connectivity or topology of the network is of the utmost importance and must exactly express the logic of the work flow. With the assembling of the network, the planning task (what should be done?) is complete, except for the refinements and modifications which are usually made to satisfy job target dates and other requirements. The need for

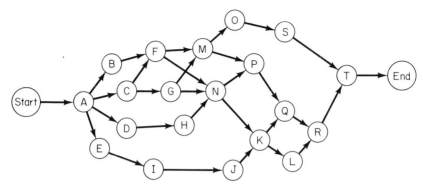

FIGURE 9.1. Completed Network.

refinements becomes apparent after the calculations on the network have been completed and the critical operations determined.

9.8 COMPUTATION OF THE CRITICAL PATH

The duration of an operation is the amount of time required to complete it. These durations are estimated by the project leader or job superintendent and are assigned to each operation in the network. An operation which represents an event in time, such as "start project," is assigned a zero duration because, in itself, it takes no time for completion. Its only function is to specify milestone events as the project develops. The critical path can be computed once all the operations have been given estimated durations.

The objectives in performing the computations are:

1. To determine the completion time for the entire project.
2. To establish which operations control total project duration. These are the critical operations which define the critical path.
3. To determine the amount of slack time or float that exists for the operations not on the critical path.

9.9 FUNDAMENTAL DEFINITIONS

Eight definitions are fundamental to the understanding of network computations, as follows:

I. ACTIVITY-ORIENTED DEFINITIONS:

1. *Earliest Start Time* (EST) of an activity is the Earliest Possible Time at which the activity can be started.

2. *Earliest Finish Time* (EFT) of an activity equals its Earliest Start Time plus its Duration and is the earliest time at which the activity can be completed.
3. *Latest Finish Time* (LFT) of an activity is the latest time at which the activity can be completed without delaying any other activity.
4. *Latest Start Time* (LST) of an activity equals its Latest Finish Time minus its Duration and is the latest time at which the activity can be started without delaying any other activity.
5. *Total Float* of an activity is the total amount of time the activity can be delayed without delaying the *entire project*.
6. *Free Float* of an activity is the total amount of time that the activity can be delayed without delaying *any other activity*.

II. EVENT-ORIENTED DEFINITIONS:

Because events have zero durations, their Earliest Start and Finish Times are equal as are their Latest Start and Finish Times. For this reason, the following two additional definitions are useful:

7. *Earliest Possible Occurrence* (EPO) of an event is the Earliest Possible Time at which an event can occur and has the same numerical value as the Earliest Start and Finish Time of the operation when considered as an activity with zero duration.
8. *Latest Possible Occurrence* (LPO) of an event is the Latest Possible Time at which an event can occur and has the same numerical value as the Latest Start and Finish Time of the operation when considered as an activity with zero duration.

In the following sections, these definitions will be given operational interpretations with the aid of an example network. Note that the last two definitions will have the same operational formulas as the Earliest Start Time and the Latest Finish Time, respectively, because an event is only an activity with zero duration.

9.9.1 Operational Definitions

Consider the network of Figure 9.2. Operation durations are given inside each node. To attach operational meaning to the definitions stated in the previous section, first consider the performance of the computations as one moves forward from Start to End; this will be called the *Early Start Time Schedule*. Afterward, the computations will be performed by moving backward from End to Start; this will be called the *Late Finish Time Schedule*.

9.9.2 Early Start Time Schedule

The Earliest Start Time of Start is defined to be zero, to indicate that the event Start occurs at the end of the zero day, which is equivalent to stating

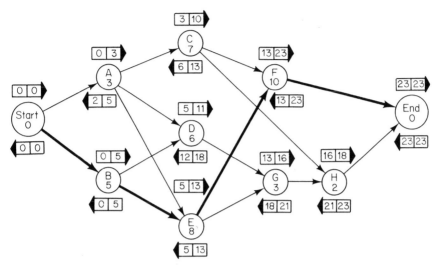

FIGURE 9.2. Sample Network.

that it occurs at the beginning of the first day. This convention will be observed throughout the remainder of this chapter. The number 0 is thereby entered in the left cell of the forward arrow, $\boxed{0\ \ }$.

Because the duration of event Start is zero, its Earliest Finish Time is $0 + 0 = 0$. This number is entered in the right cell of the forward arrow, $\boxed{0\ 0}$. Because Start is an event, its EST (Earliest Start Time) equals its EFT (Earliest Finish Time) and this quantity is called the EPO (Earliest Possible Occurrence) of event Start. Proceeding forward to Activity A, it is seen that the EST of A is equal to the EFT of Start, thus zero is entered in the left cell of the forward arrow at A. The EFT of A equals its EST plus its duration, $0 + 3 = 3$, which is entered in the right cell of the forward arrow at A. No major difficulty is encountered at activity B and its EST and EFT are entered in the appropriate cells. For activity C, the EST = 3 and the EFT $= 3 + 7 = 10$. At activity D one encounters a small difficulty; the EST of D is the largest of the EFT's of A and B, that is, 5, because D cannot start until both activities A and B have been completed and B takes the longest time to complete. Thus, the EST of D is 5 and its EFT $= 5 + 6 = 11$. Now one can give operational definitions of EST, EFT, and EPO, as follows:

1. EST (Earliest Start Time) of an activity = maximum EFT of operations immediately preceding activity in question.
2. EFT (Earliest Finish Time) of an activity = EST + Duration.
3. EPO (Earliest Possible Occurrence) of an event = EST = EFT.

Proceeding in a similar manner, one finds the EST's and EFT's of all activities and the EPO of End. These are given in the appropriate cells of all forward arrows.

Note that the EST $=$ EFT $=$ EPO of End is 23. Thus, 23 days is the *minimum* amount of time in which the project can be completed and is the duration of the *longest* path through the network.

9.9.3 Late Start Time Schedule

Refer again to Figure 9.2. If the project is not to be delayed, the Latest Finish Time of End must equal its Earliest Finish Time. Thus, the number 23 is entered in the right cell of the backward arrow, $\boxed{|23}$. Because the duration of End is zero, its Latest Start Time is $23 - 0 = 23$, and this number is entered in the left cell of the backward arrow, $\boxed{23|23}$. Since End is an event (operation with zero duration), its LFT (Latest Finish Time) $=$ LST (Latest Start Time) $=$ LPO (Latest Possible Occurrence). Proceeding backward to Activity H, one finds that the LFT of H must also be 23 because otherwise H would delay the occurrence of End. This number is again entered in the right cell of the backward arrow at H. The LST of H is $23 - 2 = 21$, i.e., LST $=$ LFT $-$ Duration, and its value is entered in the left cell of the corresponding backward arrow.

At F one finds, without difficulty, that its LFT $= 23$ and its LST $=$ LFT $-$ Duration $= 23 - 10 = 13$. At activity C, one encounters a small difficulty because C is followed immediately by two activities, F and H. Thus, its LFT must be such that C does not delay any of the operations immediately following it. Consequently, the LFT of C must be equal to the smallest of the LST's of the activities immediately ahead of it. That is, the LFT of C must be 13, which is the LST of F, since the LST of H is $21 > 13$. This figure (13) is entered in the right cell of the backward arrow at C. The LST of C is simply $13 - 7 = 6$.

Now one can state the operational definitions of LFT, LST, and LPO, as follows:

1. LFT (Latest Finish Time) of an activity $=$ Minimum LST of operations immediately following activity in question.

2. LST (Latest Start Time) of an activity $=$ LFT $-$ Duration.

3. LPO (Latest Possible Occurrence) of an event $=$ LFT $=$ LST.

Note that the definitions for LFT, LST, and LPO are exactly the opposite of those for EST, EFT, and EPO, respectively.

Proceeding in a similar manner, one finds the LFT's and LST's of all activities and the LPO of Start. These are given in the appropriate cells of all backward arrows.

9.10 ESTABLISHING THE CRITICAL PATH

At this point, one is in the position to determine which operations are critical because they must have Earliest Start Times equal to Latest Start Times or, alternatively, Earliest Finish Times equal to Latest Finish Times. This condition indicates that those operations cannot be delayed if the project is to be completed on time. Referring to Figure 9.2, one finds the critical operations to be Start, B, E, F, and End. The arrows or branches connecting them form the critical path. Those branches have been heavied in Figure 9.2.

The critical path is not unique, that is, there can be several critical paths in a network.

9.11 COMPUTATION OF FLOATS

Two types of floats will be computed; Total Float and Free Float. As expressed earlier, the Total Float of an operation is the total amount of time that the operation can be delayed without delaying the entire project and, consequently, it must equal the difference between its LST and EST or, alternatively, the difference between its LFT and LST. Thus, the operational definition of Total Float is

$$\text{Total Float} = \text{LST} - \text{EST} = \text{LFT} - \text{EFT}.$$

Consequently, all critical operations have zero Total Floats. This is possibly the best criterion for establishing the critical path. For the network of Figure 9.2, one has

Operation	Total Float
Start	0 (critical)
A	2
B	0 (critical)
C	3
D	7
E	0 (critical)
F	0 (critical)
G	5
H	5
End	0 (critical)

The Free Float of an operation is the amount of time it can be delayed without delaying any other operation. Consequently, its operational definition is as follows:

Free Float = Minimum EST of operations immediately following operation in question — EFT of operation in question.

For the network of Figure 9.2, one has

Operation		Free Float
Start	....	0 (critical)
A	....	0
B	....	0 (critical)
C	....	3
D	....	2
E	....	0 (critical)
F	....	0 (critical)
G	....	0
H	....	5
End	....	0 (critical)

It can be shown that

$$\text{Free Float} \leq \text{Total Float.} \tag{9.1}$$

Hence, each critical operation must have zero Free Float as well as zero Total Float.

Although, strictly speaking, events (operations with zero duration) have no floats, the definitions were extended to include them for operational purposes in computing the critical path.

9.12 SUMMARY OF OPERATIONAL DEFINITIONS

I. ACTIVITY-ORIENTED:

1. EST (Earliest Start Time) = Maximum EFT of operations immediately preceding activity in question.
2. EFT (Earliest Finish Time) = EST + Duration.
3. LFT (Latest Finish Time) = Minimum LST of operations immediately following activity in question.
4. LST (Latest Start Time) = LFT — Duration.
5. Free Float = Minimum EST of operations immediately following operation in question — EFT of operation in question.
6. Total Float = LST — EST = LFT — EFT.

II. EVENT-ORIENTED:

1. EPO (Earliest Possible Occurrence) = EST = EFT.
2. LPO (Latest Possible Occurrence) = LFT = LST.

9.13 CALENDAR DAY CONVERSION

All Start and Finish times are related to the zero starting day. The following convention will be used in establishing calendar dates:

1. All operations start at the beginning of the next day. Thus, the EST of C is 3; this means that C can start as early as the beginning of the 4th day of work. Similarly, the LST of C is 6, indicating that it can start as late as the beginning of the 7th day.
2. All operations finish at the end of the day. The EFT of C is 10 and, consequently, C can finish as early as the end of the 10th day of work. The LFT of C is 13, indicating that it can finish as late as the end of the 13th day.

With this convention and specifying the number of working days in the week, the dates corresponding to each EST, EFT, LFT, and LST can be computed with the aid of a calendar.

Because of conversion to calendar days, if an activity requires a total of 8 hours of work, but can only be performed in four hour durations on each of two days, its duration for the calculations on the network must be given as two days. Conversion to calendar days is performed automatically by most computer programs once the work week duration, the starting date, and the holidays have been specified.

9.14 CPM ALGORITHM

Having studied the procedure for computing the critical operations in a project, it is now convenient to express it in the form of an algorithm, as follows:

1. Define all operations (activities and events).
2. Define the operational logic.
3. Draw the network.
4. Estimate the duration of each activity and assign a zero duration to each event.
5. Calculate all EST's and EFT's.
6. Set the EFT of End equal to its LFT.
7. Calculate all LFT's and LST's.
8. Calculate Total Floats.
9. Establish the critical path by determining operations with zero Total Float.
10. Calculate Free Floats.
11. Convert to calendar days.

9.14.1   Updating

The usefulness of CPM is preserved throughout the course of a project only because of its updating capabilities. Updating can be defined as planning and programming the remaining portion of a job by introducing into the network the latest information available.

At the end of any day of work, all operations in a network must be in one of the following status categories:

1. Operations completed.
2. Operations in progress.
3. Operations not yet started.

Consider, for example, the network of Figure 9.2. If all operations had been started at their Earliest Start Times, at the end of the 15th day of work, the following would be the status of the project:

1. Operations completed: Start, A, B, C, D, and E.
2. Operations in progress:
   F has eight (8) more days to go.
   G has one (1) more day to go.
3. Operations not yet started: H and End.

The status described above could be called the "planned status." Had the job progressed according to the Early Start Time schedule, the status of the project would, in fact, coincide with the one outlined and no update would be necessary at this point. Assume, however, that, as in most cases, the job did not develop exactly as planned and that the actual project status is the following:

1. Operations completed:
   Start, A, B, C, and E.
2. Operations in progress:
   D has one more day to go, F has only 5 more days to go.
3. Operations not yet started: G, H, and End.

In addition to the above changes in estimated activity durations at the time of an update, there could also be changes in the logic (connectivity) of the network by deletion and/or addition of operations. Assuming only the duration changes reported above, one can proceed to update the network simply by recomputing the EST, EFT, LFT, LST, Free Float, and Total Float of each operation, and to locate the critical path as before but with durations and starting workday changed as follows:

1. All completed activities are given zero durations (thus strictly speaking, they become events for updating purposes).

2. All activities in progress are given durations equal to the number of days remaining for completion.
3. All activities not yet started are given durations as reported in the latest estimate.
4. To keep the time base line (reference date) unchanged, the EST of Start is increased from 0 to 15 to reflect the fact that the new computations are based on the status of the project at the end of the 15th day (at the beginning of the 16th day) of work.

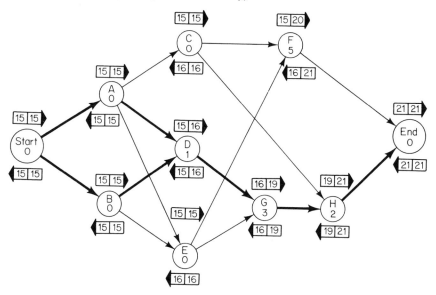

FIGURE 9.3.   Updated Network.

With-these modifications, the network calculations are updated in Figure 9.3. The Free Floats and Total Floats are:

Operation	Free Float	Total Float
Start	0	0
A	0	0
B	0	0
C	0	1
D	0	0
E	0	1
F	1	1
G	0	0
H	0	0
End	0	0

The new two-branch critical path is shown in Figure 9.3 to consist of Start,

A, B, D, G, H, and End. The overall effect of the actual progress report was a decrease in total project duration from 23 days (Figure 9.2) to 21 days.

9.14.2 Updating Frequency

Updates should normally be performed monthly or biweekly and, depending upon the character of the project, even at shorter intervals.

Shorter projects should be updated more frequently than longer ones and, since long projects become shorter toward the end, the frequency of updating should be increased as the project unfolds.

Always update following radical changes in activity durations, network logic, or both. A good practice is to update at the close of fiscal periods.

9.15 ACTIVITY ON BRANCH OR ARROW NETWORKS

Arrow networks are also in general use for CPM scheduling. Each operation in this type of network is represented by an arrow instead of a node. The tail of the arrow indicates the beginning of the activity, its head, the end of the activity.

Some examples of arrow diagrams equivalent to circle diagrams are:

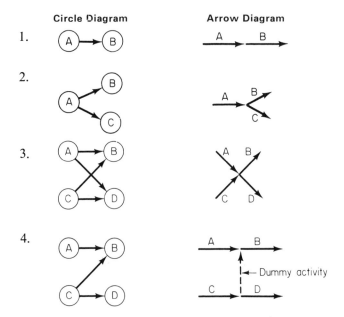

In statement 4, in order to preserve the logic: B follows A and C, but D follows only C; the arrow diagram had to include a "dummy" activity, symbolized by a dashed line. A dummy activity requires no time (zero duration) and is used merely to identify operational dependence (connectivity). Another use of "dummy" activities applies to the following case: Activities D and E can be performed simultaneously. Activity F must follow D and E. The arrow diagram is

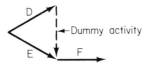

and, again, the dummy activity is used only to maintain the proper logic. In activity or branch networks, nodes are used at the heads and the tails of the connecting arrows:

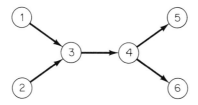

The network numbering is usually done inside these nodes. The nodes are called events and treated as such in the computations. Thus, the activity-oriented computations refer to the arrows or branches of the network, whereas the event-oriented computations refer to its nodes.

The computations are performed in a similar manner as for activity on node or circle networks. No further discussion of arrow networks will be undertaken in this chapter because the activity on node or circle networks are, in the author's opinion, more flexible and efficient and simpler to use.

9.15.1 Resource Allocation

The sequencing and scheduling of activities and events to accommodate limitations in the availability of resources is called resource allocation.

Resources are analyzed either to determine the scheduling levels required to implement the project or to alter an unacceptable work plan, in an efficient manner, to convert it into an implementable CPM project.

One calculation procedure designed to modify unacceptable networks is called the Resource Scheduling Method (RSM). The method's goal is to resequence activities such that all resource conflicts are resolved and the total project duration is increased a minimum amount. Because of the nature of the procedure, it can be implemented only after the initial network has been laid out and the calculations performed.

9.15.2 Information Requirements

The Resource Scheduling Method (RSM) requires that the level of resources necessary to effect the completion of every activity in the duration specified be known, and that the maximum availability of each resource be established. All resources for which no availability limitations exist will not cause resource conflicts because conflicts ensue only when the utilization of a given resource exceeds its availability. For example, consider the following activities and associated daily resource requirements:

Activity	Duration	Cranes	Pipe Fitters	Iron Workers
A	4	2	10	15
B	5	1	15	8
C	2	0	12	25

If only two cranes are available and if activities A and B are scheduled to be performed simultaneously according to the work plan, a resource conflict would ensue. Three cranes would be required when only two are available. Either activities A and B have to be resequenced, or one additional crane has to be procured. The decision will have to be made based on the economics of the situation at hand. RSM is a technique designed to aid the planner in formulating this type of decision.

In general, it must be stressed that the maximum amount available of any resource must be equal to or greater than the maximum amount of that resource required by a single activity in the work plan.

If no limitations exist for a particular resource, conflicts will not develop and the project planner can disregard the resource in his scheduling analysis. This could be the case with pipe fitters and iron workers, for example, where the only concern with these trades may be to keep level the manpower requirements of the project to avoid extreme fluctuations in the labor force.

9.15.3 Implementation of RSM

Assume that three activities—P, Q, and R—have the following characteristics:

Activity	Duration	Cranes	EST	EFT	LST	LFT
P	5	1	8	13	8	13
Q	4	1	7	11	9	13
R	5	1	9	14	10	15

Plot these activities in a bar graph such as the one given in Figure 9.4.

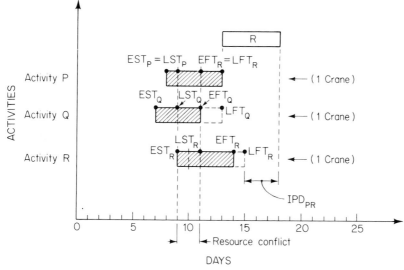

FIGURE 9.4. Activity Bar Chart.

The graph shows an Early Start Time Schedule and gives the EST, EFT, LST, and LFT of each activity. The dotted lines are, in fact, the Total Floats. Activity P has zero Total Float and for this reason it is a critical operation. Suppose that only 2 cranes are available. Under these circumstances (2 cranes available and all activities starting at their EST's), a resource conflict develops (3 cranes required, only 2 available) for days 10 and 11.

The objective of RSM is to establish a procedure for resequencing activities, *two at a time*, in order to resolve the conflict with a minimum or no increase in project duration.

In this case, it is seen that resequencing any two of the activities would resolve the conflict. The question is, Which two, and how? For instance, P could be made to follow Q or Q made to follow P; P made to follow R, or R to follow P; and so on. There are six combinations and, without performing an exhaustive search of all possibilities, one would like to establish the optimal resequencing plan.

In order to develop a criterion for optimal resequencing, assume that the best plan is to let R follow P. If this were done, the Increase in Project Duration (IPD_{PR}) can be computed as follows: Figure 9.4 shows R following P and gives a graphical measure of the Increase in Project Duration IPD_{PR}. The increase is seen to be the number of days that the Total Float of R will be exceeded. Thus,

$$IPD_{PR} = EFT_P + Duration_R - LFT_R. \qquad (9.2)$$

However,

$$Duration_R - LFT_R = -(LFT_R - Duration_R) = -LST_R. \qquad (9.3)$$

Therefore,

$$IPD_{PR} = EFT_P - LST_R. \qquad (9.4)$$

In general, then, if Activity J is resequenced to follow I, the Increase in Project Duration is

$$IPD_{IJ} = EFT_I - LST_J. \qquad (9.5)$$

Because the objective of the procedure is to *minimize* the IPD, one must select for Operation I, the activity to be followed, that with the smallest EFT, and for Operation J, the activity following, that with the largest LST. This procedure yields the smallest IPD. One must keep in mind that the resequencing must be performed two operations at a time and that the ultimate goal is to resolve the resource conflict. If the same operation simultaneously has the smallest EFT and the largest LST, a tie develops and one must consider the second best activities, both from the EFT and LST points of view, to arrive at the optimal combination.

If the Increase in Project Duration turns out to be negative, it indicates that the Total Float of the activity following has not been exceeded and, consequently, the project duration will not increase (no delay in the completion of the job).

Returning to the problem involving P, Q, and R, note that activity Q has the smallest EFT, 11, and that R has the largest LST, 10. Therefore,

$$I \Longrightarrow Q,$$

$$J \Longrightarrow R,$$

and

$$IPD_{QR} = EFT_Q - LST_R = 11 - 10 = 1 \text{ day.}$$

Therefore, if R is resequenced to follow Q, the Increase in Project Duration would be 1 day. An arrow from Q to R must be added to the network (R must follow Q) and the calculations performed again to reestablish the EST, EFT, LST, LFT, Free Float, and Total Float of each activity.

Whether or not the resequencing should be done depends upon the following factors:

1. The indirect and other costs of extending the project completion date.
2. The cost of procuring additional resources from external sources.

Suppose that, in the example, the indirect cost per day were $1,000 and that an additional crane could be rented at a daily cost of $400, then

$$\text{Cost of Renting} = \$400 \times 2 = \$800,$$
$$\text{Cost of Delay} = \$1,000.$$

The additional crane would be required during 2 days (days 10 and 11) without resequencing, and the Increase in Project Duration is 1 day if the activities are resequenced. In this case, the obvious answer is *not* to resequence but to rent the additional crane.

Other possibilities also exist. For example, activities Q and R have floats and could be displaced in time within their respective Total Floats without increasing the total job duration. This adjustment could yield a more economical resource scheduling plan. The project planner has ample opportunity to display his knowledge and to challenge his imagination when allocating resources in an optimal manner.

9.15.4 RSM Summary

1. Scan the CPM schedule, day by day, until a resource conflict is detected.
2. Resequence the operations causing the conflict, two at a time, until the conflict is resolved with a minimum Increase in Project Duration.
3. Each resequencing must be followed by the introduction of appropriate arrows to recompute all network data.
4. Cost trade-offs which result from procuring additional resources from external sources versus resequencing activities, must be evaluated to establish the most economically advantageous course of action.

9.16 TIME-COST OPTIMIZATION

The underlying assumption in the work done up to this point has been that each activity can be performed in one way only. Now CPM will be extended to include the case where, for several alternatives, a duration and a cost per method of performance can be associated to each activity in the network.

Duration entries will be used to establish the completion time of the project for each method selected, to define the critical operations for each completion time, and to compute the Total Float of each activity.

Cost entries will be used to evaluate total project cost and to ascertain which activities must be shortened or compressed to decrease the project duration at a minimum cost. The cost-duration entries lead to a series of project completion times.

9.16.1 Activity Time-Cost Trade-Off Graphs

Consider the relationship between cost and duration for a typical activity. There exists a duration with its associated cost which can be labeled the *normal duration*. This corresponds to the lowest performance cost and the longest practical completion time. As resources are increased, the time required to complete an activity decreases with a corresponding increase in cost, until a duration is reached below which it is physically impossible to complete the activity, regardless of the level, quality, and/or cost of the resources employed. This duration is called the *crash duration*. A typical Activity Time-Cost Trade-Off Graph (ATC Graph) is given in Figure 9.5.

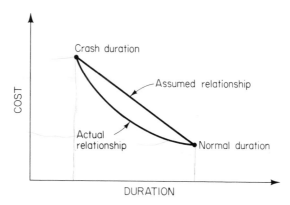

FIGURE 9.5.   Activity Time-Cost Trade-Off Graph.

Because of the uncertainties involved in developing such a graph, and to simplify the calculations, a straight line relationship will be assumed to exist between cost and time from the crash duration to the normal duration. This assumption results in an error equal to the difference in cost between the straight line and the curve. This, however, is a deviation with a high degree of uncertainty (labor costs, especially, are extremely difficult to estimate with accuracy) and, for all practical purposes, the linear relationship is as accurate as one can expect for this type of work.

One additional requirement will be imposed upon ATC graphs; the straight line joining normal with crash durations must be continuous (no time-cost discontinuities are acceptable).

9.16.2 Project Time-Cost Trade-Off Graph

The information generated by the ATC curves is used in developing the Project Time-Cost Trade-Off Graph (PTC Graph). Figure 9.6 shows a typical PTC Graph. The following observations must be made about PTC curves:

1. The PTC curve is the lower bound envelop of all the cost-duration points that can be generated by compressing (shortening) individual network activities. If, for example, a noncritical activity is shortened, the total project cost increases but the total project duration remains the same.

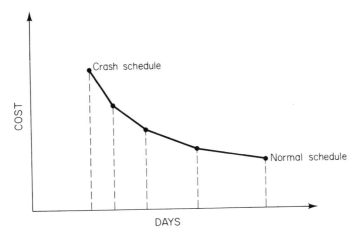

FIGURE 9.6. Project Time-Cost Trade-Off Graph.

2. From the previous observation one concludes that only critical activities must be compressed.
3. If there are several critical paths in a network, as many critical activities as there are common path branches must be simultaneously compressed. Otherwise, one or more of the paths may become noncritical, thus increasing project cost without decreasing duration. (This point will become much clearer later when the computations are performed.)
4. The *normal schedule* is obtained with all activities at their normal durations. The *crash schedule* is established when all activities on the critical path(s) have been crashed.
5. Any end point on a straight line segment is associated with a particular schedule. Other points on the straight line segment are associated with a combination of the end point schedules.
6. Each schedule is an Early Start Time schedule.
7. Because the PTC curve is the lower bound envelop of all possible schedules, each schedule on the PTC curve represents the least direct cost for completing the project in the corresponding time.
8. In summary, the PTC curve defines which method should be used with each activity in the network in order to accomplish each possible project completion time at a minimum cost.

9.16.3 Time-Cost Optimization Algorithm

The algorithm used in generating PTC curves can best be explained with the aid of an example. Consider the following network and associated information on each activity:

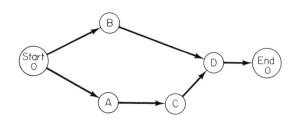

Activity	Normal Duration/Cost		Crash Duration/Cost	
A	4	$600	2	$1,000
B	6	$800	3	$1,400
C	8	$500	3	$1,200
D	7	$600	2	$1,200

The following ATC graphs are obtained from the information given above:

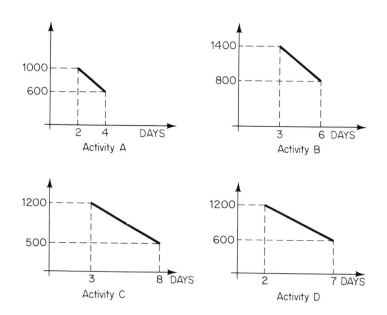

Initializing Step:

The first step in the algorithm is to compute the normal schedule. This is simply the EST schedule with all activities at their normal durations.

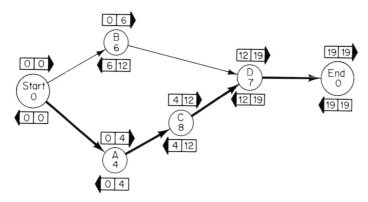

The critical path is marked with heavy arrows and the Total Floats are:

Activity	Total Float
A	0
B	6
C	0
D	0

The total project cost is computed from the ATC graphs,

$$\text{Cost} = \$600 + \$800 + \$500 + \$600 = \$2,500.$$

The first point on the PTC curve (the normal schedule) is (19 days, \$2,500).

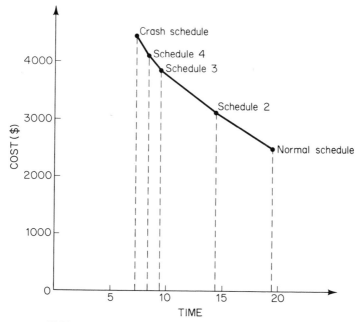

FIGURE 9.7.   **Project Time-Cost Trade-Off Graph.**

This is plotted in Figure 9.7. Now, the algorithm proceeds as follows:

Cycle 1:

STEP 1:

Determine the activities whose durations should be decreased in order to develop schedule 2.

The project completion time is controlled by the activities on the critical path. Thus, an activity whose duration is to be shortened to develop a schedule with a shorter completion time must be on the critical path. There is only *one* critical path in this cycle. Hence, only *one* of the activities will be *compressed*.

It must be noted that the maximum number of activities to be compressed can be as large as the number of critical paths.

Which activity must be compressed? The one, on the critical path, with the smallest direct cost per day of compression. The rate of increase of direct cost from compressing an activity is the slope of its ATC curve. Consequently, the activity to be compressed is the one with the least slope. The slopes are computed to be:

Activity	Slope	
A	$200/day	
B	$200/day ←————— Not on critical path.	
C	$140/day	
D	$120/day ←————— Select this one.	

Hence, compress activity D first.

STEP 2:

How much can the activity selected be compressed? There are two important limits to the compression process:

1. The Crash Limit.
2. The Total Float Limit.

These limits will be explained as they appear in the algorithm. The amount of compression is the *smallest* of these two limits. For this computation, the *Crash Limit* is the governing one. Activity D can, therefore, be compressed a maximum of 5 days, because its normal duration is 7 days and its crash duration is 2 days.

STEP 3:

Determine the EST schedule resulting from compressing activity D, 5 days.

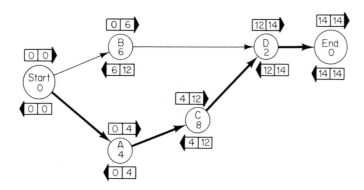

The completion time is 14 days. The Total Floats are:

Activity	Total Float
A	0
B	6
C	0
D	0

STEP 4:

The direct cost for schedule 2 equals the direct cost for schedule 1 plus 5 days compression times $120 per day = $2,500 + $600 = $3,100.

STEP 5:

Determine the PTC point for schedule 2. It is (14 days, $1,300) plotted in Figure 9.7.

STEP 6:

Determine whether schedule 2 is the crash schedule. It is *not*, because each activity on the critical path is *not* at its crash duration.

Cycle 2:

STEP 1:

Determine the next activity to be compressed on the critical path. Alternates: Activities A or C.

Activity C has the lowest cost per day, that is, the least slope = $140/day.

STEP 2:

The Crash Limit of Activity C is 3 days. Hence, operation C can be compressed 5 days.

STEP 3:

Determine the EST schedule.

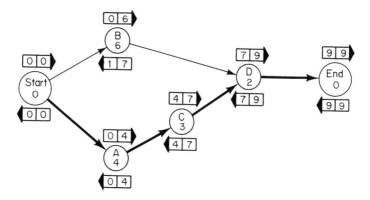

The completion time is 9 days. The Total Floats are:

Activity	Total Float
A	0
B	1
C	0
D	0

Note that the Total Float of B decreased from 3 days to 1 day.

STEP 4:

Direct Cost of schedule 3 = \$3,100 + 5 × \$140 = \$3,820.

STEP 5:

The PTC point is (9 days, \$3,820). See Figure 9.7.

STEP 6:

Schedule 3 is not the crash schedule.

Cycle 3:

STEP 1:

Only activity A on the critical path can be compressed.

STEP 2:

Activity A has a crash limit of 2 days; however, in this cycle the Crash Limit is not the least of the two limits; the *Total Float* Limit is the smallest.

The Total Float Limit is the amount of compression of an activity causing a positive Total Float for some other operation to be reduced to zero.

When Activity A is compressed 1 day, the Total Float of B is reduced to zero. Therefore, the Total Float Limit is 1 *day. To determine whether or not a Total Float Limit exists and its value if it does, compress by one day the selected activity. The smallest original value of any Total Float reduced correspondingly by one day establishes a Total Float Limit.*

When the Total Float of an operation is reduced to zero, it generates another critical path. Any further reduction in total project duration has to include this additional critical path.

STEP 3:

Determine the EST schedule resulting from compressing A by 1 day.

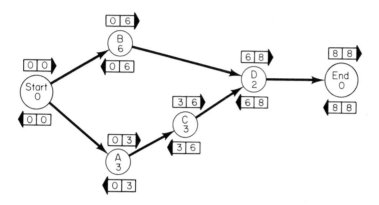

Project duration is 8 days and the entire network is critical. The Total Floats are:

Activity	Total Float
A	0
B	0
C	0
D	0

STEP 4:

Direct Cost of Schedule 4 = $3,820 + 1 × $200 = $4,020.

STEP 5:

The PTC point is (8 days, $4,020). See Figure 9.7.

STEP 6:

Schedule 4 is *not* the crash schedule.

Cycle 4:

STEP 1:

Now there are *two* critical paths. Two activities must be compressed simultaneously. Crash duration of A $\longrightarrow$ Compress 1 day. Crash duration of B ⟶ Compress 3 days. Therefore, one can compress each activity only 1 day. This is the maximum compression (if B is compressed further, it would fall out of the critical path).

STEP 2:

Compress A 1 day and B 1 day.

STEP 3:

The EST schedule is

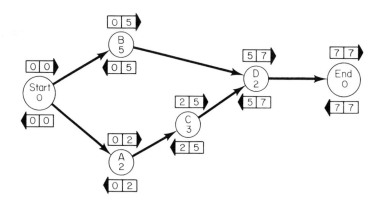

The total project completion time is 7 days.
The Total Floats are zero for all operations.
The entire network is critical.

STEP 4:

Direct Cost of crash schedule = $4,020 + 1 × $200 + 1 × $200 = $4,420.

STEP 5:

The PTC point is (7 days, $4,420).

STEP 6:

Schedule 5 is the *crash schedule*. All operations (A, C, D) on *one* critical path are at their *crash durations*.

With this last cycle, the Project Time-Cost Trade-Off curve is completely determined as given in Figure 9.7.

9.16.4 Total Cost Curves

The curve generated for the PTC graph is a direct cost curve. Indirect costs, bonus payments, and penalty costs (liquidated damages) can also be incorporated into the graph, and a total cost curve obtained by algebraically adding all the cost elements (bonus payments can be treated as negative costs) to determine the least overall cost (optimum) schedule. The procedure is

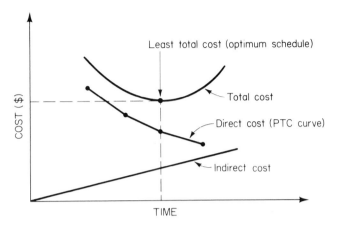

FIGURE 9.8. Optimum CPM Schedule.

depicted in Figure 9.8. This concludes the discussion of CPM. It must be pointed out that many computer programs exist to implement CPM schedules for large, complex projects with ease and efficiency.

EXERCISES

9-1. Consider the following operational dependencies:
   a. A is the first operation.
   b. E and F can be performed simultaneously and cannot be started before B is finished.
   c. I follows F, G, and H.
   d. K can start only after E and I are finished.
   e. L must follow J and K.
   f. G follows C.
   g. J depends on E.
   h. B, C, and D must follow A and can be performed simultaneously.
   i. H can begin only after D is completed.
   k. L is the last activity.
      i. Draw the CPM activity on node or circle network.
      ii. With the following durations, complete EST and LFT schedules and determine the critical path.

Activity	Duration
A	5
B	3
C	5
D	7
E	6
F	4
G	3
H	2
I	3
J	4
K	5
L	2

iii.

Activity	Cranes
E	1
F	1
G	1

Total cranes available $= 2$.

With the information supplied above on resource availability (2 cranes) and utilization, and assuming that activities start at their EST's, resequence the operations in an *optimal manner* to resolve any resource conflicts that may exist. What is the Increase in Project Duration?

9–2. Consider the following operational sequence:
  a. A, B, and C can be performed simultaneously and are the first activities in the network.
  b. F follows C and B.
  c. H must follow D.
  d. I depends on G and H.
  e. K follows I and J.
  f. L depends on M.
  g. K and L are the last operations in the network.
  h. D follows A.
  i. E cannot be started before F, B, and D are completed.
  j. E precedes G. .
  k. J must wait for completion of H.
  l. M must follow I.
      i. Draw the activity on node network for the operations given.

ii. Establish the *critical path* with the following durations:

Activity	Duration
A	4
B	3
C	4
D	5
E	4
F	10
G	12
H	3
I	7
J	8
K	14
L	13
M	9

   iii. Activity M requires 2 cranes and K also requires 2 cranes. Only 3 cranes are available. Indirect costs are estimated at $1,200/day for this job; while crane rental runs $150/day. What should be done to *resolve a resource conflict* if it develops?

9-3. a. Draw the activity on node network for the following operational logic:

    i. T follows P.

    ii. P and R can be performed simultaneously and are the first activities.

    iii. S follows T.

    iv. A is the last operation.

    v. A follows F and H.

    vi. E and K can be performed simultaneously and depend on T, L, and B.

    vii. H depends on G, K, and N.

    viii. L and M can be performed simultaneously and follow P and R.

    ix. F and G can be performed simultaneously and depend on E and S.

    x. B and N can be performed simultaneously and follow M.

b. Perform the calculations on the network and establish the critical path(s). Activity durations follow:

Activity	Duration
P	5
R	5
T	8
L	8
M	5
B	3
N	10
S	4
E	4
K	7
F	8
G	3
H	5
A	10

9–4. a. Perform the calculations for the following activity on node network. Durations are given inside the circles.

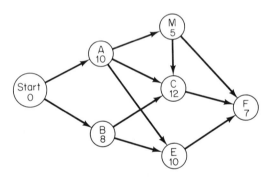

b. Resource utilization is given in the following table:

Operation	Cranes
M	2
C	1
E	1

Resource availability = 2 *cranes.*
    Determine whether or not resource conflicts develop and, if so, resolve them by *optimum* resequencing.

9–5. Develop the Project Time-Cost Trade-Off curve for the following activity on node network:

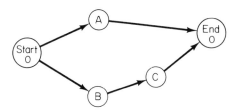

The Activity Time-Cost Trade-Off graphs are:

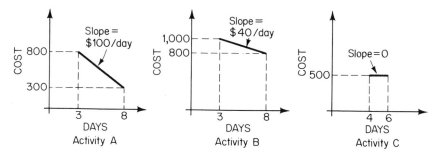

9-6. Determine the optimal completion time for the job whose CPM network is depicted below. Indirect costs are $100 per day and there is a bonus of $50 for each day that the job is completed *ahead* of day 15.

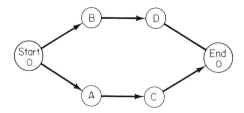

Liquidated damages are $100 per day for each day that the job is completed *after* day 15. The ATC curves for all activities are given below:

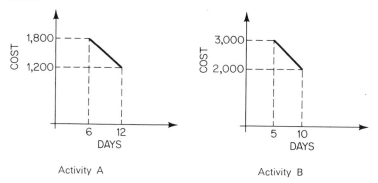

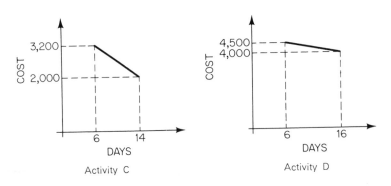

Activity C

Activity D

REFERENCES

1. Associated General Contractors of America, The, *CPM in Construction, A Manual for General Contractors*. Washington, D.C., 1965.
2. Levin, R. I. and C. A. Kirkpatrick, *Planning and Control with PERT/CPM*. New York: McGraw-Hill Book Company, 1966.
3. Peurifoy, R. L., *Construction Planning, Equipment and Methods, 2nd ed*. New York: McGraw-Hill Book Company, 1970.
4. Rubey, H. and W. W. Milner, *Construction and Professional Management, An Introduction*. New York: The Macmillan Company, 1966.
5. Shaffer, L. R., J. B. Ritter, and W. L. Meyer, *The Critical Path Method*. New York: McGraw-Hill Book Company, 1965.

CHAPTER 10

Climbing Techniques

10.1 INTRODUCTION

Climbing techniques utilize past information to generate points where the objective function achieves better values. Each climbing experiment has two main purposes:

1. To attain an improved value of the objective function—through climbing.
2. To provide information useful for locating future experiments where desirable values of the objective are likely to be found—through experimentation.

The reader will soon recognize that climbing procedures generally emcompass three distinct phases, as follows:

1. An opening game to set the stage. This is an exploratory phase.
2. Middle moves to push for advantages. These are climbing steps.
3. A final gambit to strive to reach the goal. This, again, is an exploratory phase.

The various multidimensional search schemes differ from each other only in the middle phase, for each strategy always begins and ends with an exploration. The methods that will be discussed in this chapter involve breaking a multidimensional search problem into sequences of unidimensional optimizations which are more easily resolved. This chapter sets the stage for the chapter on dynamic programming where the idea of decomposition will be pursued in much greater detail. In general, the methods presented in the following sections apply to unconstrained, continuous, or discrete objective functions. The approach followed in this chapter parallels the one given by D. J. Wilde and C. S. Beightler (see References) and their book is recommended for further study to those readers interested in pursuing this topic.

10.2 THE ALGEBRA AND GEOMETRY OF MULTIDIMENSIONAL PROBLEMS

Any ordered triplet of values (x_1, x_2, y) can be represented graphically by a point in an isometric projection. The set of points corresponding to all possible ordered triplets forms a space of three dimensions.

Biologists and statisticians, for whom the coordinate y is often the response of a living organism to environmental factors x_1 and x_2, have come to call surfaces of the type

$$y = y(x_1, x_2) \tag{10.1}$$

response surfaces.

This name has found its way into systems terminology. Response surfaces of type (10.1) are two-dimensional entities, because only two coordinates are needed to specify points on them. The difference between the number of variables and the number of independent equations relating them is called the number of *degrees of freedom*. Therefore, the space of all possible points (x_1, x_2, y) not necessarily satisfying equation (10.1) has three degrees of freedom, or dimensions, whereas the response surface has only two. The intersection of two surfaces is a curve. This is a one-dimensional entity because each surface is the representation of one equation relating the three variables. Points on the intersection belong to both surfaces and, therefore, must satisfy both equations, leaving only one degree of freedom. A point which is at the simultaneous intersection of three surfaces must be considered a zero-dimensional object, for it can have no degrees of freedom. These same considerations follow for N-dimensional spaces.

10.3 THE INITIAL MOVES— FIRST DERIVATIVE ESTIMATIONS

The first step is to locate a few experiments to determine the elevation of the initial point on the response surface and the direction of future moves in order to climb toward the maximum from the initial group of experiments.

In the case of continuous functions, their sensitivity is measured by computing first derivatives. When discrete functions are under study (functions defined at a finite number of points), the derivative formulas must be expressed in discrete terms.

Let

$$\bar{x}_0 = \begin{pmatrix} x_{01} \\ x_{02} \end{pmatrix}$$

and

$$y(\bar{x}_0) = y_0.$$

Let

$$\bar{x}_1 = \begin{pmatrix} x_{11} \\ x_{12} \end{pmatrix} = \begin{pmatrix} x_{11} \\ x_{02} \end{pmatrix},$$

that is,

$$x_{12} = x_{02},$$

and

$$y(\bar{x}_1) = y_1.$$

Keeping $x_{11} - x_{01}$ small, the first partial derivative of y with respect to the variable x_1 can be defined,

$$\left(\frac{\partial y}{\partial x_1} \right)_0 \simeq \frac{y_1 - y_0}{x_{11} - x_{01}}. \tag{10.2}$$

Now, let

$$\bar{x}_2 = \begin{pmatrix} x_{21} \\ x_{22} \end{pmatrix} = \begin{pmatrix} x_{01} \\ x_{22} \end{pmatrix},$$

that is,

$$x_{21} = x_{01},$$

and

$$y(\bar{x}_2) = y_2.$$

Again, keeping $x_{22} - x_{02}$ small, the first partial derivative of y with respect to x_2 is given by,

$$\left(\frac{\partial y}{\partial x_2} \right)_0 \simeq \frac{y_2 - y_0}{x_{22} - x_{02}}. \tag{10.3}$$

Equations (10.2) and (10.3) compute the slope of the objective function in two particular directions; parallel to the x_1 axis and parallel to the x_2 axis. The three points y_0, y_1, and y_2 on the response surface are sufficient to determine the plane approximately tangent to the surface at y_0. To determine the equation of this tangent plane, proceed as follows: In three-dimensional space, a plane satisfies the following relationship,

$$y(x_1, x_2) = m_0 + m_1 x_1 + m_2 x_2, \tag{10.4}$$

where m_0, m_1, and m_2 are constants. To evaluate them, the following set of equations must be solved:

$$y_0 = m_0 + m_1 x_{01} + m_2 x_{02},$$
$$y_1 = m_0 + m_1 x_{11} + m_2 x_{12},$$
and
$$y_2 = m_0 + m_1 x_{21} + m_2 x_{22}.$$

(10.5)

In determining the equation of the tangent plane, it is convenient to deal only with deviations of x_1, x_2, and y from the original point y_0. For experiment

$$\bar{x}_i = (x_{i1}, x_{i2}), \text{ define}$$

$$\Delta x_{i1} = x_{i1} - x_{01}, \tag{10.6}$$
$$\Delta x_{i2} = x_{i2} - x_{02}, \tag{10.7}$$
and
$$\Delta y_i = y_i - y_0. \tag{10.8}$$

This translation of coordinates eliminates the need to determine the intercept constant m_0. Take the second equation minus the first equation of (10.5) and compute

$$\Delta y_1 = m_1 \Delta x_{11} + m_2 \Delta x_{12}. \tag{10.9}$$

The third minus the first equation of (10.5) yields

$$\Delta y_2 = m_1 \Delta x_{21} + m_2 \Delta x_{22}. \tag{10.10}$$

In general,

$$\Delta y_i = m_1 \Delta x_{i1} + m_2 \Delta x_{i2}, \quad \text{for } i = 1, 2. \tag{10.11}$$

Because $\bar{x}_1$ was chosen to make $x_{12} = x_{02}$, then $\Delta x_{12} = 0$ and

$$m_1 = \frac{\Delta y_1}{\Delta x_{11}} \simeq \left(\frac{\partial y}{\partial x_1}\right)_0. \tag{10.12}$$

Also, $\bar{x}_2$ was chosen such that $x_{21} = x_{01}$, hence $\Delta x_{21} = 0$ and

$$m_2 = \frac{\Delta y_2}{\Delta x_{22}} \simeq \left(\frac{\partial y}{\partial x_2}\right)_0. \tag{10.13}$$

The general equation of the tangent plane is obtained by combining equation (10.4) with equations (10.6), (10.7), and (10.8).

$$\Delta y = m_1 \Delta x_1 + m_2 \Delta x_2 = \nabla y \Delta \bar{x}, \tag{10.14}$$

where $\nabla y = (\partial y/\partial x_1, \partial y/\partial x_2)$ is called the gradient of y and $\Delta \bar{x}$ is a column vector with components Δx_1 and Δx_2. With equation (10.14), one can estimate changes in y for any combination of small deviations in x_1 and x_2.

In choosing the three experiments $\bar{x}_0$, $\bar{x}_1$, and $\bar{x}_2$ to compute the equation of the tangent plane at y_0, the three points must not lie along the same straight line in the $x_1 - x_2$ plane for, in this case, the three equations (10.5) would not be linearly independent.

In using equation (10.14) to approximate the response surface in the neighborhood of y_0, the second-order terms in the Taylor Series Expansion of Δy are neglected, that is,

$$\Delta y = \left(\frac{\partial y}{\partial x_1}\right)_0 \Delta x_1 + \left(\frac{\partial y}{\partial x_2}\right)_0 \Delta x_2 + 0(\Delta x)^2.$$

Therefore, the deviations Δx_i, $i = 1, 2$, must be kept sufficiently small. Generalizing to N independent variables, let

$$y = y(x_1, x_2, x_3, \ldots, x_N),$$

and let $\bar{x}_0^t = (x_{01}, x_{02}, \ldots, x_{0N})$ be the original experiment and $y_0 = y(\bar{x}_0)$ be its outcome. Define

$$\Delta \bar{x}_i = \bar{x}_i - \bar{x}_0, \qquad \text{for } i = 1, 2, \ldots, N,$$

and

$$\Delta y_i = y_i - y_0 = y(\bar{x}_i) - y(\bar{x}_0).$$

Let $m_j = (\partial y / \partial x_j)_0$, be the first partial derivative of y with respect to x_j, evaluated at x_0, then

$$\Delta y = \sum_{j=1}^{N} m_j \Delta x_j = \nabla y \Delta \bar{x}. \tag{10.15}$$

To evaluate the coefficients m_j, one must perform N experiments (not including $\bar{x}_0$) and solve the N simultaneous equations

$$\Delta y_i = \sum_{j=1}^{N} m_j \Delta x_{ij} \qquad \text{for } i = 1, \ldots, N, \tag{10.16}$$

for the constants m_j, since Δy_i and Δx_{ij} are given for each experiment. These computations can be simplified by choosing $\Delta x_{ij} = 0$ for $i \neq j$, in which case equation (10.16) gives

$$\Delta y_i = m_i \Delta x_{ii},$$

from which

$$m_i = \frac{\Delta y_i}{\Delta x_{ii}}; \qquad \text{for } i = 1, 2, \ldots, N. \tag{10.17}$$

Once the m_j are known, one can state that the combinations of Δx_j giving increased y near y_0 must satisfy,

$$\nabla y \Delta \bar{x} = \sum_{j=1}^{N} m_j \Delta x_j > 0. \tag{10.18}$$

The boundary between the region of increasing y and the region below is the $N - 1$ dimensional hyperplane satisfying the equation

$$\nabla y \Delta \bar{x} = \sum_{j=1}^{N} m_j \Delta x_j = 0. \tag{10.19}$$

Note that equation (10.19) is an $N - 1$ dimensional hyperplane because it is an equation relating N variables and thus one of them is linearly dependent on the others.

10.4 MIDDLE MOVES—CLIMBING

In this section two climbing methods will be discussed in detail. The first, called *gradient search*, is a well-known procedure which has served as the basis for development of a multitude of other techniques. The second, *pattern search*, is very easy to program for computer processing. These two methods are given as typical examples of direct optimization procedures.

10.4.1 Gradient Search

This technique is called gradient search because the experiments are taken in the gradient direction. The gradient of a function y of N variables x_n; $n = 1$, ..., N, is defined to be the N-dimensional row vector of the first partial derivatives of y with respect to each independent variable x_n. Mathematically,

$$\nabla y = \left(\frac{\partial y}{\partial x_1}, \frac{\partial y}{\partial x_2}, \ldots, \frac{\partial y}{\partial x_n} \right), \tag{10.20}$$

where the symbol ∇ is the notation for gradient. In this method, the experiments are selected along the gradient, because the gradient points in the direction in which the response surface has the steepest slope. To prove this statement, consider the N-dimensional hypersphere of radius r, centered about the point $\bar{x}$. Points $\bar{x} + \partial\bar{x}$ on this sphere satisfy the equation

$$\sum_{j=1}^{N} (\partial x_j)^2 = |\partial\bar{x}|^2 = r^2. \tag{10.21}$$

The first-order approximation of y in the neighborhood of $\bar{x}$ gives the value of the objective function at various points on the sphere as

$$\Delta y = \nabla y \partial\bar{x}. \tag{10.22}$$

Suppose one wishes to find the point on the hypersphere where Δy is maximum. At this point, the following Lagrangian must be stationary:

$$L = \nabla y \partial\bar{x} - \lambda[|\partial\bar{x}|^2 - r^2]. \tag{10.23}$$

Consequently, the maximizing perturbation $\partial\bar{x}^*$ must satisfy,

$$\nabla L = \nabla y - 2\lambda\partial\bar{x}^{*t} = 0,$$

from which

$$\partial\bar{x}^{*t} = \frac{1}{2\lambda}\nabla y. \tag{10.24}$$

Since λ is a scalar, the vector of maximum perturbation, $\partial \bar{x}^*$, points in the same direction as the gradient, ∇y, thus proving the statement.

10.4.1.1 The Algorithm

The following algorithm serves to implement the gradient procedure:

STEP 1:

Select a point $\bar{x}_0$ to initialize the algorithm.

STEP 2:

Evaluate the gradient at $\bar{x}_0$.

$$\nabla y(\bar{x}_0) = \nabla y_0. \tag{10.25}$$

STEP 3:

The set of points in the gradient direction is given by

$$\Delta \bar{x}_0^t = \rho_0 \nabla y(\bar{x}_0) = \rho_0 \nabla y_0, \tag{10.26}$$

where

$$\rho_0 = \frac{1}{2\lambda}. \tag{10.27}$$

STEP 4:

Positive values of ρ_0 yield locally increasing values of y, so that the value of ρ_0 that maximizes Δy is found either by a one-dimensional search or, when possible, by direct differentiation. This latter approach will be considered here. It involves the following additional steps:

a. Substitute $\Delta \bar{x}_0 = \rho_0 \nabla y_0^t$ into y_0.
b. Differentiate with respect to ρ_0 and set this derivative to zero:

$$\frac{\partial y(\bar{x}_0 + \rho_0 \nabla y_0^t)}{\partial \rho_0} = 0.$$

c. Solve for the maximizing value ρ_0^*, by finding that value of ρ_0 which satisfies the equation

$$\left[\frac{\partial y(\bar{x}_0 + \rho_0 \nabla y_0^t)}{\partial \rho_0} \right]_{\rho_0 = \rho_0^*} = 0. \tag{10.28}$$

STEP 5:

A new point $\bar{x}_1$ is given by

$$\bar{x}_1 = \bar{x}_0 + \rho_0^* \nabla y_0^t, \tag{10.29}$$

and the procedure is iterated, thus, return to step 3 with the new set of points along the gradient direction given by

$$\Delta \bar{x}_1^t = \rho_1^* \nabla y_1. \tag{10.30}$$

10.4.1.2   Example

Apply the gradient search to the following problem:
Minimize
$$y = x_1 - 2x_2 + x_1^2 - x_1x_2 + x_2^2.$$
Using straight differentiation (chapter 3), one easily computes the minimum
$$y^* = -1,$$
and the minimizing policy,
$$\bar{x}^* = \begin{pmatrix} 0 \\ 1 \end{pmatrix}.$$

Solution by Gradient Search:

1-1. Start at $\bar{x}_0^t = (0, 0)$. Then,

2-1. $\nabla y_0 = (1, -2)$.

3-1. $\Delta \bar{x}_0^t = (x_1 - 0, x_2 - 0) = (\rho_0, -2\rho_0)$.

4-1. a. $y(\bar{x}_0 + \rho_0 \nabla y_0^t) = 5\rho_0 + 7\rho_0^2$.

     b. $\left[\dfrac{\partial y(\bar{x}_0 + \rho_0 \nabla y_0^t)}{\partial \rho_0} \right]_{\rho_0 = \rho_0^*} = 5 + 14\rho_0^* = 0.$

     c. Therefore, $\rho_0^* = -0.357$.

5-1. $\bar{x}_1^t = \bar{x}_0^t + \rho_0^* \nabla y_0 = (0, 0) + (-0.357, 0.714) = (-0.357, 0.174)$.

1-2. Now, $\bar{x}_1^t = (-0.357, 0.714)$, from which

2-2. $\nabla y_1 = (-0.428, -0.215)$.

3-2. $\Delta \bar{x}_1^t = (x_1 + 0.357, x_2 - 0.714)$
$$= (0.428\rho_1, -0.215\rho_1).$$

4-2. a.

$$y(\bar{x}_1 + \rho_1 \nabla y_1^t) = -0.357 - 0.428\rho_1$$
$$- 2(0.714 - 0.215\rho_1)$$
$$+ (-0.357 - 0.428\rho_1)^2$$
$$- (-0.357 - 0.428\rho_1)(0.714 - 0.215\rho_1)$$
$$+ (0.714 - 0.215\rho_1)^2.$$

     b. Setting

$$\left[\frac{\partial y(\bar{x}_1 + \rho_1 \nabla y_1^t)}{\partial \rho_1} \right]_{\rho_1 = \rho_1^*} = 0,$$

     c. Obtain,

$$\rho_1^* = -0.865.$$

5-2.

$$\bar{x}_2^t = \bar{x}_1^t + \rho_1^* \nabla y_1 = (0.013, 0.900).$$

Continuing the iterations, one finds that the answer approached the minimum point,

$$\bar{x}^* = \begin{pmatrix} 0 \\ 1 \end{pmatrix},$$

with some oscillation. The iterations are stopped when the differences between two consecutive answers fall within an acceptable margin of error.

10.4.2 Pattern Search

Pattern search is an easily programmed, accelerated climbing technique with ridge-following properties. The method was developed by Hooke and Jeeves in 1961 and is based on the conjecture that any set of moves that has been successful during early experiments will be worth trying again.

10.4.2.1 The Algorithm

STEP 1:

The search begins at a base point vector $\bar{b}_1$, which may be chosen arbitrarily.

STEP 2:

The experimenter chooses a step size δ_i for each independent variable $x_i, i = 1, 2, \ldots, n$. Let

$$\bar{\delta}_i = \begin{pmatrix} 0 \\ \cdot \\ \cdot \\ \cdot \\ 0 \\ \delta_1 \\ 0 \\ \cdot \\ \cdot \\ \cdot \\ 0 \end{pmatrix} \leftarrow i\text{th component.} \tag{10.31}$$

That is, $\bar{\delta}_i$ is the vector with ith component ∂_i and all the rest zero.

STEP 3:

After measuring the objective function at the initial base $\bar{b}_1$, one makes an observation at $\bar{b}_1 + \bar{\delta}_1$. If, at this new point, the objective function is better than at the base, $\bar{b}_1 + \bar{\delta}_1$ is called the temporary head $\bar{t}_{11}$, where the double subscript shows that the first pattern is being developed and that the first variable x_1 has been perturbed. If $\bar{b}_1 + \bar{\delta}_1$ is not as good as $\bar{b}_1$, one tries $\bar{b}_1 - \bar{\delta}_1$. If this new point is better than $\bar{b}_1$, it is made the temporary head; otherwise, $\bar{b}_1$ is designated temporary head. In summary, when one is minimizing,

$$\bar{t}_{11} = \begin{cases} \bar{b}_1 + \bar{\delta}_1 & \text{if } C(\bar{b}_1 + \bar{\delta}_1) < C(\bar{b}_1). \\ \bar{b}_1 - \bar{\delta}_1 & \text{if } C(\bar{b}_1 - \bar{\delta}_1) < C(\bar{b}_1) < C(\bar{b}_1 + \bar{\delta}_1). \\ \bar{b}_1 & \text{if } C(\bar{b}_1) < \min[C(\bar{b}_1 + \bar{\delta}_1), C(\bar{b}_1 - \bar{\delta}_1)]. \end{cases} \tag{10.32}$$

STEP 4:

Perturbation of x_2, the next independent variable, is similarly carried out, this time about the temporary head $\bar{t}_{11}$ instead of the original base $\bar{b}_1$. In general, the $(j+1)$th temporary head $(\bar{t}_{1,j+1})$ is obtained from the preceding one $(\bar{t}_{1,j})$ as follows:

$$t_{1,j+1} = \begin{cases} \bar{t}_{1j} + \bar{\delta}_{j+1} & \text{if } C(\bar{t}_{1j} + \bar{\delta}_{j+1}) < C(\bar{t}_{1j}), \\ \bar{t}_{1j} - \bar{\delta}_{j+1} & \text{if } C(\bar{t}_{1j} - \bar{\delta}_{j+1}) < C(\bar{t}_{1j}) < C(\bar{t}_{1j} + \bar{\delta}_{j+1}), \\ \bar{t}_{1j} & \text{if } C(\bar{t}_{1j}) < \min [C(\bar{t}_{1j} + \bar{\delta}_{j+1}), C(\bar{t}_{1j} - \bar{\delta}_{j+1})]. \end{cases} \quad (10.33)$$

This expression covers all $1 \leq j \leq n$ if the convention is adopted that

$$\bar{t}_{10} \equiv \bar{b}_1. \quad (10.34)$$

STEP 5:

After the n variables have been perturbed, the last temporary head, $\bar{t}_{1n}$, is designated the second base point $\bar{b}_2$.

$$\bar{t}_{1n} \equiv \bar{b}_2. \quad (10.35)$$

The original base point $\bar{b}_1$ and the new base point $\bar{b}_2$ establish the first pattern.

STEP 6:

Now assume that if a similar search were to be conducted from $\bar{b}_2$, the results are likely to be the same; hence, skip local excursions and double the length of the next pattern. This move establishes a new temporary head $\bar{t}_{20}$ for the second pattern based at $\bar{b}_2$, given by

$$\bar{t}_{20} = \bar{b}_1 + 2(\bar{b}_2 - \bar{b}_1) = 2\bar{b}_2 - \bar{b}_1. \quad (10.36)$$

STEP 7:

A local exploration about $\bar{t}_{20}$ is now carried out to correct the tentative second pattern, if necessary. The logical relationships establishing new temporary heads $\bar{t}_{21}, \bar{t}_{22}, \ldots, \bar{t}_{2n}$ are similar to those given by equations (10.33), with the first subscript being 2 rather than 1. The local exploration is completed after all the variables have been perturbed and the last temporary head $\bar{t}_{2n}$ has been made the third base point $\bar{b}_3$.

STEP 8:

As before, a new temporary head $\bar{t}_{30}$ is established as follows:

$$\bar{t}_{30} = \bar{b}_2 + 2(\bar{b}_3 - \bar{b}_2) = 2\bar{b}_3 - \bar{b}_2. \quad (10.37)$$

Note that repeated successes cause the pattern to grow because

$$\bar{b}_3 - \bar{b}_2 = 2(\bar{t}_{20} - \bar{b}_2) = 2(\bar{b}_2 - \bar{b}_1). \quad (10.38)$$

STEP 9:

The procedure is continued for the third and following patterns. If, for example, all the perturbations about the temporary head $\bar{t}_{i0}$ (ith pattern) fail to improve the outcome, but

$$C(\bar{t}_{i0}) < C(\bar{b}_i),\qquad\qquad(10.39)$$

then the pattern will maintain its direction and length without growth.

STEP 10:

For pattern k, assume that the temporary head $\bar{t}_{k0}$ does not produce an improvement over $\bar{b}_k$; then $\bar{b}_{k+1} = \bar{b}_k$ and the pattern is destroyed. This could mean one either is at a peak or crossing a resolution valley.

STEP 11:

Thus, a new pattern is attempted with $\bar{b}_k$ as the base point, but with the designation $\bar{b}_{k+1}$. One starts all over again by making $\bar{t}_{k+1,0} = \bar{b}_{k+1}$. If this exploration locates a better point, a new pattern can be initiated. Otherwise, the steps must be shortened in an attempt to find a better point.

STEP 12:

Finally, it must be stressed that because some components of the objective function could be step functions of the coordinates, one could easily find a *local* optimum (a better point existing in another region where these step functions abruptly drop in value). A pervasive danger is that the global optimum might be altogether ignored if the procedure converges on a local optimum with a higher cost. One way to prevent this possibility is to conduct at least two searches from randomly selected, distinct points. If both pattern searches converge to the same critical point, one's confidence in the reliability of the results is increased. Or one could search by regions. It must be remembered, however, that the search itself must be optimized (must be completed at minimum cost).

10.4.2.2 Plant Location Example

A plant structure is to be located on the site shown in Figure 10.1. Fuel lines, sewer, water, and gas lines, and a powerline must be led in from the existing tie-in locations to the industrial facility. An access road, perpendicular to the highway, must also be provided.

A subsoil exploration has revealed a stratum of stiff clay that slopes from the highway toward the river and that is overlaid by a soft deposit (Figure 10.2). Because of the soil conditions, point-bearing piles will be used.

The unit costs of lines and piles (given in Figures 10.1 and 10.2) are summarized here:

Access road	$15/ft.
Power line	$ 3/ft.
Water & Gas lines	$ 5/ft.
Sewer pipe	$ 4/ft.
Fuel lines	$12/ft.
Piles	$ 1.50/ft.

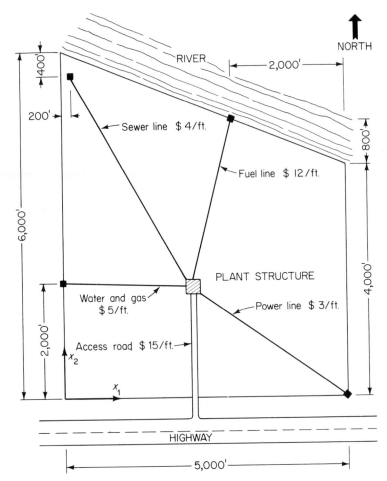

FIGURE 10.1. Plan of Site.

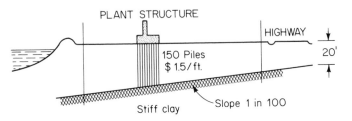

FIGURE 10.2. Subsoil Profile of Site.

The tie-in coordinates, as well as other pertinent information about the site, are given in Figure 10.1. Also, 150 point-bearing piles will be needed. Other things being equal, where must the plant structure be erected to minimize the cost of construction?

Solution by Pattern Search:

Because only incremental costs are of interest, the construction cost of the facility itself, as well as the cost of piling above 20 feet below grade, need not be considered.

The southwest corner of the site is taken to be the origin of coordinates; x_1 and x_2 are measured to the right and up from that point. The objective function, ΔC, can be written:

$$\Delta C(x_1, x_2) = 15x_2 + 3[(5,000 - x_1)^2 + x_2^2]^{\frac{1}{2}} + 5[x_1^2 + (x_2 - 2,000)^2]^{\frac{1}{2}}$$
$$+ 4[(x_1 - 200)^2 + (5,600 - x_2)^2]^{\frac{1}{2}}$$
$$+ 12[(3,000 - x_1)^2 + (4,800 - x_2)^2]^{\frac{1}{2}} + 225\left(\frac{x_2}{100}\right).$$

The geographic constraints are:

$$0 \leq x_1 \leq 5,000,$$

and

$$0 \leq x_2 \leq \frac{2x_1}{5} + 6,000.$$

The pattern search will start at the point

$$\bar{b}_1 = (2,500, 2,500).$$

With the plant structure trial sites set at multiples of 100 ft. from $\bar{b}_1$,

$$\bar{\delta}_1 = (100, 0),$$
$$\bar{\delta}_2 = (0, 100).$$

The search results are summarized below:

$$\Delta C(\bar{b}_1) = \Delta C(2,500, 2,500) = \underline{\$110,151},$$
$$\Delta C(\bar{b}_1 + \bar{\delta}_1) = \Delta C(2,600, 2,500) = \underline{\$110,443} > 110,151,$$
$$\Delta C(\bar{b}_1 - \bar{\delta}_1) = \Delta C(2,400, 2,500) = \underline{\$109,904} < 110,151.$$

Therefore,

$$\bar{t}_{11} = \bar{b}_1 - \bar{\delta}_1 = (2,400, 2,500),$$
$$\Delta C(\bar{t}_{11} + \bar{\delta}_2) = \Delta C(2,400, 2,600) = \underline{\$109,955} > 109,904,$$
$$\Delta C(\bar{t}_{11} - \bar{\delta}_2) = \Delta C(2,400, 2,400) = \underline{\$108,907} < 109,904.$$

Hence,

$$\bar{b}_2 = \bar{t}_{12} = \bar{t}_{11} - \bar{\delta}_2 = (2,400, 2,400).$$

Now,

$$\bar{t}_{20} = 2\bar{b}_2 - \bar{b}_1 = (2,300, 2,300),$$

$$\Delta C(\bar{t}_{20}) = \Delta C(2,300, 2,300) = \underline{\$108,704} < 108,907,$$

$$\Delta C(\bar{t}_{20} + \bar{\delta}_1) = \Delta C(2,400, 2,300) = \$108,882 > 108,704,$$

$$\Delta C(\bar{t}_{20} - \bar{\delta}_1) = \Delta C(2,200, 2,300) = \underline{\$108,564} < 108,704.$$

Therefore,

$$\bar{t}_{21} = \bar{t}_{20} - \bar{\delta}_1 = (2,200, 2,300),$$

$$\Delta C(\bar{t}_{21} + \bar{\delta}_2) = \Delta C(2,200, 2,400) = \$109,081 > 108,564,$$

$$\Delta C(\bar{t}_{21} - \bar{\delta}_2) = \Delta C(2,200, 2,200) = \underline{\$108,091} < 108,564.$$

Hence,

$$\bar{b}_3 = \bar{t}_{22} = \bar{t}_{21} - \bar{\delta}_2 = (2,200, 2,200),$$

from which

$$\bar{t}_{30} = 2\bar{b}_3 - \bar{b}_2 = (2,000, 2,000),$$

$$\Delta C(\bar{t}_{30}) = \Delta C(2,000, 2,000) = \underline{\$107,087} < 108,091,$$

$$\Delta C(\bar{t}_{30} + \bar{\delta}_1) = \Delta C(2,100, 2,000) = \$107,138 > 107,087,$$

$$\Delta C(\bar{t}_{30} - \bar{\delta}_1) = \Delta C(1,900, 2,000) = \$107,087 = 107,087.$$

This last point was considered to have failed, thus

$$\bar{t}_{31} = \bar{t}_{30} = (2,000, 2,000),$$

$$\Delta C(\bar{t}_{31} + \bar{\delta}_2) = \Delta C(2,000, 2,100) = \$107,506 > 107,087,$$

$$\Delta C(\bar{t}_{31} - \bar{\delta}_2) = \Delta C(2,000, 1,900) = \underline{\$106,698} < 107,087.$$

Hence,

$$\bar{b}_4 = \bar{t}_{32} = \bar{t}_{31} - \bar{\delta}_2 = (2,000, 1,900),$$

$$\bar{t}_{40} = 2\bar{b}_4 - \bar{b}_3 = (1,800, 1,600),$$

$$\Delta C(\bar{t}_{40}) = \Delta C(1,800, 1,600) = \underline{\$105,782} < 106,698,$$

$$\Delta C(\bar{t}_{40} + \bar{\delta}_1) = \Delta C(1,900, 1,600) = \underline{\$105,749} < 105,782.$$

$$\bar{t}_{41} = \bar{t}_{40} + \bar{\delta}_1 = (1,900, 1,600),$$

$$\Delta C(\bar{t}_{41} + \bar{\delta}_2) = \Delta C(1,900, 1,700) = \$106,029 > 105,749,$$

$$\Delta C(\bar{t}_{41} - \bar{\delta}_2) = \Delta C(1,900, 1,500) = \underline{\$105,512} < 105,749.$$

Consequently,

$$\bar{b}_5 = \bar{t}_{42} = \bar{t}_{41} - \bar{\delta}_2 = (1,900, 1,500),$$

$$\bar{t}_{50} = 2\bar{b}_5 - \bar{b}_4 = (1,800, 1,100),$$

$$\Delta C(\bar{t}_{50}) = \Delta C(1,800, 1,100) = \underline{\$104,952} < 105,512,$$
$$\Delta C(\bar{t}_{50} + \bar{\delta}_1) = \Delta C(1,900, 1,100) = \underline{\$104,912} < 104,952.$$
$$\bar{t}_{51} = \bar{t}_{50} + \bar{\delta}_1 = (1,900, 1,100),$$
$$\Delta C(\bar{t}_{51} + \bar{\delta}_2) = \Delta C(1,900, 1,200) = \$105,009 > 104,912,$$
$$\Delta C(\bar{t}_{51} - \bar{\delta}_2) = \Delta C(1,900, 1,000) = \$105,114 > 104,912.$$

Hence,

$$\bar{b}_6 = \bar{t}_{52} = \bar{t}_{51} = (1,900, 1,100),$$
$$\bar{t}_{60} = 2\bar{b}_6 - \bar{b}_5 = (1,900, 700),$$
$$\Delta C(\bar{t}_{60}) = \Delta C(1,900, 700) = \underline{\$104,791} < 104,912,$$
$$\Delta C(\bar{t}_{60} + \bar{\delta}_1) = \Delta C(2,000, 700) = \underline{\$104,760} < 104,791.$$
$$\bar{t}_{61} = \bar{t}_{60} + \bar{\delta}_1 = (2,000, 700),$$
$$\Delta C(\bar{t}_{61} + \bar{\delta}_2) = \Delta C(2,000, 800) = \underline{\$104,752} < 104,760,$$
$$\bar{b}_7 = \bar{t}_{62} = \bar{t}_{61} + \bar{\delta}_2 = (2,000, 800).$$

Now,

$$\bar{t}_{70} = 2\bar{b}_7 - \bar{b}_6 = (2,100, 500),$$
$$\Delta C(\bar{t}_{70}) = \Delta C(2,100, 500) = \$104,835 > 104,752,$$

and the new base head $\bar{t}_{70}$ has failed. Therefore, the pattern is destroyed and

$$\bar{t}_{80} = \bar{b}_8 = \bar{b}_7 = (2,000, 800).$$
$$\Delta C(\bar{t}_{80}) = \Delta C(2,000, 800) = \underline{104,752},$$
$$\Delta C(\bar{t}_{80} + \bar{\delta}_1) = \Delta C(2,100, 800) = \$104,762 > 104,752,$$
$$\Delta C(\bar{t}_{80} - \bar{\delta}_1) = \Delta C(1,900, 800) = \$104,782 > 104,752.$$
$$\bar{t}_{81} = \bar{t}_{80} = (2,000, 800),$$
$$\Delta C(\bar{t}_{81} + \bar{\delta}_2) = \Delta C(2,000, 900) = \$104,771 > 104,752,$$
$$\Delta C(\bar{t}_{81} - \bar{\delta}_2) = \Delta C(2,000, 700) = \$104,760 > 104,752.$$

Since all perturbations about $\bar{t}_{80}$ failed to reduce the incremental cost ΔC, one may conclude that, for the step size chosen, $\bar{t}_{80} = (2,000, 800)$ is the optimum point. The pattern behavior for the location of the plant is shown in Figure 10.3.

Very little effort was required in establishing the best point. The least incremental cost was computed to be \$104,752, whereas the incremental cost at location $\bar{b}_1 = (2,500, 2,500)$, was \$110,151. The difference is a savings of \$5,399 (approximately 4.8 % of the incremental cost of location).

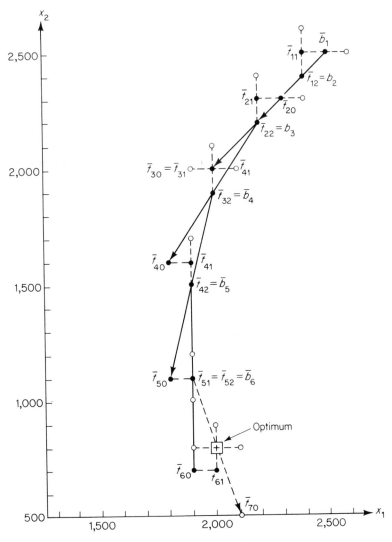

FIGURE 10.3. Pattern Behavior for Plant Location.

10.5 FINAL GAME EXPLORATION AT CRITICAL POINT

As stated in section 10.1, both at the beginning and at the end of a climb local explorations must be performed. However, the initial exploration is a simple linear study near an arbitrary point, whereas the final one is a *nonlinear* search in the neighborhood of the optimum.

In most cases, it is not only necessary to locate the optimum; one must also understand the behavior of the objective function at points nearby. Because the tangent plane is horizontal at a peak, curvature, asymmetry, and other nonlinearities are extremely important there, and the experimenter is forced to fit quadratic or higher degree expressions to the unknown function. Even when totally unconcerned about the behavior of the objective function near a critical point, the investigator would be negligent not to examine the apparent optimum carefully, for there may exist better points nearby.

The curvature of the response surface is first measured by considering only the first- and second-degree terms of the Taylor Series Expansion of the objective function in the region of interest. If the function is asymmetric, cubic terms might have to be investigated, but this is ascertained later when the quadratic predictions are checked against actual observations.

10.5.1 Example·

Suppose that for a specific function, $y(x_1, x_2)$, the maximum appears to be near the point with coordinates, $x_1 = 5.0$ and $x_2 = 9.5$, where $y(5.0, 9.5) = \$1,200/\text{day}$.

Values at four nearest points are given in Figure 10.4. Many search strategies, in addition to pattern search, yield a cruciform pattern in the vicinity of a candidate point for optimum. Consider the linear and quadratic terms

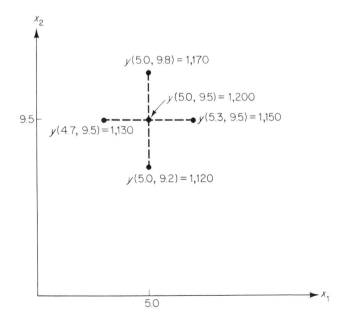

FIGURE 10.4

of the Taylor Series Expansion of the objective function's increment,

$$\Delta y = m_1 \Delta x_1 + m_2 \Delta x_2 + \tfrac{1}{2}[m_{11}(\Delta x_1)^2 + 2m_{12}\Delta x_1 \Delta x_2 + m_{22}(\Delta x_2)^2].$$
(10.40)

Five constants appear in equation (10.40) and only four extra points are available. To resolve this problem—temporarily, at least—neglect the interaction term. One has,

$$\Delta y = m_1 \Delta x_1 + m_2 \Delta x_2 + \tfrac{1}{2}[m_{11}(\Delta x_1)^2 + m_{22}(\Delta x_2)^2].$$
(10.41)

This approximation can be used to estimate the location of the true optimum. A measurement check there will reveal whether or not the approximation is good enough; if it isn't, one can use the check point to evaluate the constants when the interaction term is included. The crosslike arrangement permits great simplification in the computations.

Let

$\bar{x}_0^t = (5.0, 9.5) = (x_1, x_2)$, be the base point;

$\bar{x}_{11}^t = (5.3, 9.5) = (x_1^1, x_2^1)$, be the point to the right of $\bar{x}_0 (j = 1)$;

$\bar{x}_{12}^t = (4.7, 9.5) = (x_1^2, x_2^2)$, be the point to the left of $\bar{x}_0 (j = 2)$;

$\Delta y_{ij} = y(\bar{x}_{ij}) - y(\bar{x}_0)$;

and

$$\Delta x_{ij} = x_i^j - x_i.$$

Equation (10.41) can be written for $\bar{x}_{11}$ and $\bar{x}_{12}$ as follows:

$$\Delta y_{11} = m_1 \Delta x_{11} + m_2(0) + \tfrac{1}{2}m_{11}(\Delta x_{11})^2 + \tfrac{1}{2}m_{22}(0)$$
$$= m_1 \Delta x_{11} + \tfrac{1}{2}m_{11}(\Delta x_{11})^2,$$
(10.42)

$$\Delta y_{12} = m_1 \Delta x_{12} + m_2(0) + \tfrac{1}{2}m_{11}(\Delta x_{12})^2 + \tfrac{1}{2}m_{22}(0)$$
$$= m_1 \Delta x_{12} + \tfrac{1}{2}m_{11}(\Delta x_{12})^2.$$
(10.43)

But since $\Delta x_{12} = -\Delta x_{11}$, equation (10.43) can be rewritten

$$\Delta y_{12} = -m_1 \Delta x_{11} + \tfrac{1}{2}m_{11}(\Delta x_{11})^2,$$
(10.44)

from which,

$$m_{11} = \frac{\Delta y_{11} + \Delta y_{12}}{(\Delta x_{11})^2},$$
(10.45)

and

$$m_1 = \frac{\Delta y_{11} - \Delta y_{12}}{2\Delta x_{11}}.$$

In general, when there are N independent variables arranged according to points $\bar{x}_{ij}$ which satisfy

$$\bar{x}_{i1} = \bar{x}_0 + \Delta \bar{x}_{i1}\bar{\epsilon}_i$$

and

$$\bar{x}_{i2} = \bar{x}_0 - \Delta \bar{x}_{i1}\bar{\epsilon}_i,$$

where

$$\epsilon_i = \begin{pmatrix} 0 \\ 0 \\ \cdot \\ \cdot \\ \cdot \\ 1 \\ \cdot \\ \cdot \\ \cdot \\ 0 \end{pmatrix}, \qquad (10.46)$$

with the 1 in the ith position,
then,

$$m_i = \frac{\Delta y_{i1} - \Delta y_{i2}}{2\Delta x_{i1}}, \qquad (10.47)$$

and

$$m_{ii} = \frac{\Delta y_{i1} + \Delta y_{i2}}{(\Delta x_{i1})^2}. \qquad (10.48)$$

For the example under consideration,

$$m_1 = \frac{(1{,}150 - 1{,}200) - (1{,}130 - 1{,}200)}{2(0.3)} = 33.3,$$

$$m_{11} = \frac{(1{,}150 - 1{,}200) + (1{,}130 - 1{,}200)}{(0.3)^2} = -1{,}333.33,$$

$$m_2 = \frac{(1{,}170 - 1{,}200) - (1{,}120 - 1{,}200)}{2(0.3)} = 83.3,$$

and

$$m_{22} = \frac{(1{,}170 - 1{,}200) + (1{,}200 - 1{,}200)}{(0.3)^2} = -1{,}222.2.$$

Therefore,

$$\Delta y = 33.3\Delta x_1 + 83.3\Delta x_2 - 666.7(\Delta x_1)^2 - 611.1(\Delta x_2)^2. \qquad (10.49)$$

Differentiating partially with respect to Δx_1 and Δx_2, and setting the derivatives equal to zero, obtain the coordinate steps to the apparent optimum.

$$\left(\frac{\partial \Delta y}{\partial \Delta x_1}\right)_0 = 33.3 - 1{,}333.3\Delta x_1^0 = 0.$$

Hence, $\Delta x_1^0 = .00250.$

$$\left(\frac{\partial \Delta y}{\partial \Delta x_2}\right)_0 = 83.3 - 1{,}222.2\Delta x_2^0 = 0,$$

from which

$$\Delta x_2^0 = .00682.$$

In general, for a cruciform pattern such as the one of this example,

$$\Delta x_i^0 = -\frac{m_i}{m_{ii}} = \frac{(\Delta y_{i2} - \Delta y_{i1})x_{i1}}{2(\Delta y_{i2} + \Delta y_{i1})}. \tag{10.50}$$

Because the expression fit indicates that the optimum is indeed very close to $\bar{x}_0$, one may be tempted to conclude that the optimum point has been found. Test y at the point $(5.3, 9.8) = (x_1 + \Delta x_{11}, x_2 + \Delta x_{21})$ which lies in the same quadrant as the apparent optimum.

From equation (10.49),

$$\Delta y = 33.3(0.3) + 83.3(0.3) - 666.7(0.3)^2 - 611.1(0.3)^2,$$

or

$$\Delta y = -80.$$

Consequently, equation (10.49) predicts

$$y(5.3, 9.8) = 1,200 - 80 = \$1,120/\text{day}.$$

Suppose, however, that an evaluation of the objective function at the point $(5.3, 9.8)$ yields

$$y(5.3, 9.8) = \$925/\text{day}.$$

This disagreement indicates that the variables are interacting strongly and that, consequently, the simple model of equation (10.41) should be replaced by that of equation (10.40). Because the interaction term $2m_{12}\Delta x_1\Delta x_2$ is zero in equations (10.42) and (10.43), the values computed for m_1 and m_{11} remain valid. Similarly, m_2 and m_{22} are also correct. To compute m_{12}, one must write equation (10.40) for the new point $\bar{x}^{12}$,

$$\Delta y^{12} = -275 = 33.3(0.3) + 83.3(0.3) - 666.7(0.3)^2 - 611.1(0.3)^2 \\ + m_{12}(0.3)(0.3),$$

from which

$$m_{12} = -2,166.7,$$

clearly not a negligible value.

For the general case of N variables, there will be $N(N-1)/2$ interaction terms, each requiring a point. To evaluate the coefficients m_{ij}, it is necessary to place each new point at

$$\bar{x}^{ij} = \bar{x}_0 \pm \Delta \bar{x}_i \bar{\epsilon}_i \pm \Delta \bar{x}_j \bar{\epsilon}_j, \tag{10.51}$$

where the $(\pm)$ indicates that the sign is arbitrary. Each time equation (10.40) is written for a new point $\bar{x}^{ij}$, it will have one m_{ij} as its only unknown, for all the constants m_i and m_{ii} would have been already evaluated.

Returning to the problem in hand, one can now write

$$\Delta y = 33.3\Delta x_1 + 83.3\Delta x_2 - 666.7(\Delta x_1)^2 \\ - 2,166.7(\Delta x_1)(\Delta x_2) - 611.1(\Delta x_2)^2. \tag{10.52}$$

Next, the location of the optimum is reestimated by differentiating equation (10.52) with respect to Δx_1 and Δx_2, setting the derivatives equal to zero, and solving the two simultaneous equations:

$$\frac{\partial \Delta y}{\partial \Delta x_1} = 33.3 - 1{,}333.3\Delta x_1 - 2{,}166.7\Delta x_2 = 0,$$

$$\frac{\partial \Delta y}{\partial \Delta x_1} = 83.3 - 1{,}222.2\Delta x_2 - 2{,}166.7\Delta x_1 = 0,$$

from which

$$\Delta x_1^0 = .0456 \quad \text{and} \quad \Delta x_2^0 = -.0127,$$

and the coordinates of the optimum point turn out to be

$$x_1^0 = 5.0456 \quad \text{and} \quad x_2^0 = 9.4873.$$

A check at the new optimum point verifies that $y^0 = \$1{,}200.23/\text{day}$, a $\$0.23/\text{day}$ improvement over the previous apparent optimum.

EXERCISES

10–1. Use gradient search to obtain the minimum of the function,

$$y = 4x_1 + 2x_2 + 10x_1x_2 + 6x_1^2 + 3x_2^2.$$

Start the search at the origin of coordinates, $\bar{x}_0^t = (0, 0)$.

10–2. Use gradient search to obtain the critical point of

$$y = 10x_1^2 + 5x_2^2 - 7x_1x_2.$$

Start at $\bar{x}_0^t = (0, 0)$.

10–3. Employ gradient search to minimize the cost function of the plant location example of section 10.4.2.2. Start the search at the point $\bar{x}_0^t = (300,900)$.

10–4. Use gradient search to solve:

$$\text{Min } y(\bar{x}) = 4x_1^2 + 6x_2^2 + 7(x_1 - x_2)^2.$$

Start search at $(1, 1)$.

10–5. Use pattern search to minimize the objective of example 10.4.1.2. Start at $\bar{x}_0^t = (0, 0)$ and use a step size of 0.1.

10–6. Minimize the function of exercise 10–1 with pattern search. Start at $\bar{x}_0^t = (0, 0)$ and use a step size of 0.1.

10–7. Employ pattern search to obtain the critical point of the function in exercise 10–2. Start patterns at the origin of coordinates and use a step size of 0.1.

10–8 through 10–10. Perform final game explorations (fit functions with linear and quadratic terms) at the critical points obtained in exercises 10–4, 10–5, and 10–6. Use the final cruciform patterns developed and the original step sizes.

10–11.

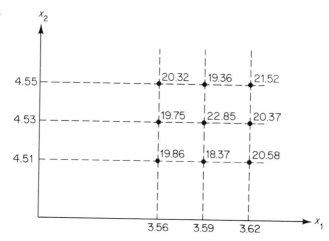

a. Develop a *quadratic* approximation for $\Delta y(\bar{x})$ in the sample space given above and estimate the location of the maximum point.
b. Determine the character of the point—that is, maximum, ridge, saddle, and so on.

10–12.

x_1	x_2	y
3.0	5.0	753
4.0	5.0	829
2.0	5.0	778
3.0	6.0	760
3.0	4.0	820
4.0	6.0	784

a. From the data given above, construct an interacting quadratic approximation.
b. What is the predicted minimum value of y?
c. Is the stationary point, in reality, a minimum?

REFERENCES

1. AGUILAR, R. J., *Optimum Location of Plant Facilities*, Division of Engineering Research, Louisiana State University, Bulletin No. 98. Baton Rouge, 1969.
2. HOOKE, R. and T. A. JEEVES, "Direct Search Solution of Numerical and Statistical Problems," *J. Assn. Comp. Mach.*, VIII, No. 2 (April 1961), 212–29.
3. WILDE, D. J. and C. S. BEIGHTLER, *Foundations of Optimization*. Englewood Cliffs, N.J.: Prentice-Hall, Inc., 1967.

CHAPTER **11**

Dynamic Programming: Optimization of Serial Systems

11.1 INTRODUCTION

Dynamic programming is applicable to the optimization of deterministic and stochastic, continuous and discontinuous, linear and nonlinear systems possessing a serial structure.

A serial structure consists of components connected head to tail with no recycle. The most common example of a serial structure is time. What happens during the month of February influences the events of March but has no effect whatsoever upon what took place in January (see Figure 11.1). There is no recycle of information in time.

Serial structures also occur in processing systems with components connected in such a way that any change in the operation of a given element influences only downstream elements. Recycle systems do not possess this property because every component within a loop is both upstream and downstream of every other component within the same loop. This does not mean, however, that systems with recycle cannot be serialized. It is necessary only to regroup the components so that the loop is redefined as a single component. Figure 11.2 shows schematically how this is accomplished. In the early 1950s, Richard Bellman postulated his so-called "Principle of Optimality"

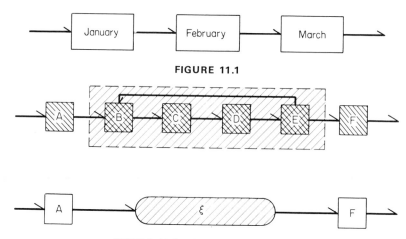

FIGURE 11.1

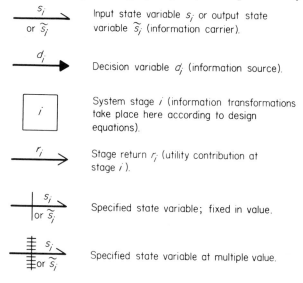

FIGURE 11.2.   Serializing a System.

and applied it to the optimization of serial systems with a strategy he called *dynamic programming*. Not surprisingly, his optimization strategy is based more on common sense than on mathematical sophistication. This is perhaps the reason for its wide acceptance.

11.2 DEFINITION OF SYMBOLS AND TERMS

Figure 11.3 gives a list of the symbols commonly used in dynamic programming problems to identify the stages, outputs, and inputs of a system. *State*

s_j or $\widetilde{s}_j$	Input state variable s_j or output state variable $\widetilde{s}_j$ (information carrier).
d_j	Decision variable d_j (information source).
j	System stage j (information transformations take place here according to design equations).
r_j	Stage return r_j (utility contribution at stage j).
s_j or $\widetilde{s}_j$	Specified state variable; fixed in value.
s_j or $\widetilde{s}_j$	Specified state variable at multiple value.

FIGURE 11.3.   Definition of Symbols and Terms.

variable, denoted by a half arrow, carries information from stage to stage and, consequently, it is an input to one stage and an output from another. The input state variable to stage i is symbolized by s_i. The output state variable from stage i is denoted by $\tilde{s}_i$. The output state variable, $\tilde{s}_i$, is related to the input state variable, s_i, and to the decision variable d_i (next paragraph), through the transition function at that stage, thus

$$\tilde{s}_i = T_i(s_i, d_i). \tag{11.1}$$

Decision variable at stage i, symbolized by a full arrow and the letter d_i, is an input variable which supplies information to the system. It can be manipulated at will by the decision maker and controls the composition of the information carried by the state variables.

System component i, denoted by a box, is the system element where transformations of information take place in accordance with design equations. The design equations relate input to output. However, their functional relationships need not be expressed mathematically and could be given by tables, computer programs, or other media.

Stage return to stage *i*, represented by an open arrow and r_i, measures the utility contribution of stage i to the system's economic health. The stage return is, of course, a function of the decision and the input state variables. This relationship, expressed by

$$r_i = R_i(s_i, d_i) \tag{11.2}$$

is the return function at stage i.

Specified state variable. Frequently, the values of one or more state variables are specified. For example, the system may be required to manufacture a certain quantity of material, or a structure required to support a certain load. The specification of the values of a variable is denoted by a vertical bar with cross lines. Decision variables may have specified values, though not frequently. This, however, *destroys* the structure of the optimization strategy at the stage where the decision variable has been specified.

11.3 TYPES OF SERIAL OPTIMIZATION PROBLEMS

1. *Initial value problem*, with input state variable specified.

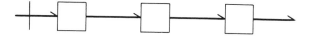

2. *Final value problem*, with output state variable specified.

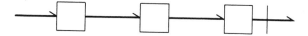

3. *Boundary value problem*, with both input and output state variables specified.

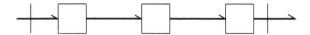

Final value problems can be transformed into initial value problems by reversing the direction of state variables through state inversion (to be discussed later in this chapter). Also, a system that exhibits a limited amount of branching still possesses a serial structure if no recycle loops are present. The following, for example, is a serial structure:

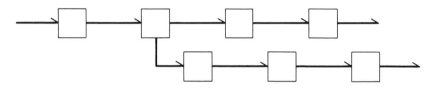

11.4 THE CONCEPT OF SUBOPTIMIZATION AND THE PRINCIPLE OF OPTIMALITY

The concept of suboptimization and the principle of optimality can be best introduced when discussed in connection with the optimization of an initial value problem.

Consider the problem of designing the most economical raised structural platform to support a specified load of w pounds per square foot of horizontal projection (see Figure 11.4). The structural system shown can be

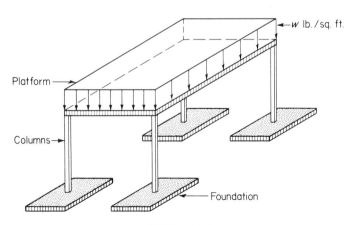

FIGURE 11.4

schematically represented as follows,

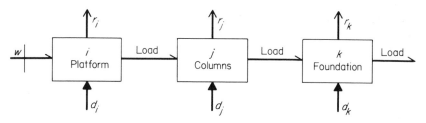

where the r's are the economic returns (cost) of the elements in the system. The objective is to minimize the total cost of the project and to design a safe structure capable of supporting the specified load w.

It would be very desirable to "tear" the system into a number of smaller pieces which could be optimized more or less individually. Unfortunately, the problem of tearing and of stage suboptimization has to be approached with a great deal of care because otherwise the procedure can lead to catastrophic results. A logical strategy is needed, and it is the principle of optimality that provides the guidelines for the optimization of serial systems.

Suppose that one isolated interior stage j (columns) and tried to suboptimize it by minimizing its individual return r_j. If the cost of steel is high, as in many countries, one may conclude that the cheapest supporting element that can be designed is a massive concrete column without reinforcement. This design may boost the dead load to such a level that the foundation cost is excessive. Thus, the suboptimization of stage j has adversely affected the economic return of the downstream stage k with the overall result of producing an uneconomical structure. The accidental suboptimization of interior stages must be dismissed as having no part in a rational strategy.

It is clear that if an interior stage in a serial structure influences all downstream stages, it cannot be suboptimized without a consideration of the effect it has downstream. The question that must be answered is, therefore, what suboptimization can be performed with full assurance that it is part of a rational optimization strategy? Obviously, the last stage in a serial structure influences no other element and, consequently, it is the only one that can be considered independently. However, the last two stages could be grouped and considered together as one larger end component. Then that larger element could be suboptimized without the danger of disrupting the contributions of any other. In fact, any number of the last stages can be systematically grouped in reverse sequential order and suboptimized. R. D. Bellman observed this fact and expressed it in his principle of optimality as follows:

> An optimal policy has the property that whatever the initial state and initial decision are, the remaining decisions must constitute an optimal policy with regard to the state resulting from the first decision.

Or, expressed in a different way: *End stages in a serial structure must be optimal with respect to the feed they receive from the upstream stages.*

Figure 11.5 shows how the principle of optimality allows the gradual inclusion of last stages into the sphere of suboptimization until the entire system is optimized. Recycle loops preclude the principle from being applied since in such a loop there exists no stage which has no other downstream stage. However, these recycle loops can be regrouped as explained before and suboptimized by cutting and matching state variables and with a method called decision inversion, to be discussed later in this chapter.

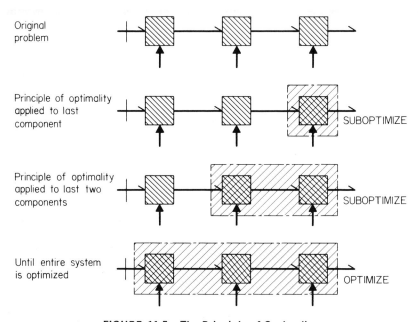

FIGURE 11.5. The Principle of Optimality.

Because the process of optimization must be carried out in reverse order, it is convenient to number the elements of the system in the same manner, thus stage i has input state variable s_i,

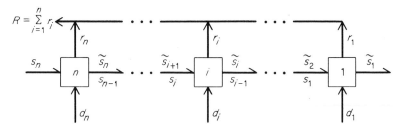

output state variable $\tilde{s}_i$, input decision variable d_i, return output r_i, transition function $T_i (s_i, d_i)$ and return function $R_i (s_i, d_i)$, giving,

$$\tilde{s}_i = T_i(s_i, d_i)$$

and

$$r_i = R_i(s_i, d_i).$$

The set of transformations carrying the same index i specifies the ith stage, and the set of all stages defines the system. Note that the input to stage i is also the output from stage $i + 1$, thus

$$\tilde{s}_{i+1} = s_i, \quad \text{for } i = 1, 2, \ldots, n - 1. \tag{11.3}$$

These relationships, called *incidence identities*, specify the flow of information through the system. A system is specified *completely* by its stages and incidence identities. In the initial value problem given, s_n is fixed, and the decision problem is to find the optimal sequence $d_1^*(s_n), d_2^*(s_n), \ldots, d_n^*(s_n)$ that maximizes (or minimizes) the total return R, defined by

$$R = \sum_{i=1}^{n} r_i = \sum_{i=1}^{n} R_i(s_i, d_i). \tag{11.4}$$

Equations (11.1) and (11.2) show that for a given input s_i, the value chosen for d_i determines not only the return r_i, but also the value of T_i which maps the input s_i into the output $\tilde{s}_i$. Although, as stated previously, a given d_i may maximize the return at stage i, it may also adversely affect the inputs to all subsequent stages, leading to a nonoptimal *total* return for the system. An optimal sequence, or policy $\{d_i^*(s_N)\}$, can be found only by taking into account the transitions coupling the stages together.

Equations (11.1) to (11.4) can be combined to express the total return as a function of the various inputs:

$$R = \sum_{i=1}^{N} R_i(s_i, d_i)$$
$$= R_1(s_1, d_1) + R_2(s_2, d_2) + \cdots + R_{N-1}(s_{N-1}, d_{N-1})$$
$$+ R_N(s_N, d_N).$$

However,

$$s_1 = \tilde{s}_2 = T_2(s_2, d_2),$$
$$s_2 = \tilde{s}_3 = T_3(s_3, d_3),$$
$$\vdots$$
$$s_{N-1} = \tilde{s}_N = T_N(s_N, d_N).$$

Therefore,

$$R_1(s_1, d_1) = R_1[T_2(s_2, d_2), d_1]$$
$$= R_1[T_2[T_3(s_3, d_3), d_2], d_1]$$
$$= R_1[T_2(T_3\{T_4 \ldots [T_N(s_N, d_N), d_{N-1}], \ldots\}, d_2), d_1].$$

In general,

$$R_i(s_i, d_i) = R_i[T_{i+1}(T_{i+2}\{T_{i+3} \cdots [T_N(s_N, d_N), d_{N-1}], \ldots\}, d_{i+1}), d_i].$$

Therefore,

$$R = \sum_{i=1}^{N} R_i(s_i, d_i) = \sum_{i=1}^{N} (T_{i+1}\{T_{i+2} \cdots [T_N(s_N, d_N), d_{N-1}], \ldots\}, d_i). \qquad (11.5)$$

This last expression indicates that for a given value of the initial input state s_N, the total return R is a function only of the decision variables d_i,

$$R = R(s_N, d_i); \; i = 1, 2, \ldots, N. \qquad (11.6)$$

The *initial value maximum return function* is $R^*(s_N)$, defined as the function which, for *all* values of $d_1, \ldots, d_N$ and for any *particular* value of s_N is such that

$$R^*(s_N) = \max \, [R(s_N, d_i)].$$

The decisions $d_i^*(s_N)$ form an *initial value optimal policy*, and because there are N such decision functions of the single state variable s_N to be specified by the policy, the problem is called an *N-decision, one-state optimization problem*. *The number of decisions plus the number of specifiable states equal the number of degrees of freedom.*

Note that s_N, not being an output from any stage, could be called a decision variable rather than a state. If s_N is such that one can choose freely among its many possible values, then s_N will be called *choice variable* and labeled c_N instead of s_N.

Multiple decision and state variables can be distinguished from each other with a second subscript (d_{i1}, d_{i2}, etc.). Similarly, a stage can have several transition functions with double subscripts where necessary to avoid confusion. Each stage has only one return *function*, and the return values are always scalars.

11.5 DECOMPOSITION OF SERIAL SYSTEMS BY APPLICATION OF THE PRINCIPLE OF OPTIMALITY

Application of Bellman's principle of optimality to a serial system permits the transformation of the N-decision, one-state problem symbolized by equation (11.6) into a set of N one-decision, one-state problems. Consider the suboptimization of stage 1, the most downstream stage.

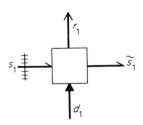

Define the maximand (for maximization, without loss of generality),

$$M_1(s_1, d_1) = R_1(s_1, d_1). \tag{11.7}$$

The maximization of equation (11.7) involves two variables, s_1 and d_1. Because the value of the input state variable s_1 is not known a priori, one must maximize (11.7) for all possible values of s_1. Define the maximized return from stage 1 as follows:

$$f_1(s_1) = \max_{d_1} M_1[(s_1, d_1)] = \max_{d_1} R_1[(s_1, d_1)]. \tag{11.8}$$

The maximum return $f_1(s_1)$ is, as indicated, a function of the input state variable s_1; that is, for each value of s_1 there exists a best decision d_1^* which maximizes the return from stage 1.

Figure 11.6 is a graphical representation of the suboptimization of stage 1. The figure shows contours of the function $M_1(s_1, d_1)$ on the $s_1 - d_1$ plane.

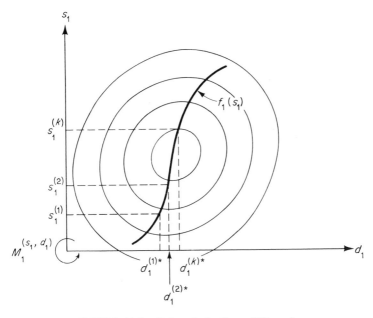

FIGURE 11.6. Suboptimization of Stage 1.

It is clearly seen that for each value of the state variable, $s_1^{(k)}$, there exists a best value of the decision variable, $d_1^{(k)*}$, which yields the maximized return $f_1(s_1)$, itself a function s_1. Now consider the simultaneous suboptimization of stages 2 and 1.

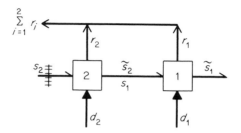

The maximand is

$$M_2(s_2, d_2) = R_2(s_2, d_2) + R_1(s_1, d_1). \tag{11.9}$$

However, the best return function for stage 1, $f_1(s_1)$, has already been computed. Substituting it in (11.9), obtain

$$M_2(s_2, d_2) = R_2(s_2, d_2) + f_1(s_1).$$

The incidence identity

$$s_1 = \tilde{s}_2$$

permits substitution of $\tilde{s}_2$ for s_1 into $f_1(s_1)$; thus,

$$M_2(s_2, d_2) = R_2(s_2, d_2) + f_1(\tilde{s}_2).$$

However, $\tilde{s}_2 = T_2(s_2, d_2)$ through the transition function at stage 2. Thus,

$$M_2(s_2, d_2) = R_2(s_2, d_2) + f_1(T_2(s_2, d_2))$$

which finally reduces to

$$M_2(s_2, d_2) = R_2(s_2, d_2) + f_1(s_2, d_2) \tag{11.10}$$

because a function of a function, $f_1(T_2(s_2, d_2))$ is simply a function of the independent variables, $f_1(s_2, d_2)$. Consequently, although, on the surface the maximand function for stages 2 and 1 appears to be a function of s_1, d_1, s_2 and d_2 in equation (11.9), because of the serial structure of the system as expressed by the incidence identity and the transition function, equation (11.10) shows that, in reality, this maximand is a function of only s_2 and d_2. Thus another one-decision, one-state problem must be maximized over the decision variable d_2 for every possible value of the state variable s_2.

Proceeding to optimize as before, obtain

$$f_2(s_2) = \max_{d_2} [M_2(s_2, d_2)]$$
$$= \max_{d_2} [R_2(s_2, d_2) + f_1(s_2, d_2)]. \tag{11.11}$$

Again, for each value of the input state variable s_2, a best value of the decision variable d_2^* yields the maximized return function $f_2(s_2)$. Figure 11.6 with the subscript $_2$ replacing $_1$ depicts the suboptimization just performed.

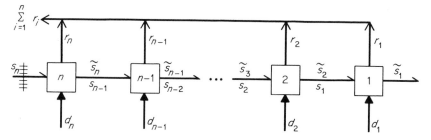

In general, at stage n one obtains the maximand

$$M_n(s_n, d_n) = R_n(s_n, d_n) + R_{n-1}(s_{n-1}, d_{n-1}) + \cdots + R_2(s_2, d_2)$$
$$+ R_1(s_1, d_1). \tag{11.12}$$

However, from the previous suboptimization one has available the maximized return, $f_{n-1}(s_{n-1})$, for the $_{n-1}$ downstream stages. Substitution of this function in equation (11.12) yields

$$M_n(s_n, d_n) = R_n(s_n, d_n) + f_{n-1}(s_{n-1}).$$

Again,

$$\tilde{s}_{n-1} = s_n$$

and

$$\tilde{s}_n = T_n(s_n, d_n),$$

therefore,

$$M_n(s_n, d_n) = R_n(s_n, d_n) + f_{n-1}[T_n(s_n, d_n)]$$

or

$$M_n(s_n, d_n) = R_n(s_n, d_n) + f_{n-1}(s_n, d_n), \tag{11.13}$$

and one has, again, a one-decision, one-state optimization problem. Proceeding as before, maximize over d_n for all values of s_n to obtain

$$f_n(s_n) = \max_{d_n} [M_n(s_n, d_n)]$$
$$= \max_{d_n} [R_n(s_n, d_n) + f_{n-1}(s_n, d_n)]. \tag{11.14}$$

As before, this expression yields the optimal decision set $\{d_n^*\}$. At the first upstream stage in an initial value problem, the input state variable s_N is fixed, and the optimization strategy yields a unique value of the maximized total return.

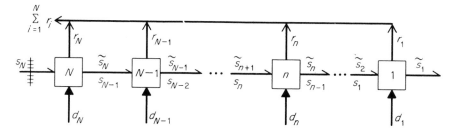

The maximand is

$$M_N(s_N, d_N) = R_N(s_N, d_N) + R_{N-1}(s_{N-1}, d_{N-1}) + \cdots + R_n(s_n, d_n)$$
$$+ \cdots + R_1(s_1, d_1), \tag{11.15}$$

but from the previous suboptimization one has the maximized return $f_{N-1}(s_{N-1})$ of the $_{N-1}$ downstream stages. Substitution of this function into equation (11.5) yields

$$M_N(s_N, d_N) = R_N(s_N, d_N) + f_{N-1}(s_{N-1}).$$

However,

$$s_{N-1} = \bar{s}_N$$

and

$$\bar{s}_N = T_N(s_N, d_N).$$

Therefore,

$$M_N(s_N, d_N) = R_N(s_N, d_N) + f_{N-1}(s_N, d_N). \tag{11.16}$$

Optimization over d_N produces the maximized return $f_N(s_N)$.

$$f_N(s_N) = \max_{d_N} \{R_N(s_N, d_N) + f_{N-1}(s_N, d_N)\}. \tag{11.17}$$

For an initial value problem, s_N is fixed, thus one needs to consider only one value of s_N, and $f_N(s_N)$ is unique. Now one only has to retrace the steps through the functions $f_n(s_n)$ generated for the entire system to gather the complete set of d_n^*, $n = 1, \ldots, N$. A simple example showing the quantitative design optimization of the platform problem introduced previously is given in the next section as a medium for transferring some of the more subtle points in the procedure.

11.6 A DISCRETE PROBLEM

Design the most economical raised platform to support a uniform load of 200 pounds per square foot of horizontal projection. The structural system consists of a rectangular surface 30 feet long and 20 feet wide; four columns, each 40 feet in length, braced as required; and an appropriate foundation to safely transfer all forces to the ground.

To simplify the problem, the number of columns was established a priori, although this could be one of the most important considerations in the optimization procedure because it affects not only the economics of the columns themselves, but that of the platform and supporting foundation. Other parameters which would normally be considered, such as various reinforced concrete strengths, reinforcing steel ratios, steel yield points, and so on, have been omitted to facilitate the understanding of the procedure.

The structural system proposed can be represented as follows:

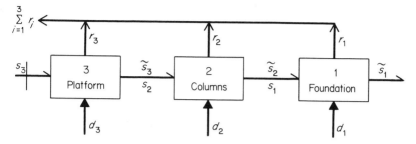

Because the intensity of the input load was set at 200 pounds per square foot, for a total of 120,000 pounds, s_3 is fixed, and the problem is an *initial value problem*.

The theories of structural analysis and design in the various materials provide design equations which yield the following tables for the system stages (values given are for illustrative purposes only):

Platform (Stage 3)

Type	s_3 (Load)	r_3 (Cost)	Weight	$\tilde{s}_3 = s_3 + Weight = s_2$
a. Solid one-way concrete slab system	120,000#	$1,100	75,000#	195,000#
b. One-way concrete pan joist system	120,000#	$1,650	37,000#	157,000#
c. Solid two-way concrete slab system	120,000#	$1,300	49,000#	169,000#
d. Concrete waffle slab	120,000#	$1,800	28,000#	148,000#
e. Steel beams and long span steel deck	120,000#	$3,000	12,000#	132,000#
f. Steel bar joist system	120,000#	$2,400	14,000#	134,000#
g. Steel beams and composite concrete deck	120,000#	$2,000	30,000#	150,000#

Columns and Bracing System (Stage 2)

Type	s_2 (Load)	r_2 (Cost)	Weight	$\tilde{s}_2 = s_2 + Weight = s_1$
a. Reinforced concrete tied columns	200,000#	$3,300	135,000#	335,000#
	170,000#	$2,500	100,000#	270,000#
	150,000#	$1,900	75,000#	225,000#
	130,000#	$1,400	60,000#	190,000#
b. Reinforced concrete spiral columns	200,000#	$3,900	115,000#	315,000#
	170,000#	$3,000	85,000#	255,000#
	150,000#	$2,200	64,000#	214,000#
	130,000#	$1,400	50,000#	180,000#
c. Structural steel	200,000#	$6,000	27,000#	227,000#
	170,000#	$5,200	23,000#	193,000#
	150,000#	$4,500	20,000#	170,000#
	130,000#	$4,000	18,000#	148,000#

Foundation (Stage 1)

Type	s_1 (Load)	r_1 (Cost)	Weight	$\tilde{s}_1 = s_1 +$ Weight
a. Spread foundation	330,000#	$1,800	120,000#	450,000#
	270,000#	$1,400	90,000#	360,000#
	220,000#	$1,000	66,000#	286,000#
	190,000#	$ 750	50,000#	240,000#
	150,000#	$ 500	34,000#	184,000#
b. Drilled concrete piles	330,000#	$1,650	105,000#	435,000#
	270,000#	$1,350	88,000#	358,000#
	220,000#	$1,100	72,000#	292,000#
	190,000#	$ 950	62,000#	252,000#
	150,000#	$ 750	49,000#	199,000#
c. Driven steel piles	330,000#	$1,400	10,000#	340,000#
	270,000#	$1,150	8,200#	278,200#
	220,000#	$ 960	6,800#	226,800#
	190,000#	$ 880	6,200#	196,200#
	150,000#	$ 780	5,500#	155,500#

Suboptimization of Stage 1 (Foundation):

Note that this is a minimization problem; accordingly, minimand functions must be defined at each stage.

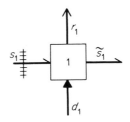

$$f_1(s_1) = \min_{d_1} (R_1(s_1, d_1)]$$

Five settings of input state variable s_1 are considered and, using the foundation design table, obtain

Summary for Stage 1:

s_1	d_1*	$f_1(s_1)$	$\tilde{s}_1$
330,000#	c	$1,400	340,000#
270,000#	c	$1,150	278,200#
220,000#	c	$ 960	226,800#
190,000#	a	$ 750	240,000#
150,000#	a	$ 500	184,000#

Suboptimization of Stages 2 and 1:

(Columns and Foundation)

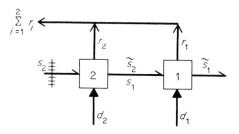

$$f_2(s_2) = \min_{d_2} [R_2(s_2, d_2) + f_1(s_2, d_2)].$$

Four settings of input variable s_2 are considered at this stage. Since this number of settings is small, the numerical values of output state variable $\tilde{s}_2$ from stage 2 will *not* necessarily coincide with tabulated values of input state variable s_1 feeding into stage 1. With a digital computer, many more settings, more closely spaced and covering a wider range, could be studied without a significant increase in labor. For the purpose of this example, numerical values will be *linearly* interpolated or extrapolated, as required. With this in mind, obtain for stages 2 and 1,

s_2	d_2	r_2	$\tilde{s}_2 = s_1$	d_1*	$f_1(s_1)$	$r_2 + f_1(s_1)$
200,000#	a	$3,300	335,000#	c	$1,420	$4,720 ←
200,000#	b	$3,900	315,000#	c	$1,340	$5,240
200,000#	c	$6,000	227,000#	c	$ 990	$6,900
170,000#	a	$2,500	270,000#	c	$1,150	$3,650 ←
170,000#	b	$3,000	255,000#	c	$1,090	$4,090
170,000#	c	$5,200	193,000#	a	$ 770	$5,970
150,000#	a	$1,900	225,000#	c	$ 980	$2,880 ←
150,000#	b	$2,200	214,000#	c	$ 920	$3,120
150,000#	c	$4,500	170,000#	a	$ 625	$5,125
130,000#	a	$1,400	190,000#	a	$ 750	$2,150
130,000#	b	$1,400	180,000#	a	$ 690	$2,090 ←
130,000#	c	$4,000	148,000#	a	$ 490	$4,490

arrows indicate the optimum solution for each setting of input state variable. Thus, obtain

Summary for Stages 2 and 1:

s_2	d_2*	$f_2(s_2)$	$\tilde{s}_2$
200,000#	a	$4,720	335,000#
170,000#	a	$3,650	270,000#
150,000#	a	$2,880	225,000#
130,000#	b	$2,090	180,000#

Optimization of Stages 3, 2, and 1:

(Platform, Columns, and Foundation)

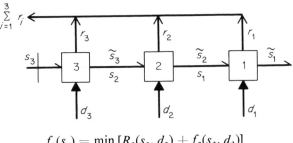

$$f_3(s_3) = \min_{d_3} [R_3(s_3, d_3) + f_2(s_3, d_3)]$$

With the information encoded in the platform design table and in the summary table for stages 2 and 1, the following is obtained:

s_3	d_3	r_3	$\tilde{s}_3 = s_2$	d_2*	$f_2(s_2)$	$r_3 + f_2(s_2)$
120,000#	a	$1,100	195,000#	a	$4,540	$5,640
120,000#	b	$1,650	157,000#	a	$3,150	$4,800
120,000#	c	$1,300	169,000#	a	$3,610	$4,910
120,000#	d	$1,800	148,000#	a	$2,800	$4,600 ⟵
120,000#	e	$3,000	132,000#	b	$2,170	$5,170
120,000#	f	$2,400	134,000#	b	$2,250	$4,650
120,000#	g	$2,000	150,000#	a	$2,880	$4,880

Therefore,

Summary for Stages 3, 2, and 1:

s_3	d_3*	$f_3(s_3)$	$\tilde{s}_3$
120,000#	d	$4,600	148,000#

Now, *retrace* the steps through the summary tables generated, to gather the complete set of d_i^*, $i = 1, 2, 3$, for the system, and find,

$$d_3^* = d,$$
$$d_2^* = a,$$
$$d_1^* = c.$$

Hence, the optimum design configuration is achieved using:

Concrete waffle slab platform
Reinforced concrete tied columns
Driven steel pile foundation

The total optimum (*minimum*) cost of the structure is $4,600.
All the cost values computed in the final stage of the procedure are numerically close together. This shows the problem to be of low sensitivity.

11.7 EFFICIENCY INVESTIGATION: COMPARISON TO DIRECT RESEARCH

The amount of work required to perform a direct search for the optimization of an n-dimensional problem can be expressed as

$$W_S = a(b)^n, \tag{11.18}$$

where a and b are suitable constants (b is usually taken as 10) and n is the number of degrees of freedom (independent variables).

With dynamic programming, an n-dimensional problem is transformed into n one-dimensional problems, and if the number of settings of the input state variable at each suboptimization stage is c, the work done at each stage, other than the last, is

$$\Delta W_i = a(b)^1 c, i = 1, 2, \ldots, n - 1. \tag{11.19}$$

At the last stage of an initial value problem $c = 1$, therefore, the work done at this stage is

$$\Delta W_n = a(b)^1. \tag{11.20}$$

The total work required, with dynamic programming, is given by

$$W_D = ab + \sum_{i=1}^{n-1} abc = [(n - 1)ac + a]b, \tag{11.21}$$

therefore, the ratio of the effort required for direct search to the effort required for dynamic programming is

$$\frac{W_D}{W_S} = \frac{[(n - 1)ac + a]b}{a(b)^n} = \frac{(n - 1)c + 1}{(b)^{n-1}}. \tag{11.22}$$

As an example, taking $a = 1, b = 10$, and $c = 10$, Table 11–1 is obtained:

TABLE 11–1. Efficiency Comparison.

n	1	5	10	100
W_S	10	$(10)^5$	$(10)^{10}$	$(10)^{100}$
W_D	10	410	910	9,910
W_D/W_S	1	$.41(10)^{-2}$	$.910(10)^{-7}$	$.991(10)^{-96}$

The plot of Figure 11.7 depicts the relationship between the work required for direct search and for dynamic programming and the number of variables.

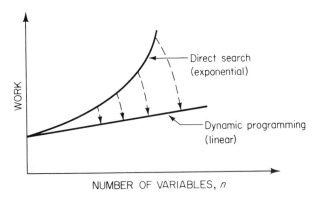

FIGURE 11.7. Comparison Between Direct Search and Dynamic Programming.

The figure shows the fundamental reason for using dynamic programming: that of linearizing the relationship between computational work and the number of variables in a problem. The same observation is applicable to optimization procedures; in general, without these procedures, multivariable problems would be next to impossible to optimize.

11.8 NETWORK ROUTING WITH DYNAMIC PROGRAMMING

Networks have been previously discussed in connection with the assignment algorithm of chapter 8 and with CPM and PERT in chapter 9. Dynamic programming is a most efficient technique for computing shortest and longest routes through networks.

Consider the network of Figure 11.8. The number on each branch represents the cost of travel between adjacent nodes. Starting at node BV, compute the least cost route to any node in zone I (the most eastern nodes). The system that symbolizes the network is shown at the bottom of the figure. Each input state variable s_i transmits information about the node of origin in the zone preceding stage i. The decisions d_i can take on only one of the following three values:

1. Go left [except for top (area A) nodes].
2. Go forward [straight ahead].
3. Go right [except for bottom (area C) nodes].

The output states $\bar{s}_i$ denote the destination node in the zone following stage i. Because it has been stipulated that the initial node is BV, this is an initial value dynamic programming problem with $s_4 = BV$, fixed. Starting the

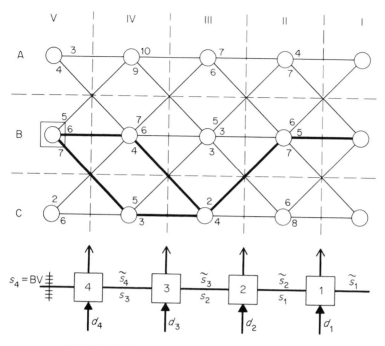

FIGURE 11.8. Network for Initial Value Problem.

optimization at stage 1, one obtains the following results:

Suboptimization of Stage 1:

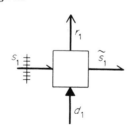

$$f_1(s_1) = \min_{d_1} [r_1(s_1, d_1)].$$

s_1	d_1	r_1	$\tilde{s}_1$
AII	F	4	AI ⟵
	R	7	BI
BII	L	6	AI
	F	5	BI ⟵
	R	7	CI
CII	L	6	BI ⟵
	F	8	CI

As before, the arrows indicate the optimum (least cost) returns and the following summary table is obtained for stage 1.

Summary for Stage 1:

s_1	d_1*	$f_1(s_1)$	$\tilde{s}_1$
AII	F	4	AI
BII	F	5	BI
CII	L	6	BI

For stages 2 and 1, obtain

Suboptimization of Stages 2 and 1:

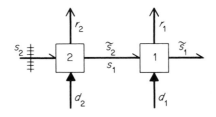

$$f_2(s_2) = \min_{d_2} [r_2(s_2, d_2) + f_1(s_2, d_2)].$$

s_2	d_2	r_2	$\tilde{s}_2 = s_1$	$f_1(s_1)$	$r_2 + f_1$
AIII	F	7	AII	4	11 ⟵
	R	6	BII	5	11 ⟵
BIII	L	5	AII	4	9
	F	3	BII	5	8 ⟵
	R	3	CII	6	9
CIII	L	2	BII	5	7 ⟵
	F	4	CII	6	10

This table shows that if $s_2 =$ AIII, then $d_2 =$ Forward and $d_2 =$ Right produce the same minimum total return of 11 for stages 2 and 1. Thus,

Summary for Stages 2 and 1:

s_2	d_2*	$f_2(s_2)$	$\tilde{s}_2$
AIII	F, R	11	AII, BII
BIII	F	8	BII
CIII	L	7	BII

The next stage is introduced in a similar manner.

Suboptimization of Stages 3, 2, and 1:

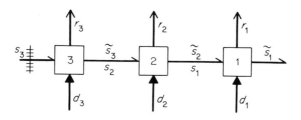

$$f_3(s_3) = \min_{d_3} [r_3(s_3, d_3) + f_2(s_3, d_3)].$$

s_3	d_3	r_3	$\tilde{s}_3 = s_2$	$f_2(s_2)$	$r_3 + f_2$
AIV	F	10	AIII	11	21
	R	9	BIII	8	17 ⟵
BIV	L	7	AIII	11	18
	F	6	BIII	8	14
	R	4	CIII	7	11 ⟵
CIV	L	5	BIII	8	13
	F	3	CIII	7	10 ⟵

Summary for Stages 3, 2, and 1:

s_3	d_3*	$f_3(s_3)$	$\tilde{s}_3$
AIV	R	17	BIII
BIV	R	11	CIII
CIV	F	10	CIII

In optimizing the entire system after introducing stage 4, one need only consider one value of input state variable s_4, namely, $s_4 = $ BV which is the fixed initial value, thus,

Optimization of Stages 4, 3, 2, and 1:

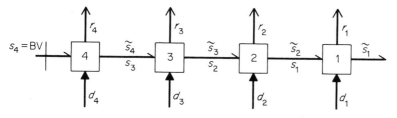

$$f_4(s_4 = \text{BV}) = \min_{d_4} [r_4(\text{BV}, d_4) + f_3(\text{BV}, d_4)].$$

s_4	d_4	r_4	$\bar{s}_4 = s_3$	$f_3(s_3)$	$r_4 + f_3$
BV	L	5	AIV	17	22
.	F	6	BIV	11	17 ←
	R	7	CIV	10	17 ←

Consequently,

Summary for Stages 4, 3, 2, and 1:

s_4	$d_4{}^*$	$f_4(s_4)$	$\bar{s}_4$
BV	F, R	17	BIV, CIV

Retracing the decisions and states through the summary tables, one finds

$d_4^* = $ Forward or Right, $\bar{s}_4 = $ BIV or CIV;
$d_3^* = $ Right from BIV and Forward from CIV, $\bar{s}_3 = $ CIII;
$d_2^* = $ Left, $\bar{s}_2 = $ BII;
$d_1^* = $ Forward, $\bar{s}_1 = $ BI;

and $f_4(s_4 = $ BV$) = 17$, the minimum cost of transportation from BV to a node in zone I which turns out to be node BI.

The least cost paths are shown with dark branches in Figure 11.8. The computations can be performed directly on the network and the numerical results of the summary tables written inside the nodes. This is done in Figure 11.9. Starting at node AII ($s_1 = $ AII), it is obvious that the best decision is to go forward ($d_2 = $ F) with a cost of 4. This minimum cost is written inside node AII. In a similar manner, one finds for nodes BII and CII that the minimum costs are, respectively, 5 and 6. Now, moving upstream to node AIII ($s_2 = $ AIII), make the following analysis: If one moves Forward ($d_2 = $ F) the cost is 7 to go to node AII plus 4, the minimum cost from AII to zone I. Thus, a decision to move forward would result in a total cost of $7 + 4 = 11$. Moving to the Right ($d_2 = $ R), one also finds a total cost of 11: 6 to get to node BII plus 5 to go at a minimum cost from BII to zone I. Thus, the minimum cost is, in fact, 11, and this number is written inside node AIII. At BIII, using the same analysis, one computes:

To the Left: Cost $= 5 + 4 = 9$;
Forward: Cost $= 3 + 5 = 8$;
To the Right: Cost $= 3 + 6 = 9$.

The minimum cost is 8 and this number is written inside node BIII. In a similar fashion, the minimum costs are computed for node CIII and for all upstream nodes, including node BV. Then simply trace downstream the

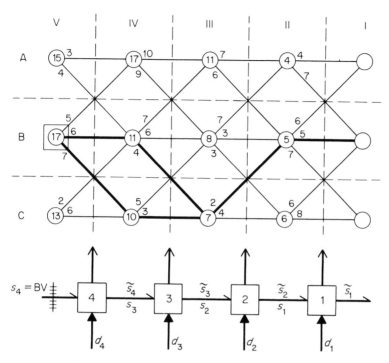

FIGURE 11.9. Computations on the Network.

node sequence with branch costs adding to the total minimum cost of 17 to obtain the least cost paths given in Figures 11.8 and 11.9. Note that with little extra effort the minimum costs can be computed for initial values $s_4 = $ AV and $s_4 = $ CV. The least cost paths for these initial conditions can be similarly scanned through the network. Thus, the minimum total returns for all values of the initial input state variable s_4 are given inside the zone V nodes in Figure 11.9.

11.9 AN ENVIRONMENTAL APPLICATION OF DYNAMIC PROGRAMMING

The following application of dynamic programming was developed by Mr. William Jacques for a report in an undergraduate course in Civil Engineering Systems at Louisiana State University.

A canning company has a plant effluent of 5,000,000 gallons per day (5 m.g.d.) at a pollutant level of 700 parts per million (p.p.m.) BOD loading. The plant operates 7 days a week. Recent federal legislation requires that industries with effluents of 0.25 m.g.d. or greater provide for minimum

TABLE 11–2. Pollution Tax Schedule.

# BOD/day	$/day
0–10	0
11–30	5
31–50	10
51–100	15
101–200	30
201–300	40
301–400	50
401–500	60
501–1,000	200
1,001–2,000	400
2,001–3,000	500
3,001–4,000	600
4,001–5,000	700
5,001–10,000	900
10,001–20,000	1,300

treatment (60% effective) of wastes with pollutant levels in excess of 100 p.p.m. In addition, industries falling into this category are required to pay a pollution tax according to the pounds of BOD discharged into the receiving stream per day. This information is summarized in Table 11–2.

The canning company wishes to initiate a pollution abatement program to meet the new standards at a minimum cost. Three types of treatment facilities are available. Information on all of them is listed in Table 11–3. Table 11–4 gives the installation costs reduced to a daily basis. These figures

TABLE 11–3. Characteristics of Treatment Facilities.

Treatment Type	BOD Removal Efficiency	Installation Cost $/m.g.d.	Daily Cost of Operation $/m.g.d.
(1) Primary	60%	200,000	116.60
(2) Secondary	90%	300,000	217.80
(3) Tertiary	99%	400,000	330.40

TABLE 11–4. Installation Costs Reduced to a Daily Basis.

Treatment Type	Installation Cost on a Daily Basis ($/m.g.d.)
(1)	$27.40
(2)	$82.20
(3)	$109.60

are, in fact, the capital recovery payments on each initial investment, at the attractive interest rate, over the expected life of each facility (see Chapter 2).

The minimum capacity for the three types of treatment is 1 m.g.d. Any one of the three types can be purchased to handle the 5 m.g.d. effluent; but only in increments of 1 m.g.d.

With the information given, the plant management must determine what facility or combination of facilities should be acquired to handle the 5 m.g.d. effluent at a minimum cost.

Solution:

The total cost per day for a given treatment type =

$$IC \times Q + OC \times Q + \text{Tax Applicable}, \qquad (11.23)$$

where,

IC = Installation cost on a daily basis (Table 11–4),

OC = Daily cost of operation (Table 11–3),

Q = Capacity of facility (1 to 5 m.g.d.).

To determine the applicable tax, one must first compute the pounds of BOD not removed for each treatment type at a given capacity. For example:

For treatment type (1) at 1 m.g.d. capacity,

$$\text{BOD load/day} = 1 \text{ m.g.d.} \times 700 \text{ p.p.m.} \times 8.34 \frac{\text{lbs.}}{\text{gal.}} \times (1 - 0.6)$$

$$= 2{,}336 \text{ lbs.}$$

Table 11–5 shows the results of these calculations.

TABLE 11–5.   BOD Load/Day (lbs.).

		TREATMENT TYPE		
		(1)	(2)	(3)
	1	2,336	584	58.4
Q	2	4,672	1,168	116.8
(m.g.d.)	3	7,008	1,752	175.2
	4	9,344	2,336	233.6
	5	11,680	2,920	292.0

With the values given in Table 11–5, the applicable tax can be determined from Table 11–2. Substitution of this value in equation (11.23) yields the total daily cost of each treatment type for a given capacity. Table 11–6 is the total daily cost summary table to be used in the optimization strategy.

TABLE 11–6. Total Daily Cost Summary Table.

		TREATMENT TYPE		
		(1)	(2)	(3)
	0	0	0	0
	1	644	500	455
Q	2	988	1,000	910
(m.g.d.)	3	1,332	1,300	1,350
	4	1,476	1,700	1,800
	5	2,020	2,000	2,240

This is an allocation problem; that is, the 5 m.g.d. of effluent must be allocated to the three treatment types. The dynamic programming model for the problem is,

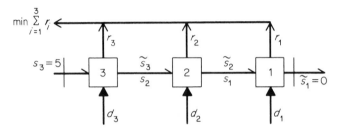

The stages represent the three types of treatment being considered. The order in which they are arranged is irrelevant. For convenience, they were placed in ascending order to coincide with the optimization strategy. In addition,

$$\tilde{s}_i = s_i - d_i,$$
$$d_i = \text{Treatment capacity (0 to 5 m.g.d.)},$$
$$r_i = \text{Cost corresponding to } d_i.$$

The calculations are given below.

Suboptimization of Stage 1 (Summary Table):

s_1	d_1*	$r_1 = f_1$	$\tilde{s}_1$
0	0	0	0
1	1	644	0
2	2	988	0
3	3	1,332	0
4	4	1,476	0
5	5	2,020	0

Note that the system symbolizes a boundary value problem because the initial input state, s_3, is fixed at 5 m.g.d. and the final output state, $\bar{s}_1$, is also fixed at 0 since all the effluent must be treated. For this reason, the optimal decision is $d_1^* = s_1$ and, consequently, $r_1 = f_1$, is the optimum return for stage 1. Thus, stage 1 is a decisionless stage. This operation is called a decision inversion, and details will be discussed later on in the chapter.

Suboptimization of Stages 2 and 1:

s_2	d_2	r_2	$\bar{s}_2 = s_1$	$f_1(s_1)$	$r_2 + f_1$
0	0	0	0	0	0
1	0	0	1	644	644
	1	500	0	0	500 ⟵
2	0	0	2	988	988 ⟵
	1	500	1	644	1,144
	2	1,000	0	0	1,000
3	0	0	3	1,332	1,332
	1	500	2	988	1,488
	2	1,000	1	644	1,644
	3	1,300	0	0	1,300 ⟵
4	0	0	4	1,476	1,476 ⟵
	1	500	3	1,332	1,832
	2	1,000	2	988	1,988
	3	1,300	1	644	1,944
	4	1,700	0	0	1,700
5	0	0	5	2,020	2,020
	1	500	4	1,476	1,976 ⟵
	2	1,000	3	1,332	2,332
	3	1,300	2	988	2,288
	4	1,700	1	644	2,344
	5	2,000	0	0	2,000

As usual, the arrows point to the minimum for each state value. The following summary table is thus obtained.

Summary for Stages 2 and 1:

s_2	d_2^*	$f_2(s_2)$	$\bar{s}_2$
0	0	0	0
1	1	500	0
2	0	988	2
3	3	1,300	0
4	0	1,476	4
5	1	1,976	4

Optimization of Stages 3, 2, and 1:

s_3	d_3	r_3	$\tilde{s}_3 = s_2$	$f_2(s_2)$	$r_3 + f_2$
5	0	0	5	1,976	1,976
	1	455	4	1,476	1,931 ⟵
	2	910	3	1,300	2,210
	3	1,350	2	988	2,338
	4	1,800	1	500	2,300
	5	2,240	0	0	2,240

The solution can now be traced through the summary tables, starting from the minimum return in the last table:

$$f_3 = \$1,931 \text{ per day,}$$
$$d_3^* = 1 \longrightarrow \tilde{s}_3 = s_2 = 4,$$
$$d_2^* = 0 \longrightarrow \tilde{s}_2 = s_1 = 4,$$
$$d_1^* = 4 \longrightarrow \tilde{s}_1 = 0.$$

Consequently, the minimum cost treatment program is to install a 4 m.g.d. primary treatment plant and a 1 m.g.d. tertiary treatment plant.

11.10 A CONTINUOUS VARIABLE PROBLEM

Use dynamic programming to prove that the optimal policy to

$$\text{Max } y = x_1 x_2 x_3$$

subject to

$$x_1 + x_2 + x_3 = L,$$

is

$$x_1^* = x_2^* = x_3^* = \frac{L}{3}.$$

This problem can also be solved by direct differentiation after substitution of the constraint into the objective function, or with Lagrange multipliers (see chapter 3). However, the dynamic programming solution possesses several interesting features. The model is given below,

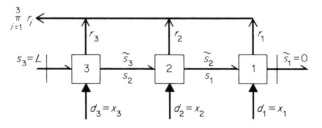

where

$$\bar{s}_i = s_i - d_i,$$

$$0 \leq d_i = x_i \leq s_i,$$

and

$$r_i = d_i, \text{ for all } i.$$

Note that in this problem the individual returns are multiplied together $\pi_{i=1}^3 r_i$ (read the π product of the r_i from $i = 1$ to 3), instead of being added to each other. This causes no particular difficulty if, as in this case, the r_i are constrained to remain non-negative.

Suboptimization of Stage 1:

$$f_1(s_1) = \max_{0 \leq d_1 = x_1 \leq s_1} (d_1) = s_1.$$

This is a decisionless stage ($d_1^* = s_1$) because $\bar{s}_1$ must equal zero.

Suboptimization of Stages 2 and 1:

$$f_2(s_2) = \max_{0 \leq d_2 = x_2 \leq s_2} [r_2 \cdot f_1(s_1)]$$

$$= \max_{0 \leq d_2 = x_2 \leq s_2} (d_2 \cdot s_1).$$

However,

$$s_1 = \bar{s}_2 = s_2 - d_2.$$

Therefore,

$$f_2(s_2) = \max_{0 \leq d_2 = x_2 \leq s_2} [d_2(s_2 - d_2)].$$

To maximize, set equal to zero the derivative of the maximand with respect to d_2,

$$\frac{d\{d_2(s_2 - d_2)\}}{d d_2} = (s_2 - d_2) - d_2 = 0,$$

therefore,

$$d_2^* = \frac{s_2}{2}$$

and

$$f_2(s_2) = \frac{s_2}{2}\left(s_2 - \frac{s_2}{2}\right) = \left(\frac{s_2}{2}\right)^2.$$

This value of d_2 falls within the feasible range, $0 \leq d_2 = x_2 \leq s_2$.

Optimization of Stages 3, 2, and 1:

$$f_3(s_3) = \max_{0 \leq d_3 = x_3 \leq s_3} [r_3 \cdot f_2(s_2)].$$

The initial input state $s_3 = L_3$, hence

$$f_3(L) = \max_{0 \leq d_3 = x_3 \leq L} \left[d_3\left(\frac{s_2}{2}\right)^2\right],$$

but

$$s_2 = \tilde{s}_3 = s_3 - d_3 = L - d_3,$$

therefore

$$f_3(L) = \max_{0 \leq d_3 = x_3 \leq L} \left[d_3 \left(\frac{L - d_3}{2} \right)^2 \right].$$

Again, set equal to zero the derivative of the maximand with respect to d_3,

$$\frac{d \left[d_3 \left(\frac{L - d_3}{2} \right)^2 \right]}{d \, d_3} = \left(\frac{L - d_3}{2} \right)^2 + d_3 \left[2 \left(\frac{L - d_3}{2} \right) \left(-\frac{1}{2} \right) \right] = 0.$$

Therefore,

$$\left(\frac{L - d_3}{2} \right)^2 - d_3 \left(\frac{L - d_3}{2} \right) = 0,$$

or

$$\frac{L - d_3}{2} - d_3 = 0 \text{ for } \left(\frac{L - d_3}{2} \right) \neq 0.$$

Hence,

$$d_3^* = \frac{L}{3}$$

and

$$f_3(L) = \frac{L}{3} \left(\frac{L - \frac{L}{3}}{2} \right)^2 = \left(\frac{L}{3} \right)^3.$$

$d_3^* = L/3$ again falls within the feasible range, $0 \leq d_3 = x_3 \leq L$. Retracing the decisions and states back to stage 1, find

$$\text{Max } y = \left(\frac{L}{3} \right)^3,$$

$$d_3^* = x_3^* = \frac{L}{3} \longrightarrow s_2 = \frac{2L}{3},$$

$$d_2^* = x_2^* = \frac{L}{3} \longrightarrow s_1 = \frac{L}{3},$$

$$d_1^* = s_1^* = \frac{L}{3} \longrightarrow \tilde{s}_1 = 0.$$

The solution can be generalized: In order to maximize $y = \pi_{i=1}^n x_i$ subject to

$$\sum_{i=1}^n x_i = L,$$

set

$$x_i = \frac{L}{n}$$

for all i, to yield Max $y = (L/n)^n$.

11.11   STATE INVERSION FOR FINAL VALUE PROBLEMS

Consider the final value problem of Figure 11.10. Assume that there exist N inverse transition functions

$$s_i = \tilde{T}_i(\bar{s}_i, d_i); \; i = 1, \ldots, N,$$

which express the input state at stage i as a function of its output state and decision variables. The interchanging of the roles of the input and output state variables is called state inversion.

Through state inversion, each return function can also be expressed in terms of the decision and output state. Thus,

$$R_i(s_i, d_i) = R_i[\tilde{T}_i(\bar{s}_i, d_i), d_i] = \tilde{R}_i(\bar{s}_i, d_i); \; i = 1, 2, \ldots, N, \qquad (11.24)$$

and the optimization problem is to find the set of decision variables $d_i^*(\bar{s}_1)$ for which the total return is an optimum:

$$\tilde{R}^*[\bar{s}_1, d_i^*(\bar{s}_1)] = \sum_{i=1}^{N} \tilde{R}_i(\bar{s}_i, d_i). \qquad (11.25)$$

State inversion reverses the flow of information through the state variables such that the optimization can now proceed from stage N downstream to stage 1 and the final value problem is transformed into an initial value problem. After each stage in the system has been reversed, the final value problem of Figure 11.10 can be put in the same notational form, with one exception, as an initial value problem by renumbering the stages in reverse order.

The exception is that in the transformed and renumbered system the tilde notation ($\sim$) is used in a different way. Thus, the input states are symbolized $\bar{s}_i$, the output states s_i, and the returns $\tilde{r}_i$. This was done in Figure 11.11.

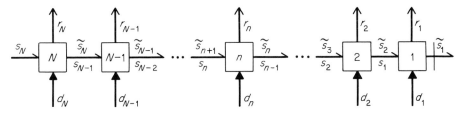

FIGURE 11.10.   A Final Value Problem.

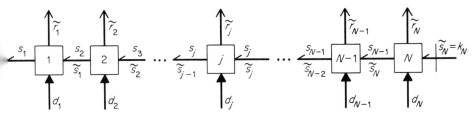

FIGURE 11.11.   Initial Value Problem Obtained from a Final Value Problem Through State Inversion.

The problem has once more become an N-decision, one-state initial value problem and its optimization is performed as usual.

11.11.1 A Final Value Allocation Problem

Consider the four-stage system

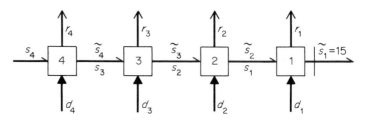

with transition and return functions defined as follows:

$$\tilde{s}_i = 4s_i + 2d_i,$$
$$r_i = 5s_i - d_i,$$
$$0 \le d_i \le s_i;\ i = 1, \ldots, 4.$$

The final output state $\tilde{s}_i$ is fixed at 15; consequently, this is a final value allocation problem.

Perform the state inversion

$$s_i = \tfrac{1}{4}(\tilde{s}_i - 2d_i) = 0.25\tilde{s}_i - 0.5d_i;\ i = 1, \ldots, 4.$$

which in turn yields,

$$\tilde{r}_i = 1.25\tilde{s}_i - 3.5d_i,$$

and

$$d_i \ge 0,$$
$$d_i \le s_i = 0.25\tilde{s}_i - 0.5d_i,$$

or

$$0 \le d_i \le \frac{0.25}{1.5}\tilde{s}_i = 1.67\tilde{s}_i;\ i = 1, \ldots, 4.$$

Now renumber the stages in reverse order to obtain the initial value problem

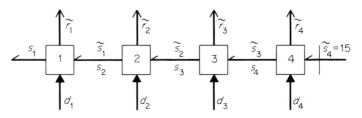

which can be optimized following the usual procedure.

Suboptimization of Stage 1:

$$f_1(\bar{s}_1) = \max_{0 \le d_1 \le 1.67\bar{s}_1} [1.25\bar{s}_1 - 3.5d_1] = 1.25\bar{s}_1 \qquad \text{with } d_1^* = 0.$$

Suboptimization of Stages 2 and 1:

$$f_2(\bar{s}_2) = \max_{0 \le d_2 \le 1.67\bar{s}_2} [1.25\bar{s}_2 - 3.5d_2 + 1.25\bar{s}_1].$$

But

$$\bar{s}_1 = s_2 = 0.25\bar{s}_2 - 0.5d_2,$$

therefore,

$$\begin{aligned} f_2(\bar{s}_2) &= \max_{0 \le d_2 \le 1.67\bar{s}_2} (1.25\bar{s}_2 - 3.5d_2 + 0.312\bar{s}_2 - 0.625d_2) \\ &= \max_{0 \le d_2 \le 1.67\bar{s}_2} (1.562\bar{s}_2 - 4.125d_2) = 1.562\bar{s}_2 \qquad \text{with } d_2^* = 0. \end{aligned}$$

Suboptimization of Stages 3, 2, and 1:

$$f_3(\bar{s}_3) = \max_{0 \le d_3 \le 1.67\bar{s}_3} (1.25\bar{s}_3 - 3.5d_3 + 1.562\bar{s}_2).$$

Again,

$$\bar{s}_2 = s_3 = 0.25\bar{s}_3 - 0.5d_3.$$

Thus,

$$\begin{aligned} f_3(\bar{s}_3) &= \max_{0 \le d_3 \le 1.67\bar{s}_3} (1.25\bar{s}_3 - 3.5d_3 + 0.39\bar{s}_3 - 0.78d_3) \\ &= \max_{0 \le d_3 \le 1.67\bar{s}_3} (1.64\bar{s}_3 - 4.28d_3) = 1.64\bar{s}_3 \qquad \text{with } d_3^* = 0. \end{aligned}$$

Optimization of Stages 4, 3, 2, and 1:

$$\begin{aligned} f_4(\bar{s}_4) = f_4(15) &= \max_{0 \le d_4 \le 25} \{18.75 - 3.5d_4 + 1.64\bar{s}_3\} \\ &= \max_{0 \le d_4 \le 25} \{18.75 - 3.5d_4 + 6.14 - 0.82d_4\} \\ &= \max_{0 \le d_4 \le 25} \{24.89 - 4.32d_4\} = 24.89 \qquad \text{with } d_4^* = 0. \end{aligned}$$

Retracing steps through the system, find

$$f_4(\bar{s}_4) = f_4(15) = 24.89$$

$$d_4^* = 0 \longrightarrow s_4 = 0.25\bar{s}_4 = 3.75 = \bar{s}_3,$$

$$d_3^* = 0 \longrightarrow s_3 = 0.25\bar{s}_3 = 0.938 = \bar{s}_2,$$

$$d_2^* = 0 \longrightarrow s_2 = 0.25\bar{s}_2 = 0.234 = \bar{s}_1,$$

and

$$d_1^* = 0 \longrightarrow s_1 = 0.25\bar{s}_1 = 0.059.$$

Returning to the original numbering system, obtain

$$f_4(\bar{s}_1) = f_4(15) = 24.89.$$

NOTE: The subscript $_4$ given to the function f indicates that this was the fourth functional suboptimization.

Also,

$$d_1^* = 0 \longrightarrow s_1 = 3.75 = \bar{s}_2,$$
$$d_2^* = 0 \longrightarrow s_2 = 0.938 = \bar{s}_3,$$
$$d_3^* = 0 \longrightarrow s_3 = 0.234 = \bar{s}_4,$$

and

$$d_4^* = 0 \longrightarrow s_4 = 0.059.$$

11.11.2 State Inversion of a Final Value Network Problem

Return to the network of Figure 11.8 but now require to find the lowest cost route from any node in zone V to node BI. This is a final value problem because the final output state s_1 is fixed (node BI). The state inversion is automatically performed by starting the optimization at zone V and proceeding downstream. This was done in Figure 11.12 with the optimum (minimum)

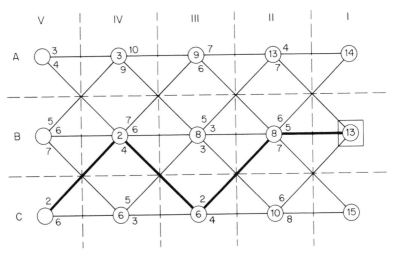

FIGURE 11.12. A Final Value Network Problem.

returns shown in the nodes themselves, as explained in section 11.8. The minimum cost route is CV − BIV − CIII − BII − BI, with a minimum cost of 13.

For completeness, the minimum costs to nodes AI and CI are also given. Although their corresponding minimum cost paths are not shown, they can be scanned as for BI.

11.12 DECISION INVERSION

Suppose that the transition function

$$\tilde{s}_1 = T_1(s_1, d_1)$$

can be solved in terms of the decision variable d_1,

$$d_1 = \hat{T}_1(s_1, \tilde{s}_1). \qquad (11.26)$$

This transformation is called a decision inversion. It permits the solution of final value and two-point boundary value problems as well as recycle loops.

Consider the solution of the two-point boundary value problem given below.

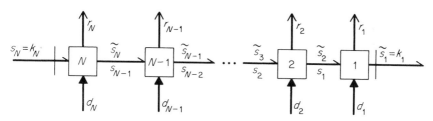

Stage 1:

A decision inversion at stage 1 yields

$$f_1(s_1, \tilde{s}_1) = R_1(\hat{T}_1(s_1, \tilde{s}_1), s_1) \qquad (11.27)$$

and, consequently, stage 1 is transformed into a decisionless stage. No optimization is needed at this stage because the decision variable d_1 was eliminated through decision inversion.

Suboptimization of Stages 2 and 1:

For stage 2 and for all upstream stages no additional decision inversions are required but each optimization at stage i, $i > 1$, has to be performed as a function of s_i and $\tilde{s}_1$ over the decision variable d_i. Thus, each suboptimization is a two-state, one-decision problem.

$$f_2(s_2, \tilde{s}_1) = \max_{d_2} [R_2(s_2, d_2) + f_1(s_1, \tilde{s}_1)] \qquad (11.28)$$

but

$$s_1 = \tilde{s}_2 = T_2(s_2, d_2),$$

therefore,

$$f_2(s_2, \tilde{s}_1) = \max_{d_2} \{R_2(s_2, d_2) + f_1[T_2(s_2, d_2), \tilde{s}_1]\},$$

or

$$f_2(s_2, \tilde{s}_1) = \max_{d_2} \{R_2(s_2, d_2) + f_1(s_2, d_2, \tilde{s}_1)\}. \qquad (11.29)$$

Suboptimization of Stages $N - 1$ through 1:

As for stage $N - 2$ through 2,

$$f_{N-1}(s_{N-1}, \bar{s}_1) = \max_{d_{N-1}} [R_{N-1}(s_{N-1}, d_{N-1}) + f_{N-2}(s_{N-1}, \bar{s}_1)] \qquad (11.30)$$

which reduces to

$$f_{N-1}(s_{N-1}, \bar{s}_1) = \max_{d_1} [R_{N-1}(s_{N-1}, d_{N-1}) + f_{N-2}(s_{N-1}, d_{N-1}, \bar{s}_1)]. \qquad (11.31)$$

Again, a two-state, one-decision optimization problem is the result.

Optimization of Stages N through 1:

At stage N, obtain

$$f_N(s_N, \bar{s}_1) = \max_{d_N} [R_N(s_N, d_N) + f_{N-1}(s_{N-1}, \bar{s}_1)], \qquad (11.32)$$

or

$$f_N(s_N, \bar{s}_1) = \max_{d_N} [R_N(s_N, d_N) + f_{N-1}(s_N, d_N, \bar{s}_1)]. \qquad (11.33)$$

At this point, $\bar{s}_1 = k_1$, and $s_N = k_N$, two fixed values and the solution is given by equation (11.33).

If, however, only $\bar{s}_1 = k_1$ is fixed, one would solve a final value problem by simply optimizing equation (11.33) over s_N, as follows:

$$f_N^*(\bar{s}_1) = \max_{d_N} [f_N(s_N, \bar{s}_1)]. \qquad (11.34)$$

This last manipulation shows that decision inversion is an alternate to state inversion in optimizing final value problems.
In summary:

1. Final value problems can be optimized either by state inversion (preferable for its simplicity) or, if state inversion is not possible, by decision inversion which transforms the one-state, N-decision problem into one decisionless stage, equation (11.27); $N - 2$, two-state, one-decision problems, equations (11.28) through (11.31); and one, one-state, two-decision problems, equations (11.32) through (11.34).
2. Two-point boundary value problems require decision inversion for solution. Decision inversion transforms the one-state, N-decision problem into one decisionless stage, equation (11.27); and $N - 1$, two-state, one-decision problems, equation (11.28) through (11.33).

11.12.1 Recycle Loops

The final output state is connected to the initial input state and a recycle loop develops. This system is optimized by cutting the recycle loop. This is called a cut state. Thus, the problem can now be transformed into a two-point

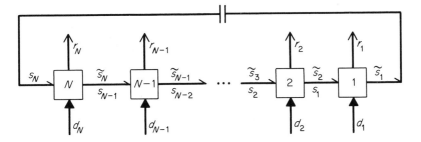

boundary value problem with matching boundaries:

$$\tilde{s}_1 = \tilde{c}_1 = s_N = c_N. \tag{11.35}$$

The $\tilde{c}_1 = c_N$ are choice variables which must be selected to maximize equation (11.33). Consequently, the maximum for the recycle loop problem is obtained from equation (11.33) as follows:

$$f_N^{**} = \max_{\tilde{s}_1 = \tilde{c}_1 = s_N = c_N} f_N(s_N, \tilde{s}_1). \tag{11.36}$$

The practical procedure for optimizing equation (11.36) is:

1. Select a value of $\tilde{c}_1 = c_N$ and solve a two-point boundary value problem through one decision inversion at stage 1.
2. Select different values of $\tilde{c}_1 = c_N$ until one is obtained for which equation (11.36) achieves its maximum value.

11.12.2 Problems with Converging and Diverging Branches

With decision and state inversions, it is possible to use dynamic programming to optimize systems with converging and diverging branches. The interested reader is referred to the list of books at the end of the chapter for additional information on this subject.

11.12.3 Networks with Recycle Loops

The allocation problems given in sections 11.9 and 11.10 illustrate the solution of boundary value problems with final output state $\tilde{s}_1$ required to vanish. In optimizing those problems, a decision inversion was performed almost automatically at stage 1.

The instructional value of the network problem with recycle given in this section is attributed to the fact that its solution involves the optimization of several two-point boundary value systems. Consider the network of Figure 11.13 and determine the least cost route "around the world," that is, the least cost route that originates and terminates at the same city.

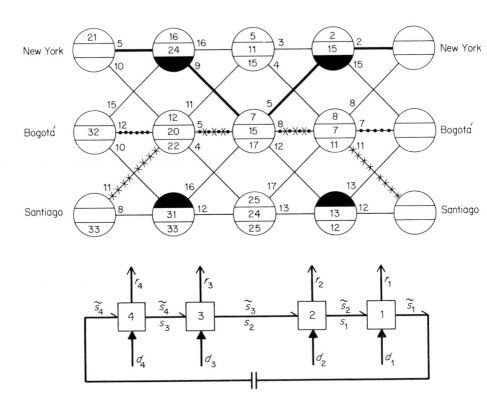

FIGURE 11.13.   Network with Recycle Loop.

1. Fix the final output state and the initial input state at New York.

$$\tilde{s}_1 = s_4 = \text{New York}.$$

Perform a decision inversion at stage 1 and proceed upstream to stage 4. The results of the computations are given in the upper cell of each node. The least cost is 21 and the least cost route is shown with heavy lines.

2. Now fix,

$$\tilde{s}_1 = s_4 = \text{Bogotá}.$$

The same procedure yields the costs shown in the middle cell of each node. The minimum cost is 32 and the minimum route is marked with heavy dots.

3. Now fix the boundary values at Santiago.

$$\tilde{s}_1 = s_4 = \text{Santiago}.$$

The computations appear in the node's lower cells. The minimum cost route is marked with x's and the minimum cost is 33.

4. Comparing the three minima, it is seen that the global minimum is

$$f_4^{**} = 21.$$

It is obtained by setting

$$\bar{s}_1 = s_4 = \text{New York.}$$

Similar procedures can be used for other boundary value and cycling problems.

EXERCISES

11–1. Use dynamic programming to compute the *longest* path from *every* node in region V to any node in region I.

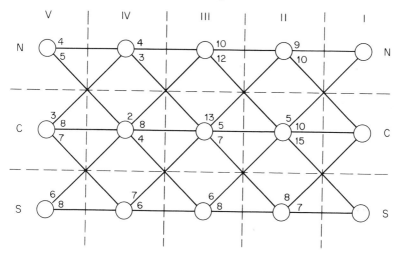

11–2. Compute the *shortest* path from node NV to node SI in the network of Problem 11–1.

11–3. Use dynamic programming to determine the cheapest route from Boston to the West Coast.

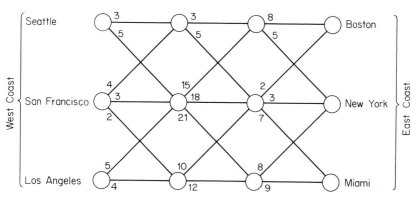

11–4. Three types of kits are to be packed in a space capsule. Their weights and relative values are given below:

Kit Type	Weight	Value
1	10#	20
2	6#	11
3	3#	7

The total capsule payload is 35#. No more than 3 of each kit type can be used. However, any kit type can be excluded if necessary.

Determine the number of units of each type of kit that maximizes *total value* without exceeding payload capacity.

11–5.

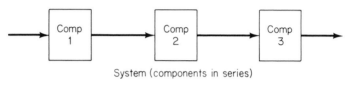

System (components in series)

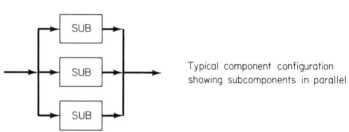

Typical component configuration showing subcomponents in parallel

Components in a space capsule are linked in series. Each component is, in fact, a set of subcomponents connected in parallel. The weights (w_i) and reliabilities (p_i) of the subcomponents are given below.

Subcomp. for Comp. i	Subcomp. Weight w_i	Subcomp. Reliab. p_i
1	2 lbs.	.95
2	1 lbs.	.90
3	3 lbs.	.97

Total payload capacity is 10 lbs., and each component must have at least one subcomponent but not more than three. Determine the system configuration that maximizes its total reliability.

NOTE:
1. Reliability of each component $R_i = 1 - (1 - p_i)^n$, $i = 1, 2, 3$ where $n =$ no. of subcomponents in parallel.
2. Total reliability of system, $R = \pi_{i=1}^{3} R_i$.

11–6. Four water projects must be allocated from limited construction funds in a small district. The alternatives are identified as projects 1, 2, 3, and 4. They will produce the net independent returns shown in the table. With dynamic programming and the grid intervals given, determine the optimal allocation of $200,000 worth of construction funds. Combine alternatives in the sequence indicated.

NET RETURN

	1	2	3	4
$ 0	$ 0	$ 0	$ 0	$ 0
20,000	−4,000	2,000	6,000	800
40,000	0	3,000	12,000	1,000
60,000	6,000	4,000	16,000	6,000
80,000	9,000	5,000	25,000	16,000
100,000	10,000	6,000	26,000	38,000
120,000	14,000	7,000	18,000	38,000
140,000	21,000	8,000	16,000	41,000
160,000	23,000	9,000	6,000	43,000
180,000	24,000	10,000	−4,000	44,000
200,000	20,000	11,000	−10,000	45,000

11–7. The owner of a new car dealership wishes to invest no more than $20,000 in the purchase of three new models:
Model A costs $2,000 and sells for $2,250.
Model B costs $4,000 and sells for $4,900.
Model C costs $3,000 and sells for $3,600.
The owner wishes to have at least *one* and no more than *three* of *each* model. Determine the buying strategy that maximizes the profit.

11–8. Develop the functional maximization strategy for the boundary value problem with recycle shown below.

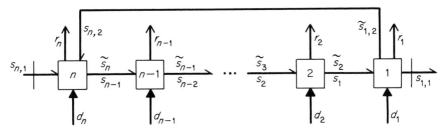

REFERENCES

1. ARIS, R., *Discrete Dynamic Programming*. New York: Blaisdell Publishing Company, 1964.

2. BELLMAN, R. E., *Dynamic Programming*. Princeton: Princeton University Press, 1957.

3. NEMHAUSER, G. L., *Introduction to Dynamic Programming*. New York: John Wiley & Sons, Inc., 1966.

4. WILDE, D. J. and C. S. BEIGHTLER, *Foundations of Optimization*. Englewood Cliffs, N.J.: Prentice-Hall, Inc., 1967.

PART **III**

Stochastic
Systems

CHAPTER 12

Stochastic Linear Programming

12.1 CLASSIFICATION

Stochastic linear programming problems can be classified in two general categories, as follows:

1. Those problems with one or more random parameters with known probability distributions but with constraints which are required to hold with probability 1. This type is usually called statistical linear programming.
2. Linear programming problems involving random parameters with known probability distributions and feasible regions defined in such a way that one or more of the constraints can be violated with a small probability. This type is called chance constrained programming.

12.2 STATISTICAL LINEAR PROGRAMMING

12.2.1 Introduction

Many problems in the planning and operation of physical facilities are influenced by the occurrence of random events and have a linear structure with regions of feasibilities defined by constraint sets which cannot be violated.

The solutions to these types of problems fall under the statistical linear programming (type 1) category.

When the parameters of the system are subject to fluctuations with known probability distributions, then, for a parameter a, the function $f(a)$ appearing in the equation

$$dp = f(a)da \tag{12.1}$$

is the probability density function of the random parameter a, and dp is the probability that the random variable a will take on a value between a and $a + da$.

These types of problems can be handled with simplicity and ease by determining expected values $E(a)$ for each parameter and using the expected values in the computations.

With this procedure, the statistical problem is replaced by a related deterministic problem and its optimal solution, $z^*[E(\bar{a})]$, is interpreted in terms of the expectation. However, in reality, one wishes to obtain $E[z^*(\bar{a})]$ for, in general, in the case of minimization, the minimum of the expected value is greater than or equal to the expected value of the minimum

$$\min z[E(\bar{a})] \geq E[\min z(\bar{a})], \tag{12.2}$$

because the deterministic problem, not involving random parameters, is more constrained than the stochastic problem. The opposite is true for maximization. In linear programming, when the cost coefficients, c_j, are stochastic but the structural coefficients, a_{ij}, and the stipulations, b_i, are deterministic, one is justified in replacing c_j by its expected value, $E(c_j)$, because it is clear that in this case the minimum of the expected value equals the expected value of the minimum. Therefore,

$$\min z[E(\bar{a})] = E[\min z(\bar{a})]. \tag{12.3}$$

However, if the a_{ij} and/or the b_i are stochastic, the problem is much more complex. In fact, no general procedure exists to handle cases such as these because the nature of each problem under consideration dictates its model formulation.

When the b_i are subject to uncertainty, a method developed by Dantzig is applicable (the reference is given at the end of the chapter).

A simplified housing construction case with *uncertain* demand will be used to illustrate the application of the method which can be applied to general linear programming problems as well, whenever the concepts on which the theory is based make physical sense in the problem.

12.2.2 A Housing Problem

The problem presented here is very simplified and, consequently, artificial but the model development applies to complex, physically realistic situations with minor or no modifications.

Suppose that to maximize the profit from the sale of homes during the following twelve-month period, a builder who owns land in Alexandria, Baton Rouge, and New Orleans, Louisiana, wishes to determine the best allocation of capital for the construction of two-, three-, and four-bedroom homes.

The unit profits in $1,000 are given in Table 12.1.

TABLE 12–1.   Unit Sale Profits in $1,000.

	(1) Alex.	(2) B.R.	(3) N.O.
(1) 2-bedroom	3.0	4.0	5.0
(2) 3-bedroom	3.5	4.8	5.7
(3) 4-bedroom	2.8	4.1	4.9

The developer has funds for building a total of 20 two-bedroom, 40 three-bedroom, and 30 four-bedroom units. If he cannot sell a house within a twelve-month period, he risks a loss of $1,000 for every house not sold in Alexandria, $1,200 for every unsold house in Baton Rouge, and $1,100 for every house not sold in New Orleans. However, he will eventually sell every house. The housing market, as well as can be determined, yielded the following demand data for *his* operation:

Location	Demand Distribution	Mean	Standard Deviation
Alexandria	Normal	40 units	10 units
Baton Rouge	Normal	40 units	5 units
New Orleans	Normal	60 units	15 units

Because demands are normally distributed, a grid of eight intervals will be used to break down the density curves, as shown in Figure 12.1. The

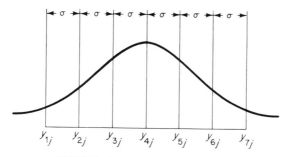

FIGURE 12.1.   Eight-Interval Grid.

model for the problem can be expressed thus:

$$\text{Max}\left[\sum_{i=1}^{3}\sum_{j=1}^{3}c_{ij}x_{ij} - \sum_{j=1}^{3}f_j(c_j, y_j)\right],\tag{12.4}$$

subject to

$$\sum_{j=1}^{3}x_{ij} = a_i \longrightarrow \text{deterministic production,}\tag{12.5}$$

$$\sum_{j=1}^{3}x_{ij} = y_i \longrightarrow \text{stochastic demand,}\tag{12.6}$$

$$\left.\begin{array}{ll}x_{ij} \geq 0 & \text{for all } i, j,\\ y_j \geq 0 & \text{for all } i, j.\end{array}\right\}\tag{12.7}$$

Where

$x_{ij} =$ number of type i homes in city j,

$i = 1 \longrightarrow$ 2-bedroom home,

$i = 2 \longrightarrow$ 3-bedroom home,

$i = 3 \longrightarrow$ 4-bedroom home,

$j = 1 \longrightarrow$ Alexandria,

$j = 2 \longrightarrow$ Baton Rouge,

$j = 3 \longrightarrow$ New Orleans,

$a_i =$ Number of type i homes built,

$c_{ij} =$ Unit profit resulting from the sale of type i home in city j,

$c_j =$ Unit cost of *not* selling home in city j during the next 12-month period,

$y_j =$ Stochastic demand in city j,

$f_j(c_j, y_j) =$ Unknown function that describes the expected cost of homes unsold in city j during the next 12-month period.

Demand information is summarized in Table 12.2.

TABLE 12.2. Demand Information by Ranges.

Probability of Demand in This Range	0	.02	.14	.34	.34	.14	.02	0
Alexandria (1)	0–10	10–20	20–30	30–40	40–50	50–60	60–70	> 70
Baton Rouge (2)	0–25	25–30	30–35	35–40	40–45	45–50	50–55	> 55
New Orleans (3)	0–15	15–30	30–45	45–60	60–75	75–90	90–105	> 105

12.2.3 Statistical Considerations

As indicated earlier, the statistical variable y_j could be replaced by the constant $E(y_j)$, $f_j(c_j, y_j)$ could be expressed in terms of c_j and $E(y_j)$ and the resulting standard problem solved. A more valuable approach is described in this section.

In Figure 12.1, the range of y_j was divided into 8 intervals and the boundary between interval k and interval $k + 1$ was denoted by y_{kj}. The last interval will be assumed to extend from $y_{7,j}$ to ∞.

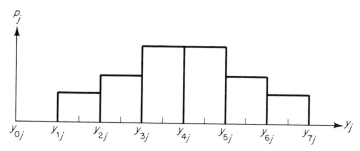

FIGURE 12.2. Histogram for Probability Distribution of y_j.

The histogram of Figure 12.2 depicts the probability p_{kj} that the demand at city j will lie in the kth interval. This is a graph of the information given in Table 12–2.

The intervals can be chosen in any manner and need not be uniform, but the assumption will be made that $p_{8j} = 0$.

If the builder constructs y_{1j} homes in city j, the probability that all of these units will be sold within the next 12 months is equal to the probability that the demand at j is equal to or exceeds y_{1j}. This probability is

$$p_{2j} + p_{3j} + p_{4j} + \cdots + p_{8j}.$$

On the other hand, the probability that the first unit out of the y_{1j} units will be sold is $p_{1j} + p_{2j} + \cdots + p_{8j} = 1$. Hence, it can be said that the average probability of selling any unit (up to y_{1j} units) is $\frac{1}{2}p_{1j} + p_{2j} + \cdots + p_{8j}$. However, in this example, c_j is the cost of *not* selling a house in city j. Therefore, the average probability of *not* selling any unit up to y_{1j} units is

$$1 - (\tfrac{1}{2}p_{1j} + \sum_{k=2}^{8} p_{kj}), \tag{12.8}$$

and the cost of *not* selling up to y_{1j} units is

$$E(C_j) = \psi_{1j}y_j; \qquad 0 \leq y_j \leq y_{1j}, \tag{12.9}$$

where

$$\psi_{1j} = c_j\left[1 - \left(\tfrac{1}{2}p_{1j} + \sum_{k=2}^{8} p_{kj}\right)\right]. \tag{12.10}$$

Now consider the interval $y_{1j} \leq y_j \leq y_{2j}$. The expected cost of *not* selling y_{1j} units is given by equation (12.9) with $y_j = y_{1j}$, so one must determine the additional expected cost of *not* selling the incremental units in excess of y_{1j}. If a total of y_{2j} homes are built in city j, the total expected cost of not selling between y_{1j} and y_{2j} units is given by

$$E(C_j) = \psi_{1j}y_{1j} + \psi_{2j}(y_j - y_{1j}); \qquad y_{1j} \leq y_j \leq y_{2j}, \qquad (12.11)$$

where

$$\psi_{2j} = c_j\left[1 - \left(\tfrac{1}{2}p_{2j} + \sum_{k=3}^{8} p_{kj}\right)\right]. \qquad (12.12)$$

In general, if the number of houses built in city j lies in the kth range, the expected cost of not selling between $y_{k-1,j}$ and $y_{k,j}$ units within the next 12-month period is

$$E(C_j) = \sum_{s=1}^{k-1} \psi_{sj}(y_{sj} - y_{s-1,j}) + \psi_{kj}(y_j - y_{k-1,j}); \qquad y_{k-1,j} \leq y_j \leq y_{kj}, \qquad (12.13)$$

where

$$\psi_{kj} = c_j\left[1 - \left(\tfrac{1}{2}p_{kj} + \sum_{s=k+1}^{8} p_{sj}\right)\right], \qquad (12.14)$$

with the definition $y_{0j} = 0$ (the origin).

12.2.4 Linear Programming Formulation

Now let y_j and $E(C_j)$ be represented in the following form:

$$y_j = \Delta_{1j} + \Delta_{2j} + \cdots + \Delta_{8j}, \qquad (12.15)$$

and

$$E(C_j) = \psi_{1j}\Delta_{1j} + \psi_{2j}\Delta_{2j} + \cdots + \psi_{8j}\Delta_{8j}, \qquad (12.16)$$

where the Δ_{kj} are *new* non-negative variables constrained as follows:

$$0 \leq \Delta_{kj} \leq y_{kj} - y_{k-1,j} \equiv U_{kj}, \qquad \text{for } k = 1, 2, \ldots, 7, \qquad (12.17)$$
$$j = 1, 2, 3.$$

The variables Δ_{kj} represent increments of housing units built in city j. They measure the degree to which the different intervals of demand are filled. If $\Delta_{kj} = 0$, the kth interval is empty, and if $\Delta_{kj} = U_{kj}$, the interval is filled to capacity. The last increment, Δ_{8j}, is *not* upper bounded. For $k = k_j$,

$$\Delta_{kj} = U_{kj} \qquad \text{for } k = 1, 2, \ldots, k_{j-1},$$
$$0 \leq \Delta_{kj} \leq U_{kj} \qquad \text{for } k = k_j, \qquad (12.18)$$

and

$$\Delta_{kj} = 0 \qquad \text{for } k = k_{j+1}, k_{j+2}, \ldots, 8.$$

This means that there exists an interval k_j, such that all intervals below it are filled to capacity and all intervals above it are empty. Substitution of equations (12.15) and (12.16) into (12.4) and (12.6) yields,

$$\text{Max}\left(\sum_{i=1}^{3} \sum_{j=1}^{3} c_{ij}x_{ij} - \sum_{j=1}^{3} \sum_{k=1}^{8} \psi_{kj}\Delta_{kj}\right) \tag{12.19}$$

subject to

$$x_{ij} \geq 0 \qquad \text{for all } i, j, \tag{12.20}$$

$$0 \leq \Delta_{kj} \leq U_{kj}; \qquad k \neq 8 \quad \text{and} \quad \Delta_{8j} \geq 0, \tag{12.21}$$

$$\sum_{j=1}^{3} x_{ij} = a_i; \qquad i = 1, 2, 3. \tag{12.22}$$

$$\sum_{i=1}^{3} x_{ij} - \sum_{k=1}^{8} \Delta_{kj} = 0; \qquad j = 1, 2, 3. \tag{12.23}$$

This model is in standard linear programming format and can be readily solved. For the specific example under consideration it becomes:

FOR ALEXANDRIA, $j = 1$:

$[U_{kj} = y_{kj} - y_{k-1,j},$ from equation (12.17)]

$\psi_{11} = 1,000\{1-[\frac{1}{2}(0)+.02+.14+.34+.34+.14+.02]\} = \quad 0; U_{11} = 10,$

$\psi_{21} = 1,000\{1-[\frac{1}{2}(.02)+.14+.34+.34+.14+.02]\} \quad = \quad 10; U_{21} = 10,$

$\psi_{31} = 1,000\{1-[\frac{1}{2}(.14)+.34+.34+.14+.02]\} \quad = \quad 90; U_{31} = 10,$

$\psi_{41} = 1,000\{1-[\frac{1}{2}(.34)+.34+.14+.02]\} \quad = \quad 330; U_{41} = 10,$

$\psi_{51} = 1,000\{1-[\frac{1}{2}(.34)+1.4+.02]\} \quad = \quad 670; U_{51} = 10,$

$\psi_{61} = 1,000\{1-[\frac{1}{2}(.14)+.02]\} \quad = \quad 910; U_{61} = 10,$

$\psi_{71} = 1,000\{1-[\frac{1}{2}(.02)]\} \quad = \quad 990; U_{71} = 10,$

$\psi_{81} = 1,000\{1-0\} \quad = 1,000 \underline{\quad\quad}.$

FOR BATON ROUGE, $j = 2$:

$\psi_{12} = 1,200\{ 0 \} = \quad 0; U_{12} = 25,$

$\psi_{22} = 1,200\{.01\} = \quad 12; U_{22} = \quad 5,$

$\psi_{32} = 1,200\{.09\} = \quad 108; U_{32} = \quad 5,$

$\psi_{42} = 1,200\{.33\} = \quad 396; U_{42} = \quad 5,$

$\psi_{52} = 1,200\{.67\} = \quad 804; U_{52} = \quad 5,$

$\psi_{62} = 1,200\{.91\} = 1,092; U_{62} = \quad 5,$

$\psi_{72} = 1,200\{.99\} = 1,188; U_{72} = \quad 5,$

$\psi_{82} = 1,200\{ 1 \} = 1,200; \underline{\quad\quad}.$

FOR NEW ORLEANS, $j = 3$:

$\psi_{13} = 1,100\{ 0 \} = \quad 0; U_{13} = 15,$

$\psi_{23} = 1,100\{.01\} = \quad 11; U_{23} = 15,$

$\psi_{33} = 1,100\{.09\} = \quad 99; U_{33} = 15,$

$\psi_{43} = 1,100\{.33\} = \quad 363; U_{43} = 15,$

$\psi_{53} = 1,100\{.67\} = \quad 737; U_{53} = 15,$

$\psi_{63} = 1,100\{.91\} = 1,001; U_{63} = 15,$

$\psi_{73} = 1,100\{.99\} = 1,089; U_{73} = 15,$

$\psi_{83} = 1,100\{ 1 \} = 1,100; \quad\text{———}.$

The model from equations (12.19) through (12.23) is,

$$\text{Max } z = 3,000x_{11} + 4,000x_{12} + 5,000x_{13} + 3,500x_{21} + 4,800x_{22}$$
$$+ 5,700x_{23} + 2,800x_{31} + 4,100x_{32} + 4,900x_{33} - \{10\Delta_{21}$$
$$+ 90\Delta_{31} + 330\Delta_{41} + 670\Delta_{51} + 910\Delta_{61} + 990\Delta_{71} + 1,000\Delta_{81}$$
$$+ 12\Delta_{22} + 108\Delta_{32} + 396\Delta_{42} + 804\Delta_{52} + 1,092\Delta_{62} + 1,188\Delta_{72}$$
$$+ 1,200\Delta_{82} + 11\Delta_{23} + 99\Delta_{33} + 363\Delta_{43} + 737\Delta_{53} + 1,001\Delta_{63}$$
$$+ 1,089\Delta_{73} + 1,100\Delta_{83}\},$$

subject to

$$x_{ij} \geq 0 \quad \text{for } i = 1, 2, 3; j = 1, 2, 3;$$

$$\Delta_{kj} \geq 0 \quad \text{for } k = 1, 2, \ldots, 8; j = 1, 2, 3;$$

$$\Delta_{k1} \leq 10 \quad \text{for } k = 1, 2, \ldots, 7;$$

$$\Delta_{12} \leq 25; \Delta_{k2} \leq 5 \quad \text{for } k = 2, 3, \ldots, 7;$$

$$\Delta_{k3} \leq 15 \quad \text{for } k = 1, 2, \ldots, 7;$$

$$x_{11} + x_{12} + x_{13} = 20;$$

$$x_{21} + x_{22} + x_{23} = 40;$$

$$x_{31} + x_{32} + x_{33} = 30;$$

and

$$\sum_{i=1}^{3} x_{ij} - \sum_{k=1}^{8} \Delta_{kj} = 0, \quad \text{for } j = 1, 2, 3.$$

This model can be maximized immediately with the simplex algorithm studied in chapter 5. Linear programming programs are available at most computer centers. The numerical solution to the problem is of no interest for it follows the standard methods already discussed.

12.3 CHANCE CONSTRAINED PROGRAMMING

12.3.1 Assumptions

Chance constrained programming, developed by Charnes and Cooper, is a type of statistical linear programming which allows a small probability that the constraints can be violated. The following assumptions are standard for the types of problems belonging to this category:

1. The structural coefficients, a_{ij}, are constant parameters.
2. The stipulations, b_i, have known multivariate normal distributions.
3. The cost coefficients, c_j, have known distributions and are statistically independent of the stipulations, b_i.
4. The variables, x_j, must de determined before the values taken by any of the random parameters are known.

12.3.2 The Model

Consider the problem

$$\text{Max } E(z) = \sum_{j=1}^{n} E(c_j)x_j \tag{12.24}$$

subject to

$$x_j \geq 0 \qquad \text{for } j = 1, \ldots, n, \tag{12.25}$$

and

$$P\left(\sum_{j=1}^{n} a_{ij}x_j \leq b_i\right) \geq \alpha_i \qquad \text{for } i = 1, \ldots, m, \tag{12.26}$$

where expression (12.26) must be interpreted to mean that the probability that $\sum_{j=1}^{n} a_{ij}x_j \leq b_i$ is greater than or equal to α_i, with

$$0 \leq \alpha_i \leq 1 \qquad \text{for } i = 1, \ldots, m. \tag{12.27}$$

The α_i's are generally chosen very near unity. The probability $1 - \alpha_i$, $i = 1, \ldots, m$, represents the risk that the random variables will take on values such that the constraints are violated, that is,

$$\sum_{j=1}^{n} a_{ij}x_j > b_i, \qquad i = 1, \ldots, m. \tag{12.28}$$

The problem's objective is to determine the optimal non-negative solution vector which will "probably" satisfy each of the constraints when the random parameters take on their values.

The form of the constraints is, at this point, not in standard linear programming format. Consequently, the aim of the procedure is to convert the probabilistically structured constraints into deterministic ones which fit linear programming theory and which permit the simplex algorithm to be applied.

The assumptions stated previously make this a relatively easy task. Let $E(b_i)$ and σ_{b_i} respectively, be the expected value and the standard deviation of the random variable b_i.

Because the b_i's were assumed normally distributed (assumption 2), that is,

$$b_i \Longrightarrow \text{Normal } (E(b_i), \sigma_{b_i}), \tag{12.29}$$

the random variable

$$\frac{b_i - E(b_i)}{\sigma_{b_i}} \tag{12.30}$$

is also normally distributed with zero mean and standard deviation of one,

$$\frac{b_i - E(b_i)}{\sigma_{b_i}} \Longrightarrow \text{Normal } (0, 1). \tag{12.31}$$

The probabilistic constraints given by expression (12.26) can be written,

$$P\left(\sum_{j=1}^{n} a_{ij}x_j \leq b_i\right) = P\left(\frac{\sum_{j=1}^{n} a_{ij}x_j - E(b_i)}{\sigma_{b_i}} \leq \frac{b_i - E(b_i)}{\sigma_{b_i}}\right) \geq \alpha_i$$

$$\text{for } i = 1, \ldots, m. \tag{12.32}$$

Let Z be a normally distributed random variable with zero mean and standard deviation of one, that is,

$$Z \Longrightarrow \text{Normal } (0, 1). \tag{12.33}$$

Then,

$$\frac{b_i - E(b_i)}{\sigma_{b_i}} = Z. \tag{12.34}$$

A table of values of areas under the normal curve Z is given in Appendix B. This table gives the probability that the random variable Z take on values less than or equal to a number of standard deviations to the left or to the right of the mean. For example,

$$P(Z \leq +2.000) = 0.9773.$$

This is the area under the curve from $-\infty$ to $+2.000$ standard deviations to the right of the mean. By the same token, the probability that Z take on values greater than $+2.000$ standard deviations to the right of the mean is given by the difference between 1 and 0.9773. Thus,

$$P(Z \geq 2.000) = 1 - 0.9773 = 0.0227.$$

This is the area under the curve from $+2.000$ standard deviations to the right of the mean to $+\infty$.

In general,

$$P(Z \geq K_\alpha) = \alpha, \tag{12.35}$$

where,

$K_\alpha =$ Number of standard deviations to the left or to the right of zero mean,

$\alpha =$ Probability that the random variable Z will lie to the right of K_α.

Other examples are:

$$P(Z \geq 0) = 0.5000,$$
$$P(Z \geq +1.000) = 1.000 - 0.8413 = 0.1587,$$
$$P(Z \geq -1.000) = 1.000 - 0.1587 = 0.8413.$$

From this discussion it follows that

$$P\left[K_{\alpha_i} \leq \frac{b_i - E(b_i)}{\sigma_{b_i}}\right] = P\left[\frac{b_i - E(b_i)}{\sigma_{b_i}} \geq K_{\alpha_i}\right] = \alpha_i. \tag{12.36}$$

If K_{α_i} were decreased in value, α_i would increase, and if K_{α_i} were increased, α_i would decrease. Consequently, the expression

$$P\left[\frac{\sum\limits_{j=1}^{n} a_{ij}x_j - E(b_i)}{\sigma_{b_i}} \leq \frac{b_i - E(b_i)}{\sigma_{b_i}}\right] \geq \alpha_i$$

holds for a given solution if and only if

$$\frac{\sum\limits_{j=1}^{n} a_{ij}x_j - E(b_i)}{\sigma_{b_i}} \leq K_{\alpha_i}. \tag{12.37}$$

Hence, one concludes that

$$P\left[\sum\limits_{j=1}^{n} a_{ij}x_j \leq b_i\right] \geq \alpha_i \tag{12.38}$$

holds if and only if

$$\sum\limits_{j=1}^{n} a_{ij}x_j \leq E(b_i) + K_{\alpha_i}\sigma_{b_i}, \tag{12.39}$$

and the stochastic constraint (12.38) can be replaced by the standard linear programming constraint given by (12.39). The model becomes,

$$\text{Max } E(z) = \sum\limits_{j=1}^{n} E(c_j)x_j, \tag{12.40}$$

subject to

$$x_j \geq 0; j = 1, \ldots, n, \tag{12.41}$$

and

$$\sum_{j=1}^{n} a_{ij}x_j \leq E(b_i) + K_{\alpha_i}\sigma_{b_i}; \qquad i = 1, \ldots, m. \qquad (12.42)$$

12.3.3 Example

Consider the following problem:

$$\text{Max } E(z) = E(c_1)x_1 + E(c_2)x_2,$$

subject to

$$x_i \geq 0 \qquad \text{for } i = 1, 2,$$

and

$$P\{2x_1 + 3x_2 \leq b_1\} \geq 0.9772,$$

where c_1, c_2, and b_1 are normally distributed random variables with means and standard deviations as follows:

$$c_1 \implies \text{Normal (3, 2)},$$
$$c_2 \implies \text{Normal (2, 1)},$$
$$b_1 \implies \text{Normal (8, 2)}.$$

Solution:

The objective function is

$$\text{Max } E(z) = 3x_1 + 2x_2$$

because the expected values of c_1 and c_2 are, respectively, $E(c_1) = 3$ and $E(c_2) = 2$. For the constraint to hold with probability greater than or equal to 0.9772, consider

$$P(2x_1 + 3x_2 \leq b_1) = P\left[\frac{2x_1 + 3x_2 - E(b_1)}{\sigma_{b_1}} \leq \frac{b_1 - E(b_1)}{\sigma_{b_1}}\right] \geq 0.9772.$$

But $E(b_1) = 8$ and $\sigma_{b_1} = 2$, therefore,

$$P(2x_1 + 3x_2 \leq b_1) = P\left(\frac{2x_1 + 3x_2 - 8}{2} \leq \frac{b_1 - 8}{2}\right) \geq 0.9772.$$

The random variable $(b_1 - 8)/2$ is normally distributed with zero mean and standard deviation of one. Consequently,

$$Z = \frac{b_1 - 8}{2}.$$

Thus,

$$P(2x_1 + 3x_2 \leq b_1) = P\left(\frac{2x_1 + 3x_2 - 8}{2} \leq Z\right) \geq 0.9772,$$

and the question can be phrased: How many standard deviations to the left or to the right of the zero mean will the random variable Z be with probability 0.9772 or greater? Mathematically,

$$P(Z \geq K_{0.9772}) \geq 0.9772.$$

From Table B–I, find $K_{0.9772} = -2.000$ because the area under the curve from $-\infty$ to -2.000 standard deviations (to the left of the zero mean) is 0.0228, therefore, the area from -2.000 standard deviations to $+\infty$ is $1.000 - 0.0228 = 0.9772$.

Thus, a necessary and sufficient condition for

$$P(2x_1 + 3x_2 \leq b_1) = P\left(\frac{2x_1 + 3x_2 - 8}{2} \leq Z\right) \geq 0.9772$$

is that

$$\frac{2x_1 + 3x_2 - 8}{2} \leq -2.000$$

or

$$2x_1 + 3x_2 \leq 8 - 2(2) = 4.$$

This last expression corresponds to inequality (12.39). The model is, then,

$$\text{Max } E(z) = 3x_1 + 2x_2$$

subject to

$$x_i \geq 0 \qquad \text{for } i = 1, 2,$$

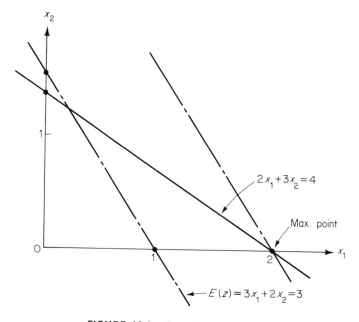

FIGURE 12.3. Graphical Solution.

and

$$2x_1 + 3x_2 \leq 4.$$

This is a two-dimensional problem. Its graphical solution is given in Figure 12.3.

The maximum point is seen to be $x_1^* = 2$, $x_2^* = 0$ which yields $E(z)^* = 6$. This solution is guaranteed not to violate the stochastic constraint with probability 0.9772 or larger.

EXERCISES

12-1. Solve the housing problem of section 12.2.2 by eliminating Alexandria from consideration because, according to Table 12-1, this city yields the lowest unit profits.

12-2. Restructure the model of the housing problem presented in section 12.2.2 with the following modifications:

a. Funds available have increased to permit construction of a maximum of 30 two-bedroom, 50 three-bedroom, and 40 four-bedroom units.

b. The market has expanded 20% such that the mean for Alexandria is 48 units; for Baton Rouge, 48 units; and for New Orleans, 72 units.

c. The market demand distributions remain normal but the standard deviations (risks) have all increased 10%.

d. All other elements in the problem remain the same.

12-3. The management of an industrial plant producing three types of heavy construction equipment is in the process of developing an optimal production policy. The monthly demands for each type of equipment are summarized below:

Equipment Type	Demand Distribution	Mean	Standard Deviation
1	Normal	150 units	15 units
2	Normal	200 units	25 units
3	Normal	180 units	20 units

The production belt capacities restrict the number of type 1 units to a maximum of 300 per month, type 2 units to a maximum of 400 per month, and type 3, to no more than 350 per month.

The unit profits from the sale of each type of equipment are $2,000 for type 1, $3,000 for type 2, and $2,500 for each type 3 unit sold. Unsold units constitute a loss to the company because of inventory and overhead costs. The estimated monthly costs incurred

from overstocking are \$400 for each type 1 unit not sold, \$500 for each type 2 unit not sold, and \$350 for each unsold type 3 unit.
a. Develop the model to maximize the expected monthly profit for the company.
b. Solve using the simplex method.

12-4. Solve the problem:

$$\text{Max } E(z) = E(c_1)x_1 + E(c_2)x_2$$

subject to

$$x_i \geq 0 \quad \text{for } i = 1, 2,$$
$$P(3x_1 + 2x_2 \leq b_1) \geq 0.9772,$$

and

$$P(x_1 + 4x_2 \leq b_2) \geq 0.9938,$$

where c_1, c_2, b_1, and b_2 are normally distributed random variables with means and standard deviations as follows:

$$c_1 \longrightarrow \text{Normal } (3, 1),$$
$$c_2 \longrightarrow \text{Normal } (2, 1),$$
$$b_1 \longrightarrow \text{Normal } (7, 3),$$
$$b_2 \longrightarrow \text{Normal } (8, 1).$$

12-5. Solve the problem,

$$\text{Max } E(z) = E(c_1)x_1 + E(c_2)x_2 + E(c_3)x_3$$

Subject to

$$x_i \geq 0 \quad \text{for } i = 1, 2, 3,$$
$$P(2x_1 + x_2 + 3x_3 \leq b_1) \geq 0.9713,$$
$$P(x_1 + 4x_2 + 2x_3 \leq b_2) \geq 0.9974,$$

where c_1, c_2, c_3, b_1, and b_2 are normally distributed random variables with the following means and standard deviations:

$$c_1 \longrightarrow \text{Normal } (4, 1),$$
$$c_2 \longrightarrow \text{Normal } (3, 1),$$
$$c_3 \longrightarrow \text{Normal } (5, 2),$$
$$b_1 \longrightarrow \text{Normal } (9, 3),$$
$$b_2 \longrightarrow \text{Normal } (7, 2).$$

12-6. In the industrial park development problem of section 6.2 (chapter 6), assume that the available labor force is normally distributed with means equal to the values shown under the heading "Max. Number," in Table 6-4, and with standard deviations equal to 10% of the means for each labor type.

Develop the optimization model under the assumption that the available labor force for each type must not be exceeded more than 5% of the time, that is, the labor requirements for each class of laborer must be less than or equal to the available number, with probability greater than or equal to 0.95.

REFERENCES

1. AGUILAR, R. J., "Architectural Optimization Under Conditions of Uncertainty; Stochastic Models," *Proceedings, Computer Applications to Environmental Design*, School of Architecture, University of Kentucky. Lexington, April 1970.

2. CHARNES, A. and W. W. COOPER, "Chance-Constrained Programming," *Management Science*, VI (1959), 73–80.

3. DANTZIG, G. B., "Linear Programming Under Uncertainty," *Management Science*, I (1955), 197–206.

CHAPTER 13

Stochastic
Dynamic Programming

13.1 INTRODUCTION

In chapter 11 it was stated that dynamic programming can be applied to the optimization of deterministic and stochastic, continuous and discontinuous, linear and nonlinear systems possessing a serial structure. All the topics discussed in chapter 11 apply equally well to the contents of this chapter; however, the theory will now be expanded to include the consideration of random parameters in the return and transition functions. The student should be thoroughly familiar with the contents of chapter 11 before proceeding with the study of this chapter. Consider a stochastic return function

$$r_i = R_i(s_i, d_i, k_i) \tag{13.1}$$

where s_i is the input state variable to stage i, d_i is the decision variable, and k_i is a random variable. The probability distribution of k_i is given by $p_i(k_i)$.

For a fixed value of s_i and d_i, one would expect to receive an average (expected) return of

$$\bar{r}_i(s_i, d_i) = \sum_k p_i(k_i) r_i(s_i, d_i, k_i) \tag{13.2}$$

if k_i is discrete, and of

$$\bar{r}_i(s_i, d_i) = \int p_i(k_i) r_i(s_i, d_i, k_i) dk_i \qquad (13.3)$$

if k_i is a continuous random variable.

13.2 OPTIMALITY CRITERION

For a fixed input state s_i, the decision policy d_i^* will be considered globally optimal (maximal) if and only if

$$\bar{r}_i(s_i, d_i^*) \geq \bar{r}_i(s_i, d_i) \qquad (13.4)$$

for all feasible values of the decision variable d_i.

This criterion, in fact, states that the decision policy is maximal when the *expected value* of the return function is a maximum. The maximization of expected value under risk may appear to disregard the importance of the variation of the random variable as measured by its variance or standard deviation. Consider the following situation: For a fixed value of s_i, the probable outcomes of two possible decisions are,

	$d_i^{(1)}$		$d_i^{(2)}$	
$p_i(k_i)$	$\frac{29}{30}$	$\frac{1}{30}$	$\frac{1}{3}$	$\frac{2}{3}$
$r_i(s_i, d_i, k_i)$	0	100	1	4

Consequently,

$$\bar{r}_i(s_i, d_i^{(1)}) = \tfrac{29}{30}(0) + \tfrac{1}{30}(100) = 3\tfrac{1}{3},$$
$$\bar{r}_i(s_i, d_i^{(2)}) = \tfrac{1}{3}(1) + \tfrac{2}{3}(4) = 3.$$

Under the maximum expected value criterion, $d_i^{(1)}$ (the first decision) would be selected. This does not consider the much higher risk involved in the first decision where the return is zero twenty-nine out of thirty times. With the second decision, however, one is certain to obtain at least one, $33\tfrac{1}{3}\%$ of the time, and four, $66\tfrac{2}{3}\%$ of the time. Expected value theory does *not*, in itself, provide a measure of the risk involved. Would it be preferable to accept a slightly lesser expected return from $d_i^{(2)}$ (the second decision) to avoid the much larger risk of having no return from $d_i^{(1)}$? If this were true, it would merely indicate that proper weights have not been assigned to the returns. In evaluating returns, it is necessary to consider all the factors involved. Thus, to justify choosing $d_i^{(2)}$ over $d_i^{(1)}$ one could argue that an outcome that produces a zero return would endanger the economic survival of the system and, consequently, one would assign a large negative value to this otherwise zero return. In general, all objections to the expected value criterion can be overcome by properly measuring the value of the return. When the returns have been measured correctly, expected value is a valid criterion for comparing decision policies under conditions of risk. This is the criterion adopted in this chapter for the discussion of all topics in stochastic dynamic programming.

13.3 MULTISTAGE DECISION PROCESSES UNDER RISK

Consider the serial system shown in Figure 13.1. There is only one basic difference between this system and the ones studied in chapter 11; a stochastic variable k_i is an input to each stage i in the chain. This stochastic variable introduces risk into the decision-making process. The returns are given by equation (13.1) as

$$r_i = R_i(s_i, d_i, k_i) \qquad \text{for } i = 1, \ldots, N,$$

and the transition functions can be similarly expressed by the following equation relating the output state variable to the input state variable, the decision variable, and the random parameter,

$$\tilde{s}_i = T_i(s_i, d_i, k_i) \qquad \text{for } i = 1, \ldots, N. \qquad (13.5)$$

It is assumed that the random variables k_i, $i = 1, \ldots, N$, are independently distributed with respective probability distributions $p_i(k_i)$, $i = 1, \ldots, N$.

As in previous discussions of dynamic programming, the total system return is assumed to be the sum of the stage returns. Therefore,

$$R(s_N, \ldots, s_1, d_N, \ldots, d_1, k_N, \ldots, k_1) = \sum_{i=1}^{N} r_i(s_i, d_i, k_i). \qquad (13.6)$$

The incidence identities are given, as usual, by

$$\tilde{s}_i = s_{i-1}, \qquad i = 2, \ldots, N. \qquad (13.7)$$

Because the input state variable to stage i is a function of all upstream stage

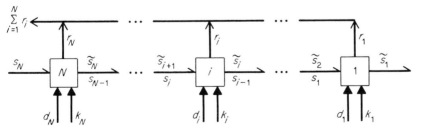

FIGURE 13.1. Serial System.

variables, the return at stage i depends not only on the random variable k_i but also on the random variables $k_{i+1}, k_{i+2}, \ldots, k_N$ affecting the upstream stages. Therefore, for the stochastic serial system, the input state variables depend on the previously observed random variables as well as on the upstream decision variables and the initial input state s_N. For this reason, even when a decision policy has been formulated, the input to stage i, $i \neq N$, is not known before specific values of the random variables, $k_{i+1}, k_{i+2}, \ldots, k_N$, affecting the upstream stages, have been realized.

In order to compute the expected value of R, one must know how to compute the expected value of a function of several random variables.

Consider the function $G(k_1, k_2, \ldots, k_N)$ of the random variables $k_1, k_2, \ldots, k_N$, with probability distributions $p_1(k_1), p_2(k_2), \ldots, p_N(k_N)$. If the k_i are discrete, the expected value of G is computed as follows:

$$E(G) = \bar{G} = \sum_{k_1} \cdots \sum_{k_N} \prod_{i=1}^{N} p_i(k_i) G(k_1, k_2, \ldots, k_N). \tag{13.8}$$

If the random variables are continuous, one has,

$$E(G) = \bar{G} = \int_{k_1} \cdots \int_{k_N} \prod_{i=1}^{N} p_i(k_i) G(k_1, k_2, \ldots, k_N) dk_1, \ldots, dk_N. \tag{13.9}$$

In what follows, the random variables are assumed to be discrete without loss of generality.

The expected value of the total system return R, equation (13.6), can thus be computed as follows:

$$\bar{R}(s_N, d_N, \ldots, d_1) = \sum_{k} \cdots \sum_{k_N} \left[\prod_{i=1}^{N} p_i(k_i) \sum_{i=1}^{N} r_i(s_i, d_i, k_i) \right]$$

$$= \sum_{k_1} \cdots \sum_{k_N} \left[\prod_{i=1}^{N} p_i(k_i) r_N(s_N, d_N, k_N) \right] + \cdots$$

$$+ \sum_{k_1} \cdots \sum_{k_N} \left[\prod_{i=1}^{N} p_i(k_i) r_1(s_1, d_1, k_1) \right]. \tag{13.10}$$

However, the ith stage return is independent of the random variables $k_{i-1}, \ldots, k_1$, occurring downstream.

Therefore,

$$\bar{R}(s_N, d_N, \ldots, d_1)$$

$$= \sum_{k_N} [p_N(k_N) r_N(s_N, d_N, k_N) \sum_{k_{N-1}} \cdots \sum_{k_1} \prod_{i=1}^{N-1} p_i(k_i)]$$

$$+ \sum_{k_N} \{ p_N(k_N) \sum_{k_{N-1}} [p_{N-1}(k_{N-1}) r_{N-1}(s_{N-1}, d_{N-1}, k_{N-1}) \sum_{k_{N-2}} \cdots \sum_{k_1} \prod_{i=1}^{N-2} p_i(k_i)] \}$$

$$+ \cdots + \sum_{k_N} (p_N(k_N) \sum_{k_{N-1}} \{ p_{N-1}(k_{N-1}) \cdots \sum_{k_1} [p_1(k_1) r_1(s_1, d_1, k_1)] \cdots \}). \tag{13.11}$$

Consequently,

$$\bar{R}(s_N, d_N, \ldots, d_1) = \sum_{k_N} (p_N(k_N)r_N(s_N, d_N, k_N) \sum_{k_{N-1}} \{p_{N-1}(k_{N-1}) \cdots$$
$$\sum_{k_2} [p_2(k_2) \sum_k p_1(k_1)] \cdots \})$$
$$+ \sum_{k_N} [p_N(k_N) \sum_{k_{N-1}} (p_{N-1}(k_{N-1})r_{N-1}(s_{N-1}, d_{N-1}, k_{N-1})$$
$$\sum_{k_{N-2}} \{p_{N-2}(k_{N-2}) \cdots \sum_{k_2} [p_2(k_2) \sum_{k_1} p_1(k_1)] \cdots \})] + \cdots$$
$$+ \sum_{k_N} \{p_N(k_N) \sum_{k_{N-1}} [p_{N-1}(k_{N-1}) \sum_{k_{N-2}} (p_{N-2}(k_{N-2})$$
$$\cdots \sum_{k_2} \{p_2(k_2) \sum_k [p_1(k_1)r_1(s_1, d_1, k_1)]\} \cdots)]\}. \quad (13.12)$$

However,

$$\sum_{k_i} p_i(k_i) = 1 \quad \text{for } i = 1, \ldots, N.$$

Therefore,

$$\bar{R}(s_N, d_N, \ldots, d_1) = \sum_{k_N} [p_N(k_N)r_N(s_N, d_N, k_N)]$$
$$+ \sum_{k_N} \{p_N(k_N) \sum_{k_{N-1}} [p_{N-1}(k_{N-1})r_{N-1}(s_{N-1}, d_{N-1}, k_{N-1})]\} \cdots$$
$$+ \sum_{k_N} \{p_N(k_N) \sum_{k_{N-1}} [p_{N-1}(k_{N-1}) \cdots$$
$$\cdots \sum_{k_2} (p_2(k_2) \sum_{k_1} [p_1(k_1)r_1(s_1, d_1, k_1)])]\}. \quad (13.13)$$

subject to the transition functions

$$\tilde{s}_i = T_i(s_i, d_i, k_i), \quad i = 1, \ldots, N,$$

and the incidence identities

$$\tilde{s}_i = s_{i-1}, \quad i = 2, \ldots, N.$$

Without loss of generality, assume that the total return function is to be maximized. Let $\bar{f}_N(s_N)$ be the maximum expected return as a function of the initial input state s_N. This notation parallels the one used in the development of chapter 11. Hence,

$$\bar{f}_N(s_N) = \underset{d_N, d_{N-1}, \ldots, d_1}{\text{Max}} \bar{R}(s_N, d_N, \ldots, d_1). \quad (13.14)$$

Substitution of equation (13.13) into (13.14) yields

$$\bar{f}_N(s_N) = \underset{d_N, d_{N-1}, \ldots, d_1}{\text{Max}} [\sum_{k_N} [p_N(k_N)r_N(s_N, d_N, k_N)]$$
$$+ \sum_{k_N} \{p_N(k_N) \sum_{k_{N-1}} [p_{N-1}(k_{N-1})r_{N-1}(s_{N-1}, d_{N-1}, k_{N-1})]\}$$
$$+ \cdots + \sum_{k_N} (p_N(k_N) \sum_{k_{N-1}} \{p_{N-1}(k_{N-1})$$
$$\cdots \sum_{k_1} [p_1(k_1)r_1(s_1, d_1, k_1)] \cdots \})]. \quad (13.15)$$

The sum $\sum_{k_N} p_N(k_N)$ is common to every term. Factoring it out, obtain

$$
\begin{aligned}
\bar{f}_N(s_N) = \ &\underset{d_N, d_{N-1}, \ldots, d_1}{\text{Max}} \sum_{k_N} p_N(k_N)[r_N(s_N, d_N, k_N) \\
&+ \sum_{k_{N-1}} [p_{N-1}(k_{N-1})r_{N-1}(s_{N-1}, d_{N-1}, k_{N-1})] + \cdots \\
&+ \sum_{k_{N-1}} (p_{N-1}(k_{N-1}) \sum_{k_{N-2}} \{p_{N-2}(k_{N-2}) \\
&\cdots \sum_{k_1} [p_1(k_1)r_1(s_1, d_1, k_1)] \cdots \})],
\end{aligned}
\tag{13.16}
$$

subject to

$$
\bar{s}_i = T_i(s_i, d_i, k_i), \qquad i = 1, \ldots, N
$$

and

$$
\bar{s}_i = s_{i-1}, \qquad i = 2, \ldots, N.
$$

Following the same line of thought used in the derivation of the deterministic recursive equations in chapter 11, one can write,

$$
\underset{d_N, d_{N-}, \ldots, d_1}{\text{Max}} [\bar{R}(s_N, d_N, \ldots, d_1)] = \underset{d_N}{\text{Max}} \{ \underset{d_{N-1}, \ldots, d_1}{\text{Max}} [\bar{R}(s_N, d_N, \ldots, d_1)] \}
\tag{13.17}
$$

and thus remove the Nth stage return from the inner maximization process, because this return is not a function of the downstream decision variables $d_{N-1}, \ldots, d_1$. Obtain,

$$
\begin{aligned}
\bar{f}_N(s_N) = \ &\underset{d_N}{\text{Max}} \{ \sum_{k_N} p_N(k_N)\{r_N(s_N, d_N, k_N) \\
&+ \underset{d_{N-1}, \ldots, d_1}{\text{Max}} \{ \sum [p_{N-1}(k_{N-1})r_{N-1}(s_{N-1}, d_{N-1}, k_{N-1})] \\
&+ \sum_{k_{N-1}} \{p_{N-1}(k_{N-1}) \sum_{k_{N-2}} [p_{N-2}(k_{N-2})r_{N-2}(s_{N-2}, d_{N-2}, k_{N-2})]\} \\
&+ \cdots + \sum_{k_{N-1}} [p_{N-1}(k_{N-1}) \sum_{k_{N-2}} [p_{N-2}(k_{N-2}) \\
&\cdots \sum_{k_1} [p_1(k_1)r_1(s_1, d_1, k_1)] \cdots]]\}\}\}.
\end{aligned}
\tag{13.18}
$$

But

$$
\bar{f}_{N-1}(s_{N-1}) = \bar{f}_{N-1}[T_N(s_N, d_N, k_N)].
\tag{13.19}
$$

Therefore,

$$
\begin{aligned}
\bar{f}_{N-1}(s_{N-1}) = \ &\underset{d_{N-1}, \ldots, d_1}{\text{Max}} (\sum_{k_{N-1}} [p_{N-1}(k_{N-1})r_{N-1}(s_{N-1}, d_{N-1}, k_{N-1})] + \\
&\cdots + \sum_{k_{N-1}} \{p_{N-1}(k_{N-1}) \sum_{k_{N-2}} [p_{N-2}(k_{N-2}) \\
&\cdots \sum_{k_1} [p_1(k_1)r_1(s_1, d_1, k_1)] \cdots]\}).
\end{aligned}
\tag{13.20}
$$

Substitution of equation (13.20) into equation (13.18) results in

$$
\bar{f}_N(s_N) = \underset{d_N}{\text{Max}} \sum_{k_N} p_N(k_N)[r_N(s_N, d_N, k_N) + \bar{f}_{N-1}(s_{N-1})].
\tag{13.21}
$$

However, because

$$
s_{N-1} = \bar{s}_N
\tag{13.22}
$$

and

$$\tilde{s}_N = T_N(s_N, d_N, k_N), \tag{13.23}$$

one has that

$$\tilde{f}_N(s_N) = \operatorname*{Max}_{d_N} \sum_{k_N} p_N(k_N)\{r_N(s_N, d_N, k_N) + \tilde{f}_{N-1}[T_N(s_N, d_N, k_N)]\}. \tag{13.24}$$

The expression $r_N(s_N, d_N, k_N) + \tilde{f}_{N-1}[T_N(s_N, d_N, k_N)]$ corresponds to the maximand expression defined in chapter 11 for the deterministic case. Following the same procedure, let the stochastic maximand be defined as

$$M_N(s_N, d_N, k_N) = r_N(s_N, d_N, k_N) + \tilde{f}_{N-1}[T_N(s_N, d_N, k_N)]. \tag{13.25}$$

Then, equation (13.24) becomes

$$\tilde{f}_N(s_N) = \operatorname*{Max}_{d_N} \sum_{k_N} p_N(k_N)M_N(s_N, d_N, k_N). \tag{13.26}$$

By induction, as for the deterministic case, one has that for any stage i, $1 \le i \le N$, the fundamental stochastic recursion equations are

$$\tilde{f}_i(s_i) = \operatorname*{Max}_{d_i} \sum_{k_i} p_i(k_i)M_i(s_i, d_i, k_i) \tag{13.27}$$

where

$$M_i(s_i, d_i, k_i) = r_i(s_i, d_i, k_i) + \tilde{f}_{i-1}[T_i(s_i, d_i, k_i)] \qquad \text{for } i = 2, \ldots, N, \tag{13.28}$$

and

$$M_1(s_1, d_1, k_1) = r_1(s_1, d_1, k_1). \tag{13.29}$$

It must be noted that risk causes no increase in state variables and that because the maximand is a function of only one random variable, only one random parameter at a time is introduced into the optimization procedure, thus overcoming the bulk of the formidable difficulties normally encountered when one attempts to optimize functions of several random variables.

13.4 MULTIPLICATION OF STATE RETURNS

When the total system return is defined to be the product of the individual stage returns, the total return function is expressed, thus,

$$R(s_N, \ldots, s_1, d_N, \ldots, d_1, k_N, \ldots, k_1) = \prod_{i=1}^{N} r_i(s_i, d_i, k_i), \tag{13.30}$$

subject to

$$\tilde{s}_i = T_i(s_i, d_i, k_i), \qquad i = 1, \ldots, N$$

and

$$\tilde{s}_i = s_{i-1}, \qquad i = 2, \ldots, N.$$

Following the same line of thought as in the previous section, one concludes that for a typical stage i,

$$f_i = \underset{d_i}{\text{Max}} \sum_{k_i} p_i(k_i) M_i(s_i, d_i, k_i) \qquad \text{for } 1 \leq i \leq N, \qquad (13.31)$$

where the maximand is now defined as a product.

$$M_i(s_i, d_i, k_i) = r_i(s_i, d_i, k_i) \tilde{f}_{i-1}[T_i(s_i, d_i, k_i)] \qquad \text{for } 2 \leq i \leq N, \qquad (13.32)$$

and

$$M_1(s_1, d_1, k_1) = r_1(s_1, d_1, k_1). \qquad (13.33)$$

As for the deterministic case, return composition by multiplication requires non-negativity of the individual return functions for maximization; thus one needs

$$r_i(s_i, d_i, k_i) \geq 0 \qquad \text{for all } s_i, d_i, k_i, \qquad \text{for } i = 1, \ldots, N, \qquad (13.34)$$

for the recursive expression to yield a maximum value.

13.5 THE STOCHASTIC CHARACTER OF OPTIMAL DECISION POLICIES

Stochastic dynamic programming yields an optimal decision policy which is itself stochastic. This is not to be construed as a deficiency of dynamic programming; rather, it is an intrinsic property of the stochastic multistage decision system itself.

To clarify this statement, consider the optimal (maximal) decision at the initial upstream stage N. This decision is

$$d_N^*(s_N). \qquad (13.35)$$

Substitution of $d_N^*(s_N)$ into the transition function at stage N yields,

$$\tilde{s}_N^* = T_N[s_N, d_N^*(s_N), k_N]. \qquad (13.36)$$

Consequently, the optimal value of the output state variable $\tilde{s}_N^*$ is known only probabilistically because of the presence of the random variable k_N in expression (13.36).

From the incidence identity linking stage N to stage $N - 1$,

$$\tilde{s}_N^* = s_{N-1}, \qquad (13.37)$$

one finds that the decision set at stage $N - 1$ depends also upon k_N, because,

$$d_{N-1}(s_{N-1}) = d_{N-1}(\tilde{s}_N^*) = d_{N-1}[T_N(s_N, d_N^*(s_N), k_N]. \qquad (13.38)$$

Thus, an N-stage, stochastic multistage optimization process yields incomplete results; only the first decision is obtained from the solution procedure.

The remaining optimal decisions for the downstream stages are determined, one by one, as the stochastic process unfolds. This is, in fact, a basic concept in the behavior of adaptive systems and for this reason dynamic programming points the way into the study of adaptive optimization processes. A final remark must be made regarding the assumption of statistical independence among the random variables. As the degree of dependence among the random variables increases, the number of state variables, required to account for that lack of independence, increases, making the recursion equations progressively more unwieldy. In the most general and worst case where k_i is dependent on all the upstream random variables $k_N, k_{N-1}, \ldots, k_{i+1}$, the ith recursion equation has $N - (i - 1)$ state variables, thus,

$$\tilde{f}_i(s_i, k_N, k_{N-1}, \ldots, k_i) = \underset{d_i}{\text{Max}} \sum_{k_i} p_i(k_N, k_{N-1}, \ldots, k_i) M_i(s_i, d_i, k_i).$$

$$(13.39)$$

This type of formulation also hints of adaptive optimization processes, which are repeating stochastic systems in which the information transmitted by the preceding stages generates the probability distribution for the next, downstream, stochastic stage.

13.6 AN EXAMPLE PROBLEM

13.6.1 Problem Statement

To illustrate the theory presented, consider the following investment situation:

A businessman with a capital of $150,000 is analyzing three investment packages. He can invest in any, all, or none of the packages in incremental amounts of $50,000. Tables 13.1, 13.2 and 13.3 show the net returns on investments for each package, and the probability that each return will be realized. Note that many of the returns are negative, indicating a net loss to the investor.

TABLE 13–1. Investment Package No. 1.

Investment	$50,000			$100,000			$150,000		
Net Return	$ 80,000	$ 30,000	0	$ 50,000	$ 20,000	$ 10,000	$100,000	$ 80,000	$ 50,000
Probability of Return	0.15	0.50	0.35	0.20	0.60	0.20	0.10	0.50	0.40

Determine the maximum expected net return for the optimal investment plan.

TABLE 13–2. Investment Package No. 2.

Investment	$50,000			$100,000			$150,000		
Net Return	$ 30,000	$ 10,000	–$ 30,000	$200,000	$ 60,000	–$ 80,000	$200,000	$ 70,000	–$ 50,000
Probability of Return	0.20	0.60	0.20	0.25	0.60	0.15	0.10	0.70	0.20

TABLE 13–3. Investment Package No. 3.

Investment	$50,000			$100,000			$150,000		
Net Return	$ 80,000	$ 20,000	0	$120,000	$ 40,000	–$ 50,000	$170,000	$120,000	–$ 70,000
Probability of Return	0.10	0.70	0.20	0.05	0.75	0.20	0.20	0.50	0.30

13.6.2 Solution

The dynamic programming model is simply a three-stage decision system with a fixed initial input state of $150,000, the total capital available for investment purposes.

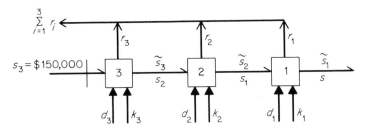

For each stage, the transition function is given by

$$\bar{s}_i = s_i - d_i.$$

Because this is an initial value problem, start the stage-wise maximization at stage 1.

STAGE 1:

For this stage, from equations (13.27) and (13.29),

$$\tilde{f}_1(s_1) = \underset{d_1}{\text{Max}} \sum_{k_1} p_1(k_1) M_1(s_1, d_1, k_1)$$

where

$$M_1(s_1, d_1, k_1) = r_1(s_1, d_1, k_1).$$

The suboptimization table for this stage follows:

Suboptimization Table for Stage 1

s_1	d_1	$\sum p_1 M_1$
$150,000	0	0
$150,000	$ 50,000	0.15(80,000) + 0.50(30,000) + 0.35(0) = $27,000
$150,000	$100,000	0.20(50,000) + 0.60(20,000) + 0.20(−10,000) = $20,000
$150,000	$150,000	0.10(100,000) + 0.50(80,000) + 0.40(−50,000) = $30,000 ←
$100,000	0	0
$100,000	$ 50,000	$27,000 ←
$100,000	$100,000	$20,000
$ 50,000	0	0
$ 50,000	$ 50,000	$27,000 ←
0	0	0 ←

Note: The arrows indicate the maximum expected net returns for each value of the input state variable s_1.

A summary table for this stage is prepared as discussed in chapter 11.

Summary Table for Stage 1

s_1	d_1^*	$\tilde{f}_1$	$\bar{s}_1$
$150,000	$150,000	$30,000	0
$100,000	$ 50,000	$27,000	$50,000
$ 50,000	$ 50,000	$27,000	0
0	0	0	0

STAGES 2 AND 1:

From equations (13.27) and (13.28), obtain

$$\tilde{f}_2(s_2) = \underset{d_2}{\text{Max}} \sum_{k_2} p_2(k_2) M_2(s_2, d_2, k_2)$$

where

$$M_2(s_2, d_2, k_2) = r_2(s_2, d_2, k_2) + \tilde{f}_1[T_2(s_2, d_2, k_2)].$$

The suboptimization table for stages 2 and 1 follows:

Suboptimization Table for Stages 2 and 1

s_2	d_2	$\bar{s}_2 = s_1$	$\bar{f}_1$	$\sum p_2[r_2 + \bar{f}_1]$
$150,000	0	$15,0000	$30,000	$30,000
$150,000	$ 50,000	$100,000	$27,000	0.20(30,000 + 27,000) + 0.60(10,000 + 27,000) + 0.20(−30,000 + 27,000) = $33,000
$150,000	$100,000	$ 50,000	$27,000	0.25(200,000) + 0.60(60,000) + 0.15(−80,000) + 27,000 = $101,000 ←
$150,000	$150,000	0	0	0.10(200,000) + 0.70(70,000) + 0.20(−50,000) = $59,000
$100,000	0	$100,000	$27,000	$27,000
$100,000	$ 50,000	$ 50,000	$27,000	$33,000
$100,000	$100,000	0	0	$74,000 ←
$ 50,000	0	$ 50,000	$27,000	$27,000 ←
$ 50,000	$ 50,000	0	0	$ 6,000
0	0	0	0	0 ←

The summary table for stages 2 and 1 is as follows:

Summary Table for Stage 2 and 1

s_2	d_2^*	$\bar{f}_2$	$\bar{s}_2 = s_1$
$150,000	$100,000	$101,000	$50,000
$100,000	$100,000	$ 74,000	0
$ 50,000	$ 50,000	$ 27,000	0
0	0	0	0

STAGES 3, 2, AND 1:

The recursive equation for stages 3, 2, and 1 is

$$\bar{f}_3(s_3) = \bar{f}_3(\$150,000) = \underset{d_3}{\text{Max}} \sum_{k_3} p_3(k_3)M_3(s_3, d_3, k_3),$$

where

$$M_3(s_3, d_3, k_3) = r_3(s_3, d_3, k_3) + \bar{f}_2[T_3(s_3, d_3, k_3)].$$

The suboptimization table for stages 3, 2, and 1 follows:

Suboptimization Table for Stages 3, 2, and 1

s_3	d_3	$\bar{s}_3 = s_2$	$\bar{f}_2$	$\sum p_3[r_3 + \bar{f}_2]$
$150,000	0	$150,000	$101,000	$101,000 ←
$150,000	$ 50,000	$100,000	$ 74,000	0.10(80,000) + 0.70(20,000) + 0.20(0) + 74,000 = $96,000
$150,000	$100,000	$ 50,000	$ 27,000	0.05(120,000) + 0.75(40,000) + 0.20(−50,000) + 27,000 = $53,000
$150,000	$150,000	0	0	0.20(170,000) + 0.50(120,000) + 0.30(−70,000) = $73,000

Summary Table for Stages 3, 2, and 1

s_3	d_3^*	$\bar{f}_3$	$\bar{s}_3 = s_2$
$150,000	0	$101,000	$150,000

Therefore, the total expected net return is $101,000 which represents a 67% profit on the investment of $150,000.

It must be reemphasized that the true return on investment will be known only after the outcome of the events whose probabilities are given is itself known. This element of risk is characteristic of stochastic systems.

EXERCISES

13-1. What would be the maximum expected net return in the example problem of section 13.6 if the net returns from Investment Package No. 2 *increased* 20% each? (In the case of a negative return—say, −$30,000—the new value would be −$30,000 + $6,000 = −$24,000.)

13-2. Recompute the maximum expected net return in the example problem of section 13.6 by assuming that the net returns from Investment Package No. 1 increased 50%. (For a negative return of −$10,000, the new value would be −$10,000 + $5,000 = −$5,000.)

13-3. A piece of equipment has a four-year life, at the end of which it must be replaced. The equipment can be replaced at the end of any year, however, at a cost of $150,000. If it is *not* replaced and it breaks down during the following year it causes the company a loss of $100,000 in revenues. The probability of a breakdown increases with equipment age, as follows:

New equipment, $p = 0.00$,
1-year-old equipment, $p = 0.20$,
2-year-old equipment, $p = 0.50$,
3-year-old equipment, $p = 0.80$.

When a breakdown occurs the equipment must be replaced. The cost of replacement must be added to the company's loss of revenues. Compute the minimum expected cost of a typical four-year life cycle.

13-4. Rework Problem 13-3 with the following new set of breakdown probabilities:

New equipment, $p = 0.05$,
1-year-old equipment, $p = 0.15$,
2-year-old equipment, $p = 0.30$,
3-year-old equipment, $p = 0.50$.

REFERENCES

1. HOWARD, R. A., *Dynamic Programming and Markov Processes.* New York: John Wiley & Sons, Inc., 1960.
2. NEMHOUSER, G. L., *Introduction to Dynamic Programming.* New York: John Wiley & Sons, Inc., 1966.

CHAPTER 14

Introduction
to the
Theory of Games

14.1 GROUP DECISON MAKING

The involvement of more than one decision maker in policy formulation and development results in group decision making. When all of the decision makers contribute to the solution of the same problem or set of problems without conflicts of interest, group decision making can be an extremely creative and successful method for generating alternative solution procedures. This is what frequently occurs during "brainstorming" sessions when ideas are tossed about, discussed, and selected according to preestablished criteria.

When the decision makers work against each other under conditions of competition, a game situation develops. The quantitative study of decision making under uncertainty is commonly referred to as game theory.

14.2 GAME THEORY DEFINITIONS

In game theory the competitors are called *players*. A *strategy* is a plan listing all the possible courses of action that the players can take independently. When a player chooses an independent course of action, a *play* results. The outcome of the play determines the *payoff*. This is simply the consequence of the particular strategy independently selected by the player.

14.3 TWO-PERSON, ZERO-SUM GAMES

The simplest games to analyze and the only ones treated in this book are two-person, zero-sum games; two players of equal intelligence are assumed to be engaged in the game, which is said to be zero sum because the gains of one player exactly equal the losses of the other. Thereby, the net loss or gain in every play is zero.

If a two-person, zero-sum game is played under conditions of certainty, each course of action leads to a specific outcome. This type of strategy is called a *pure strategy*.

A *mixed strategy* results when the game is played under conditions of uncertainty. In this case each action leads to a set of possible outcomes and yields a probability distribution of the outcomes. The purpose of game theory is the establishment of criteria for selecting optimal strategies to play each game. A game has a *solution* if a set of optimal strategies exists for both players such that the payoff expected by each is invariably realized when the strategies are activated. The payoff resulting from playing the optimal strategies is called the *value* of the game.

Although a two-person game may appear too restrictive, it must be pointed out that a player may think of the set of all opponents as only one contender. In bidding strategy for construction, for example, the set of all other contractors can be thought of as the only other player (the opponent). Furthermore, bidding is a zero-sum, winner-takes-all type of game; what one wins, the opposition loses (all other contractors involved in bidding the job), and vice-versa.

14.4 PAYOFF MATRICES

Suppose that two players, A and B, are engaged in a zero-sum game. Say that each player has n courses of action or plays and that each play leads to a payoff c_{ij} for A and $-c_{ij}$ for B. The results of the game can be depicted with the aid of two payoff matrices, as follows:

1. Payoff Matrix for A.

B Plays

A Plays

$$
\begin{array}{c|ccccc}
 & 1 & 2 & 3 & & n \\
\hline
1 & c_{11} & c_{12} & c_{13} & \cdots & c_{1n} \\
2 & c_{21} & c_{22} & c_{23} & \cdots & c_{2n} \\
3 & c_{31} & c_{32} & c_{33} & \cdots & c_{3n} \\
\cdot & \cdot & \cdot & \cdot & \cdots & \cdot \\
\cdot & \cdot & \cdot & \cdot & \cdots & \cdot \\
\cdot & \cdot & \cdot & \cdot & \cdots & \cdot \\
n & c_{n1} & c_{n2} & c_{n3} & \cdots & c_{nn}
\end{array}
$$

2. Payoff Matrix for B.

B Plays

A Plays

$$
\begin{array}{c|ccccc}
 & 1 & 2 & 3 & & n \\
\hline
1 & -c_{11} & -c_{12} & -c_{13} & \cdots & -c_{1n} \\
2 & -c_{21} & -c_{22} & -c_{23} & \cdots & -c_{2n} \\
3 & -c_{31} & -c_{32} & -c_{33} & \cdots & -c_{3n} \\
\cdot & \cdot & \cdot & \cdot & \cdots & \cdot \\
\cdot & \cdot & \cdot & \cdot & \cdots & \cdot \\
\cdot & \cdot & \cdot & \cdot & \cdots & \cdot \\
n & -c_{n1} & -c_{n2} & -c_{n3} & \cdots & -c_{nn}
\end{array}
$$

Since the game is zero-sum, each element in the payoff matrix for A is equal and opposite in sign to the corresponding element in the payoff matrix for B. Consequently, it is unnecessary to show both matrices in order to play the game. If only the payoff matrix for player A is shown, then for a given play of B, player A would select the row with the largest payoff and, for a given play of A, player B would select the column with the smallest payoff. That is, player A's goal is to maximize his own payoff, whereas player B's is to minimize the opposition's. By following this course of action on the payoff matrix for A, player B automatically maximizes his payoff on his own payoff matrix. In the following discussion, only the payoff matrix for player A (the one playing the rows) will be given, and the optimization strategy will be: The row player's goal is to maximize his payoff on this matrix; the column player's goal is to minimize the opposition's payoff (thereby maximizing his own).

14.5 PURE STRATEGIES

Consider the payoff matrix for the row player. If there is an element in this payoff matrix which is simultaneously the smallest element on its row and the largest on its column, the game is said to have a saddle point and a pure strategy exists, that is, both players will select this same payoff. For example, in the matrix given below, element $c_{11} = 2$ is saddle point.

B

		1	2	3	4
	1	②	5	7	8
A	2	−1	−3	3	9
	3	−3	4	8	−4
	4	−2	7	3	1

The strategy of the game will invariably lead to this payoff:

1. Because the goal of player B is to minimize the opposition's payoff, he will select the play (column) containing the smallest payoff (-4). Thus, B plays column 4.
2. On the other hand, A wishes to maximize his own payoff. Thus, given that B selected to play column 4, A plays row 2.
3. Now, in view of this move by A, B plays column 2 which contains the smallest element in row 2.
4. To which A responds by playing row 4.
5. B plays column 1.
6. A plays row 1.
7. At this point, the best strategy for B is to play column 1, and the game ends because c_{11} yields the best payoff for each player under this competitive situation.

Note that the strategy can be summarized as follows:

Player A finally selects the row with the maximum-minimum element; player B, the column with the minimum-maximum element. Thus, the strategy is said to be maxi-min for A and mini-max for B.

The value of the game is given by the saddle point element. For the payoff matrix of the example, the game value is therefore 2.

If must pointed out that a game can have more than one saddle point. For example, in the payoff matrix given below,

B

		1	2	3
	1	−3	5	7
A	2	①	3	2
	3	①	4	2

c_{12} and c_{13} are both saddle points. In either case the value of the game is 1.

14.5.1   Eliminating Rows and Columns

It is possible to simplify a game by eliminating rows and/or columns from the payoff matrix if dominant and recessive rows and/or columns exist.

1. Row i is said to be dominant and row j recessive if every element in row i is *larger* than the corresponding element in row j, that is, $a_{ik} > a_{jk}$, $k = 1, 2, \ldots, n$. The recessive row j can be deleted.
2. Column i is said to be dominant and column j recessive if every element of column i is smaller than the corresponding element in column j, that is, $a_{ki} < a_{kj}$, $k = 1, 2, \ldots, n$.

The recessive column j can be deleted. For example, consider the payoff matrix

$$B$$

		1	2	3
	1	3	5	7
A	2	−1	4	−6
	3	−3	2	8

Every element in row 1 is larger than the corresponding element in row 2. Row 2 is recessive and can be deleted. Every element in column 2 is larger than the corresponding element in column 1. Column 2 is recessive and can be deleted.

The payoff matrix reduces to

	1	3
1	③	7
3	−3	8

Furthermore, $c_{11} = 3$ is a saddle point. The game has a pure strategy and a game value of 3.

A final observation that can be of value as an aid in simplifying payoff matrices is that the addition of the same positive constant to every element of a payoff matrix does *not* alter the optimal strategies although the value of the game is increased by the constant.

14.6   GAMES WITH MIXED STRATEGIES

If a payoff matrix does not have a saddle point, a pure strategy is not available and the two-person, zero-sum game is said to possess a mixed strategy.

Consider the following payoff matrix as an example of a game with a mixed strategy.

B

		1	2	3
	1	3	5	7
A	2	-1	4	6
	3	7	-6	-5

No element in this matrix meets the criterion for saddle point; no element is simultaneously the smallest on its row and the largest on its column. If A chooses to play the strategy for the maxi-min element over the rows, A plays row 1 (the maximum of the minimum elements over the rows is 3). Upon seeing this move of A, B plays column 1. This makes A switch to row 3; whereby B plays column 2; A now plays row 1 again, and so on.

A linear programming model can be developed to determine the optimal mixed strategies for both players. First, one must rid the payoff matrix of all negative elements. For the example matrix, add 6 to every element. The optimal strategy has *not* been altered and the new value of the game, v, is now assured to be positive. The true value is, of course, $v - 6$.

The new payoff matrix is given below:

B

		1	2	3	
	1	9	11	13	x_1
A	2	5	10	12	x_2
	3	13	0	1	x_3
		y_1	y_2	y_3	

Let x_i, $i = 1, 2, 3$, be the probability that player A selects strategy i. Consequently,

$$\sum_{i=1}^{n} x_i = 1. \tag{14.1}$$

Similarly, let y_j, $j = 1, 2, 3$, be the probability that player B chooses strategy j. Hence,

$$\sum_{j=1}^{n} y_j = 1. \tag{14.2}$$

Player A wishes to maximize the value of the game, v, while player B wishes to minimize it. This is because the matrix gives the payoffs for A, and the game is zero-sum (what A wins, B loses and vice-versa).

The optimal mixed strategies for A and B are formulated in the following two sections.

14.6.1 Optimal Mixed Strategy for Player B

To establish the optimal mixed strategy for B, one must consider all possible moves of A. Player A wishes to maximize the value of the game, v. However, regardless of the strategy A chooses to play, his winnings cannot exceed v. That is, for every one of his plays, A's winning must be less than or, at most, equal to v. Therefore,

1. If A plays row 1, his expected payoff is

$$9y_1 + 11y_2 + 13y_3 \leq v. \tag{14.3}$$

This relationship simply states that when A plays row 1, he will win 9 units with probability y_1, plus 11 units with probability y_2, plus 13 units with probability y_3 and that this sum (the expected payoff) cannot exceed the value of the game. The y_j are, of course, the strategies for player B, that is, the probabilities that B will play columns $j = 1, 2, 3$.

2. If A plays row 2, his expected payoff is

$$5y_1 + 10y_2 + 12y_3 \leq v. \tag{14.4}$$

3. Similarly for row 3, the expected payoff turns out to be

$$13y_1 + 0y_2 + 1y_3 \leq v. \tag{14.5}$$

From equation (14.2) the following relationship also holds:

$$y_1 + y_2 + y_3 = 1. \tag{14.6}$$

Dividing expressions (14.3) through (14.6) by v and defining

$$\hat{y}_j = \frac{y_j}{v}, \qquad \text{for all } j, \tag{14.7}$$

obtain the set,

$$\left. \begin{aligned} 9\hat{y}_1 + 11\hat{y}_2 + 13\hat{y}_3 &\leq 1, \\ 5\hat{y}_1 + 10\hat{y}_2 + 12\hat{y}_3 &\leq 1, \\ 13\hat{y}_1 + 0\hat{y}_2 + 1\hat{y}_3 &\leq 1, \\ \\ \hat{y}_1 + \hat{y}_2 + \hat{y}_3 &= \frac{1}{v}. \end{aligned} \right\} \tag{14.8}$$

and

The objective of B, the player associated with the y_j probabilities, is to minimize v, or, equivalently, to maximize $1/v$. Consequently, one can formulate the following linear programming model for the optimal mixed strategy for player B.

$$\text{Max} \frac{1}{v} = \hat{y}_1 + \hat{y}_2 + \hat{y}_3, \tag{14.9}$$

subject to $\hat{y}_j \geq 0$ for all j and

$$\left. \begin{array}{l} 9\hat{y}_1 + 11\hat{y}_2 + 13\hat{y}_3 \leq 1, \\ 5\hat{y}_1 + 10\hat{y}_2 + 12\hat{y}_3 \leq 1, \\ 13\hat{y}_1 + 0\hat{y}_2 + 1\hat{y}_3 \leq 1. \end{array} \right\} \tag{14.10}$$

Solution of this linear program together with equation (14.7) yields the probabilities y_j which minimize the value of the game for B. By definition, this is the optimal mixed strategy for player B. The value of the game is $v - 6$.

14.6.2 Optimal Mixed Strategy for Player A

The optimal mixed strategy for A is developed similarly by considering all possible moves of player B. Because B wishes to minimize A's winnings, regardless of the strategy B chooses to play, A's winnings will be greater or at worst equal to the value of the game, v.

1. If B plays column 1, the payoff is

$$9x_1 + 5x_2 + 13x_3 \geq v. \tag{14.11}$$

The x_i are the strategies for player A; that is, the probabilities that A will play rows $i = 1, 2, 3$.

2. If B chooses column 2, the payoff is

$$11x_1 + 10x_2 + 0x_3 \geq v. \tag{14.12}$$

3. Finally, if B plays column 3, obtain

$$13x_1 + 12x_2 + 1x_3 \geq v. \tag{14.13}$$

Also, from equation (14.1),

$$x_1 + x_2 + x_3 = 1. \tag{14.14}$$

Dividing expressions (14.11) through (14.14) by v and defining

$$\hat{x}_i = \frac{x_i}{v}, \qquad \text{for all } i, \tag{14.15}$$

one obtains the set,

$$\left. \begin{array}{l} 9\hat{x}_1 + 5\hat{x}_2 + 13\hat{x}_3 \geq 1, \\ 11\hat{x}_1 + 10\hat{x}_2 + 0\hat{x}_3 \geq 1, \\ 13\hat{x}_1 + 12\hat{x}_2 + 1\hat{x}_3 \geq 1, \\ \hat{x}_1 + \hat{x}_2 + \hat{x}_3 = \frac{1}{v}. \end{array} \right\} \tag{14.16}$$

The objective of A is to maximize v or, equivalently, to minimize $1/v$. Therefore, one can again formulate a linear programming problem for the optimal mixed strategy for player A, as follows:

$$\text{Min } \frac{1}{v} = \hat{x}_1 + \hat{x}_2 + \hat{x}_3, \tag{14.17}$$

subject to

$$\hat{x}_i \geq 0 \qquad \text{for all } i$$

and

$$\left.\begin{array}{l} 9\hat{x}_1 + 5\hat{x}_2 + 13\hat{x}_3 \geq 1, \\ 11\hat{x}_1 + 10\hat{x}_2 + 0\hat{x}_3 \geq 1, \\ 13\hat{x}_1 + 12\hat{x}_2 + 1\hat{x}_3 \geq 1. \end{array}\right\} \tag{14.18}$$

Solution of this linear program and equation (14.15) yields the probabilities x_i which maximize the value of the game for A. By definition, this is the optimal mixed strategy for player A.

The formulation presented can be easily extended to $m \times n$ two-person, zero-sum games with $m \geq 1$ and $n \geq 1$.

14.7 A SIMPLIFIED TWO-DIMENSIONAL SOLUTION

The solution can be greatly simplified for a payoff matrix with only two rows and two columns.

Consider the payoff matrix of section 14.6. Because every element in row 1 is larger than the corresponding element in row 2, row 2 is recessive and can be deleted. Similarly, every element in column 2 is smaller than the corresponding element in column 3; therefore, column 3 is recessive and can also be deleted.

The simplified payoff matrix becomes

		B		
		1	2	
A	1	3	5	x_1
	3	7	-6	x_3
		y_1	y_2	

Again, no saddle point exists and the game does not have a pure strategy but, since now each player has only two possibilities, regardless of which strategy each plays, the payoff (value of the game) must be the same.

Consequently, for player B one obtains,

1. If A plays row 1: $3y_1 + 5y_2 = v'$.
2. If A plays row 2: $7y_1 - 6y_2 = v'$.
   Also, $y_1 + y_2 = 1$.

In these expressions, $v' = v - 6$.
   Solving simultaneously, obtain

$$y_1 = \tfrac{11}{15}; \; y_2 = \tfrac{4}{15}; \; v' = \tfrac{53}{15}.$$

For player A, obtain,

1. If B plays column 1: $3x_1 + 7x_3 = v'$.
2. If B plays column 2: $5x_1 - 6x_3 = v'$.
   Also, $x_1 + x_3 = 1$.

Again, solving simultaneously, one finds

$$x_1 = \tfrac{13}{15}; \; x_3 = \tfrac{2}{15}; \; v' = \tfrac{53}{15}.$$

In summary, the solution for the larger 3×3 payoff matrix is

$$x_1 = \tfrac{13}{15}, \; x_2 = 0, \; x_3 = \tfrac{2}{15};$$
$$y_1 = \tfrac{11}{15}, \; y_2 = \tfrac{4}{15}, \; y_3 = 0;$$

and

$$v' = \tfrac{53}{15}.$$

This simply states that player A should play row 1, $\tfrac{13}{15}$ of the time; never play row 2; and play row 3, $\tfrac{2}{15}$ of the time. Player B should play column 1, $\tfrac{11}{15}$ of the time; column 2, $\tfrac{4}{15}$ of the time; and never play column 3. By following this strategy, the value of the game over many plays will approach $\tfrac{53}{15}$.

The simplified solution was possible only because the original payoff matrix could be reduced to a 2×2 matrix; for larger dimensions the linear programming method of solution must be used.

14.8 GRAPHICAL SOLUTION FOR A 2 × 2 PAYOFF MATRIX

A graphical solution method can also be employed with 2×2 payoff matrices. The approach consists of solving graphically the system of three equations in three unknowns developed for each player.

Consider again the reduced payoff matrix of section 14.7.

$$B$$

		1	2	
	1	3	5	x_1
A	3	7	-6	x_3
		y_1	y_2	

To develop the graphical solution for player B, plot y along the horizontal axis and the game's payoff along the vertical axis as in Figure 14.1. Because $y_1 + y_2 = 1$, knowledge of the value of y_1 is sufficient to compute y_2.

Now consider the payoffs that result from each of the moves player A can make.

If A plays row 1 the payoff is

$$3y_1 + 5y_2 = 3y_1 + 5(1 - y_1). \qquad (14.19)$$

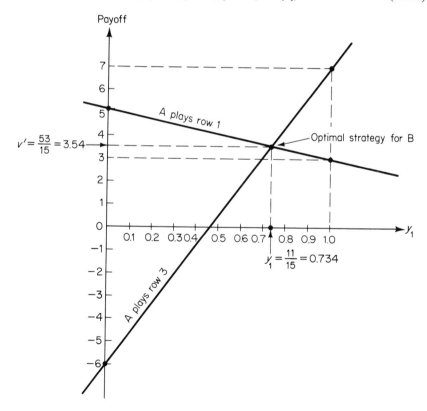

FIGURE 14.1. Mixed Strategy for Player B.

This equation of a straight line plots as shown in Figure 14.1. Its end points are obtained as follows: When $y_1 = 0$, the payoff is 5; when $y_1 = 1$, the payoff is 3.

If A plays row 3 the payoff is

$$7y_1 - 6y_2 = 7y_1 - 6(1 - y_1). \tag{14.20}$$

The straight line given by equation (14.20) is similarly plotted in Figure 14.1. The intersection of these two straight lines gives the same solution as the one obtained analytically before, that is,

$$y_1 = \tfrac{11}{15}; y_2 = 1 - y_1 = \tfrac{4}{15}; v' = \tfrac{53}{15}.$$

The graphical solution for player A is similarly developed in Figure 14.2 which shows a plot of x_1 versus the game's payoff.

If B plays column 1 the payoff is

$$3x_1 + 7x_3 = 3x_1 + 7(1 - x_1). \tag{14.21}$$

Equation (14.21) is plotted in Figure 14.2.

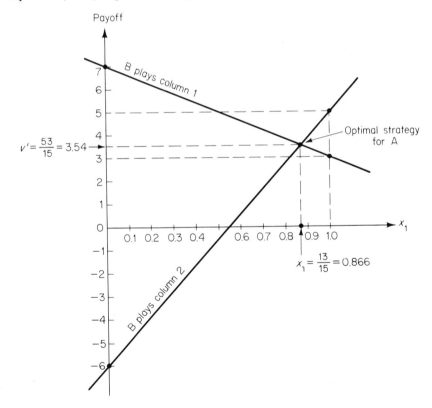

FIGURE 14.2. Mixed Strategy for Player A.

If B plays column 2 the payoff is

$$5x_1 - 6x_3 = 5x_1 - 6(1 - x_1).   \text{(14.22)}$$

Equation (14.22) is similarly plotted and the intersection of the two straight lines yields the same solution as the one obtained analytically:

$$x_1 = \tfrac{13}{15}; \, x_3 = 1 - x_1 = \tfrac{2}{15}; \, v' = \tfrac{53}{15}.$$

A wealth of information on game theory exists in the literature; the interested reader is referred to the texts listed at the end of the chapter for additional study in this area.

EXERCISES

14–1 through 14–4.

Determine the values of all the games with payoff matrices shown below.

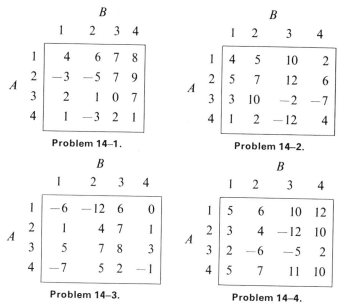

Problem 14–1.

Problem 14–2.

Problem 14–3.

Problem 14–4.

14–5. Use analytical and graphical methods to determine the value of the game and the player strategies for the payoff matrix given below.

$$
\begin{array}{c|cc}
 & \multicolumn{2}{c}{B} \\
 & 1 & 2 \\
\hline
A \quad 1 & 5 & -3 \\
2 & -9 & 7 \\
\end{array}
$$

14–6. Use a graphical procedure to determine player strategies and game value for the payoff matrix that follows:

$$
\begin{array}{c c}
 & B \\
 & \begin{array}{ccc} 1 & 2 & 3 \end{array} \\
A \begin{array}{c} 1 \\ 2 \\ 3 \end{array} &
\begin{array}{|ccc|}
\hline
4 & 5 & -7 \\
10 & 7 & 2 \\
-12 & 6 & 3 \\
\hline
\end{array}
\end{array}
$$

14–7 through 14–9. Use linear programming to determine the values of the games and the strategies for players A and B for the payoff matrices shown.

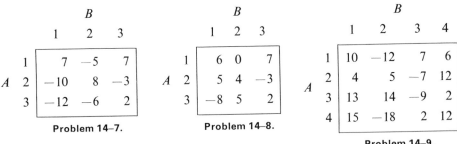

Problem 14–7.   Problem 14–8.

Problem 14–9.

REFERENCES

1. BIERMAN, H., L. E. FOURAKER, and R. K. JAEDICKE, *Quantitative Analysis for Business Decisions.* Homewood, Ill.: Richard D. Irwin, Inc., 1961.

2. KARLIN, S., *Math Methods and Theory in Games, Programming and Economics.* Reading, Mass.: Addison-Wesley Publishing Company, 1959.

3. LEVIN, R. I. and C. A. KIRKPATRICK, *Quantitative Approaches to Management.* New York: McGraw-Hill Book Company, 1956.

4. LUCE, R. D. and H. RAIFFA, *Games and Decisions: Introduction and Critical Survey.* New York: John Wiley & Sons, Inc., 1957.

5. McKINSEY, J. C. C., *Introduction to the Theory of Games.* New York: McGraw-Hill Book Company, 1952.

6. VON NUEMANN, J. and O. MORGENSTERN, *Theory of Games and Economic Decisions,* 3rd ed. 1953. Princeton: Princeton University Press, 1953.

7. WILLIAMS, J. D., *The Compleat Strategyst.* New York: McGraw-Hill Book Company, 1954.

A

Compound Interest Tables

TABLE A–1. 1% Compound Interest Factors

	SINGLE PAYMENT		UNIFORM SERIES				
	Compound Amount Factor	Present Worth Factor	Sinking Fund Factor	Capital Recovery Factor	Compound Amount Factor	Present Worth Factor	
n	$F \longrightarrow P$	$P \longrightarrow F$	$A \longrightarrow F$	$A \longrightarrow P$	$F \longrightarrow A$	$P \longrightarrow A$	n
1	1.0100	0.9901	1.00000	1.01000	1.000	0.990	1
2	1.0201	0.9803	0.49751	0.50751	2.010	1.970	2
3	1.0303	0.9706	0.33002	0.34002	3.030	2.941	3
4	1.0406	0.9610	0.24628	0.25628	4.060	3.902	4
5	1.0510	0.9515	0.19604	0.20604	5.101	4.853	5
6	1.0615	0.9420	0.16255	0.17255	6.152	5.795	6
7	1.0721	0.9327	0.13863	0.14863	7.214	6.728	7
8	1.0829	0.9235	0.12069	0.13069	8.286	7.652	8
9	1.0937	0.9143	0.10674	0.11674	9.369	8.566	9
10	1.1046	0.9053	0.09558	0.10558	10.462	9.471	10
11	1.1157	0.8963	0.08645	0.09645	11.567	10.368	11
12	1.1268	0.8874	0.07885	0.08885	12.683	11.255	12
13	1.1381	0.8787	0.07241	0.08241	13.809	12.134	13
14	1.1495	0.8700	0.06690	0.07690	14.947	13.004	14
15	1.1610	0.8613	0.06212	0.07212	16.097	13.865	15
16	1.1726	0.8528	0.05794	0.06794	17.258	14.718	16
17	1.1843	0.8444	0.05426	0.06426	18.430	15.562	17
18	1.1961	0.8360	0.05098	0.06098	19.615	16.398	18
19	1.2081	0.8277	0.04805	0.05805	20.811	17.226	19
20	1.2202	0.8195	0.04542	0.05542	22.019	18.046	20
21	1.2324	0.8114	0.04303	0.05303	23.239	18.857	21
22	1.2447	0.8034	0.04086	0.05086	24.472	19.660	22
23	1.2572	0.7954	0.03889	0.04889	25.716	20.456	23
24	1.2697	0.7876	0.03707	0.04707	26.973	21.243	24
25	1.2824	0.7798	0.03541	0.04541	28.243	22.023	25
26	1.2953	0.7720	0.03387	0.04387	29.526	22.795	26
27	1.3082	0.7644	0.03245	0.04245	30.821	23.560	27
28	1.3213	0.7568	0.03112	0.04112	32.129	24.316	28
29	1.3345	0.7493	0.02990	0.03990	33.450	25.066	29
30	1.3478	0.7419	0.02875	0.03875	34.785	25.808	30
31	1.3613	0.7346	0.02768	0.03768	36.133	26.542	31
32	1.3749	0.7273	0.02667	0.03667	37.494	27.270	32
33	1.3887	0.7201	0.02573	0.03573	38.869	27.990	33
34	1.4026	0.7130	0.02484	0.03484	40.258	28.703	34
35	1.4166	0.7059	0.02400	0.03400	41.660	29.409	35
40	1.4889	0.6717	0.02046	0.03046	48.886	32.835	40
45	1.5648	0.6391	0.01771	0.02771	56.481	36.095	45
50	1.6446	0.6080	0.01551	0.02551	64.463	39.196	50
55	1.7285	0.5785	0.01373	0.02373	72.852	42.147	55
60	1.8167	0.5504	0.01224	0.02224	81.670	44.955	60
65	1.9094	0.5237	0.01100	0.02100	90.937	47.627	65
70	2.0068	0.4983	0.00993	0.01993	100.676	50.169	70
75	2.1091	0.4741	0.00902	0.01902	110.913	52.587	75
80	2.2167	0.4511	0.00822	0.01822	121.672	54.888	80
85	2.3298	0.4292	0.00752	0.01752	132.979	57.078	85
90	2.4486	0.4084	0.00690	0.01690	144.863	59.161	90
95	2.5735	0.3886	0.00636	0.01636	157.354	61.143	95
100	2.7048	0.3697	0.00587	0.01587	170.481	63.029	100

TABLE A–2. 1¼% Compound Interest Factors

	SINGLE PAYMENT		UNIFORM SERIES				
	Compound Amount Factor	Present Worth Factor	Sinking Fund Factor	Capital Recovery Factor	Compound Amount Factor	Present Worth Factor	
n	F → P	P → F	A → F	A → P	F → A	P → A	n
1	1.0125	0.9877	1.00000	1.01250	1.000	0.988	1
2	1.0252	0.9755	0.49689	0.50939	2.012	1.963	2
3	1.0380	0.9634	0.32920	0.34170	3.038	2.927	3
4	1.0509	0.9515	0.24536	0.25786	4.076	3.878	4
5	1.0641	0.9398	0.19506	0.20756	5.127	4.818	5
6	1.0774	0.9282	0.16153	0.17403	6.191	5.746	6
7	1.0909	0.9167	0.13759	0.15009	7.268	6.663	7
8	1.1045	0.9054	0.11963	0.13213	8.359	7.568	8
9	1.1183	0.8942	0.10567	0.11817	9.463	8.462	9
10	1.1323	0.8832	0.09450	0.10700	10.582	9.346	10
11	1.1464	0.8723	0.08537	0.09787	11.714	10.218	11
12	1.1608	0.8615	0.07776	0.09026	12.860	11.079	12
13	1.1753	0.8509	0.07132	0.08382	14.021	11.930	13
14	1.1900	0.8404	0.06581	0.07831	15.196	12.771	14
15	1.2048	0.8300	0.06103	0.07353	16.386	13.601	15
16	1.2199	0.8197	0.05685	0.06935	17.591	14.420	16
17	1.2351	0.8096	0.05316	0.06566	18.811	15.230	17
18	1.2506	0.7996	0.04988	0.06238	20.046	16.030	18
19	1.2662	0.7898	0.04696	0.05946	21.297	16.819	19
20	1.2820	0.7800	0.04432	0.05682	22.563	17.599	20
21	1.2981	0.7704	0.04194	0.05444	23.845	18.370	21
22	1.3143	0.7609	0.03977	0.05227	25.143	19.131	22
23	1.3307	0.7515	0.03780	0.05030	26.457	19.882	23
24	1.3474	0.7422	0.03599	0.04849	27.788	20.624	24
25	1.3642	0.7330	0.03432	0.04682	29.135	21.357	25
26	1.3812	0.7240	0.03279	0.04529	30.500	22.081	26
27	1.3985	0.7150	0.03137	0.04387	31.881	22.796	27
28	1.4160	0.7062	0.03005	0.04255	33.279	23.503	28
29	1.4337	0.6975	0.02882	0.04132	34.695	24.200	29
30	1.4516	0.6889	0.02768	0.04018	36.129	24.889	30
31	1.4698	0.6804	0.02661	0.03911	37.581	25.569	31
32	1.4881	0.6720	0.02561	0.03811	39.050	26.241	32
33	1.5067	0.6637	0.02467	0.03717	40.539	26.905	33
34	1.5256	0.6555	0.02378	0.03628	42.045	27.560	34
35	1.5446	0.6474	0.02295	0.03545	43.571	28.208	35
40	1.6436	0.6084	0.01942	0.03192	51.490	31.327	40
45	1.7489	0.5718	0.01669	0.02919	59.916	34.258	45
50	1.8610	0.5373	0.01452	0.02702	68.882	37.013	50
55	1.9803	0.5050	0.01275	0.02525	78.422	39.602	55
60	2.1072	0.4746	0.01129	0.02379	88.575	42.035	60
65	2.2422	0.4460	0.01006	0.02256	99.377	44.321	65
70	2.3859	0.4191	0.00902	0.02152	110.872	46.470	70
75	2.5388	0.3939	0.00812	0.02062	123.103	48.489	75
80	2.7015	0.3702	0.00735	0.01985	136.119	50.387	80
85	2.8746	0.3479	0.00667	0.01917	149.968	52.170	85
90	3.0588	0.3269	0.00607	0.01857	164.705	53.846	90
95	3.2548	0.3072	0.00554	0.01804	180.386	55.421	95
100	3.4634	0.2887	0.00507	0.01757	197.072	56.901	100

TABLE A–3. 1½% Compound Interest Factors

	SINGLE PAYMENT		UNIFORM SERIES				
	Compound Amount Factor	Present Worth Factor	Sinking Fund Factor	Capital Recovery Factor	Compound Amount Factor	Present Worth Factor	
n	F → P	P → F	A → F	A → P	F → A	P → A	n
1	1.0150	0.9852	1.00000	1.01500	1.000	0.985	1
2	1.0302	0.9707	0.49628	0.51128	2.015	1.956	2
3	1.0457	0.9563	0.32838	0.34338	3.045	2.912	3
4	1.0614	0.9422	0.24444	0.25944	4.091	3.854	4
5	1.0773	0.9283	0.19409	0.20909	5.152	4.783	5
6	1.0934	0.9145	0.16053	0.17553	6.230	5.697	6
7	1.1098	0.9010	0.13656	0.15156	7.323	6.598	7
8	1.1265	0.8877	0.11858	0.13358	8.433	7.486	8
9	1.1434	0.8746	0.10461	0.11961	9.559	8.361	9
10	1.1605	0.8617	0.09343	0.10843	10.703	9.222	10
11	1.1779	0.8489	0.08429	0.09929	11.863	10.071	11
12	1.1956	0.8364	0.07668	0.09168	13.041	10.908	12
13	1.2136	0.8240	0.07024	0.08524	14.237	11.732	13
14	1.2318	0.8118	0.06472	0.07972	15.450	12.543	14
15	1.2502	0.7999	0.05994	0.07494	16.682	13.343	15
16	1.2690	0.7880	0.05577	0.07077	17.932	14.131	16
17	1.2880	0.7764	0.05208	0.06708	19.201	14.908	17
18	1.3073	0.7649	0.04881	0.06381	20.489	15.673	18
19	1.3270	0.7536	0.04588	0.06088	21.797	16.426	19
20	1.3469	0.7425	0.04325	0.05825	23.124	17.169	20
21	1.3671	0.7315	0.04087	0.05587	24.470	17.900	21
22	1.3876	0.7207	0.03870	0.05370	25.838	18.621	22
23	1.4084	0.7100	0.03673	0.05173	27.225	19.331	23
24	1.4300	0.6995	0.03492	0.04992	28.634	20.030	24
25	1.4509	0.6892	0.03326	0.04826	30.063	20.720	25
26	1.4727	0.6790	0.03173	0.04673	31.514	21.399	26
27	1.4948	0.6690	0.03032	0.04532	32.987	22.068	27
28	1.5172	0.6591	0.02900	0.04400	34.481	22.727	28
29	1.5400	0.6494	0.02778	0.04278	35.999	23.376	29
30	1.5631	0.6398	0.02664	0.04164	37.539	24.016	30
31	1.5865	0.6303	0.02557	0.04057	39.102	24.646	31
32	1.6103	0.6210	0.02458	0.03958	40.688	25.267	32
33	1.6345	0.6118	0.02364	0.03864	42.299	25.879	33
34	1.6590	0.6028	0.02276	0.03776	43.933	26.482	34
35	1.6839	0.5939	0.02193	0.03693	45.592	27.076	35
40	1.8140	0.5513	0.01843	0.03343	54.268	29.916	40
45	1.9542	0.5117	0.01572	0.03072	63.614	32.552	45
50	2.1052	0.4750	0.01357	0.02857	73.683	35.000	50
55	2.2679	0.4409	0.01183	0.02683	84.530	37.271	55
60	2.4432	0.4093	0.01039	0.02539	96.215	39.380	60
65	2.6320	0.3799	0.00919	0.02419	108.803	41.338	65
70	2.8355	0.3527	0.00817	0.02317	122.364	43.155	70
75	3.0546	0.3274	0.00730	0.02230	136.973	44.842	75
80	3.2907	0.3039	0.00655	0.02155	152.711	46.407	80
85	3.5450	0.2821	0.00589	0.02089	169.665	47.861	85
90	3.8189	0.2619	0.00532	0.02022	187.930	49.210	90
95	4.1141	0.2431	0.00482	0.01982	207.606	50.462	95
100	4.4320	0.2256	0.00437	0.01937	228.803	51.625	100

TABLE A–4. 1¾% Compound Interest Factors

	SINGLE PAYMENT		UNIFORM SERIES				
	Compound Amount Factor	Present Worth Factor	Sinking Fund Factor	Capital Recovery Factor	Compound Amount Factor	Present Worth Factor	
n	$F \rightarrow P$	$P \rightarrow F$	$A \rightarrow F$	$A \rightarrow P$	$F \rightarrow A$	$P \rightarrow A$	n
1	1.0175	0.9828	1.00000	1.01750	1.000	0.983	1
2	1.0353	0.9695	0.49566	0.51316	2.018	1.949	2
3	1.0534	0.9493	0.32757	0.34507	3.053	2.898	3
4	1.0719	0.9330	0.24353	0.26103	4.106	3.831	4
5	1.0906	0.9169	0.19312	0.21062	5.178	4.748	5
6	1.1097	0.9011	0.15952	0.17702	6.269	5.649	6
7	1.1291	0.8856	0.13553	0.15303	7.378	6.535	7
8	1.1489	0.8704	0.11754	0.13504	8.508	7.405	8
9	1.1690	0.8554	0.10356	0.12106	9.656	8.260	9
10	1.1894	0.8407	0.09238	0.10988	10.825	9.101	10
11	1.2103	0.8263	0.08323	0.10073	12.015	9.927	11
12	1.2314	0.8121	0.07561	0.09311	13.225	10.740	12
13	1.2530	0.7981	0.06917	0.08667	14.457	11.538	13
14	1.2749	0.7844	0.06366	0.08116	15.710	12.322	14
15	1.2972	0.7709	0.05888	0.07638	16.984	13.093	15
16	1.3199	0.7576	0.05470	0.07220	18.282	13.850	16
17	1.3430	0.7446	0.05102	0.06852	19.602	14.595	17
18	1.3665	0.7318	0.04774	0.06524	20.945	15.327	18
19	1.3904	0.7192	0.04482	0.06232	22.311	16.046	19
20	1.4148	0.7068	0.04219	0.05969	23.702	16.753	20
21	1.4395	0.6947	0.03981	0.05731	25.116	17.448	21
22	1.4647	0.6827	0.03766	0.05516	26.556	18.130	22
23	1.4904	0.6710	0.03569	0.05319	28.021	18.801	23
24	1.5164	0.6594	0.03389	0.05139	29.511	19.461	24
25	1.5430	0.6481	0.03223	0.04973	31.027	20.109	25
26	1.5700	0.6369	0.03070	0.04820	32.570	20.746	26
27	1.5975	0.6260	0.02929	0.04679	34.140	21.372	27
28	1.6254	0.6152	0.02798	0.04548	35.738	21.987	28
29	1.6539	0.6046	0.02676	0.04426	37.363	22.592	29
30	1.6828	0.5942	0.02563	0.04313	39.017	23.186	30
31	1.7122	0.5840	0.02457	0.04207	40.700	23.770	31
32	1.7422	0.5740	0.02358	0.04108	42.412	24.344	32
33	1.7727	0.5641	0.02265	0.04015	44.154	24.908	33
34	1.8037	0.5544	0.02177	0.03927	45.927	25.462	34
35	1.8353	0.5449	0.02095	0.03845	47.731	26.007	35
40	2.0016	0.4996	0.01747	0.03497	57.234	28.594	40
45	2.1830	0.4581	0.01479	0.03229	67.599	30.966	45
50	2.3808	0.4200	0.01267	0.03017	78.902	33.141	50
55	2.5965	0.3851	0.01096	0.02846	91.230	35.135	55
60	2.8318	0.3531	0.00955	0.02705	104.675	36.964	60
65	3.0884	0.3238	0.00838	0.02588	119.339	38.641	65
70	3.3683	0.2969	0.00739	0.02489	135.331	40.178	70
75	3.6735	0.2722	0.00655	0.02405	152.772	41.587	75
80	4.0064	0.2496	0.00582	0.02332	171.794	42.880	80
85	4.3694	0.2289	0.00519	0.02269	192.539	44.065	85
90	4.7654	0.2098	0.00465	0.02215	215.165	45.152	90
95	5.1972	0.1924	0.00417	0.02167	239.840	46.148	95
100	5.6682	0.1764	0.00375	0.02125	266.752	47.061	100

TABLE A–5. 2% Compound Interest Factors

	SINGLE PAYMENT		UNIFORM SERIES				
	Compound Amount Factor	Present Worth Factor	Sinking Fund Factor	Capital Recovery Factor	Compound Amount Factor	Present Worth Factor	
n	F → P	P → F	A → F	A → P	F → A	P → A	n
1	1.0200	0.9804	1.00000	1.02000	1.000	0.980	1
2	1.0404	0.9612	0.49505	0.51505	2.020	1.942	2
3	1.0612	0.9423	0.32675	0.34675	3.060	2.884	3
4	1.0824	0.9238	0.24262	0.26262	4.122	3.808	4
5	1.1041	0.9057	0.19216	0.21216	5.204	4.713	5
6	1.1262	0.8880	0.15853	0.17853	6.308	5.601	6
7	1.1487	0.8706	0.13451	0.15451	7.434	6.472	7
8	1.1717	0.8535	0.11651	0.13651	8.583	7.325	8
9	1.1951	0.8368	0.10252	0.12252	9.755	8.162	9
10	1.2190	0.8203	0.09133	0.11133	10.950	8.983	10
11	1.2434	0.8043	0.08218	0.10218	12.169	9.787	11
12	1.2682	0.7885	0.07456	0.09456	13.412	10.575	12
13	1.2936	0.7730	0.06812	0.08812	14.680	11.348	13
14	1.3195	0.7579	0.06260	0.08260	15.974	12.106	14
15	1.3459	0.7430	0.05783	0.07783	17.293	12.849	15
16	1.3728	0.7284	0.05365	0.07365	18.639	13.578	16
17	1.4002	0.7142	0.04997	0.06997	20.012	14.292	17
18	1.4282	0.7002	0.04670	0.06670	11.412	14.992	18
19	1.4568	0.6864	0.04378	0.06378	22.841	15.678	19
20	1.4859	0.6730	0.04116	0.06116	24.297	16.351	20
21	1.5157	0.6598	0.03878	0.05878	25.783	17.011	21
22	1.5460	0.6468	0.03663	0.05663	27.299	17.658	22
23	1.5769	0.6342	0.03467	0.05467	28.845	18.292	23
24	1.6084	0.6217	0.03287	0.05287	30.422	18.914	24
25	1.6406	0.6095	0.03122	0.05122	32.030	19.523	25
26	1.6734	0.5976	0.02970	0.04970	33.671	20.121	26
27	1.7069	0.5859	0.02829	0.04829	35.344	20.707	27
28	1.7410	0.5744	0.02699	0.04699	37.051	21.281	28
29	1.7758	0.5631	0.02578	0.04578	38.792	21.844	29
30	1.8114	0.5521	0.02465	0.04465	40.568	22.396	30
31	1.8476	0.5412	0.02360	0.04360	42.379	22.938	31
32	1.8845	0.5306	0.02261	0.04261	44.227	23.468	32
33	1.9222	0.5202	0.02169	0.04169	46.112	23.989	33
34	1.9607	0.5100	0.02082	0.04082	48.034	24.499	34
35	1.9999	0.5000	0.02000	0.04000	49.994	24.999	35
40	2.2080	0.4529	0.01656	0.03656	60.402	27.355	40
45	2.4379	0.4102	0.01391	0.03391	71.893	29.490	45
50	2.6916	0.3715	0.01182	0.03182	84.579	31.424	50
55	2.9717	0.3365	0.01014	0.03014	98.587	33.175	55
60	3.2810	0.3048	0.00877	0.02877	114.052	34.761	60
65	3.6225	0.2761	0.00763	0.02763	131.126	36.197	65
70	3.9996	0.2500	0.00667	0.02667	149.978	37.499	70
75	4.4158	0.2265	0.00586	0.02586	170.792	38.677	75
80	4.8754	0.2051	0.00516	0.02516	193.772	39.745	80
85	5.3829	0.1858	0.00456	0.02456	219.144	40.711	85
90	5.9431	0.1683	0.00405	0.02405	247.157	41.587	90
95	6.5617	0.1524	0.00360	0.02360	278.085	42.380	95
100	7.2446	0.1380	0.00320	0.02320	312.232	43.098	100

TABLE A-6. $2\frac{1}{2}$% Compound Interest Factors

	SINGLE PAYMENT		UNIFORM SERIES				
	Compound Amount Factor	Present Worth Factor	Sinking Fund Factor	Capital Recovery Factor	Compound Amount Factor	Present Worth Factor	
n	$F \rightarrow P$	$P \rightarrow F$	$A \rightarrow F$	$A \rightarrow P$	$F \rightarrow A$	$P \rightarrow A$	n
1	1.0250	0.9756	1.00000	0.02500	1.000	0.976	1
2	1.0506	0.9518	0.49383	0.51883	2.025	1.927	2
3	1.0769	0.9286	0.32514	0.35014	3.076	2.856	3
4	1.1038	0.9060	0.24082	0.26582	4.153	3.762	4
5	1.1314	0.8839	0.19025	0.21525	5.256	4.646	5
6	1 1597	0.8623	0.15655	0.18155	6.388	5.508	6
7	1.1887	0.8413	0.13250	0.15750	7.547	6.349	7
8	1.2184	0 8207	0.11447	0.13947	8.736	7.170	8
9	1.2489	0.8007	0.10046	0.12546	9.955	7.971	9
10	1.2801	0.7812	0.08926	0.11426	11.203	8.752	10
11	1.3121	0.7621	0.08011	0.10511	12.483	9.514	11
12	1.3449	0.7436	0.07249	0.09749	13.796	10.258	12
13	1.3785	0.7254	0.06605	0.09105	15.140	10.983	13
14	1.4130	0.7077	0.06054	0.08554	16.519	11.691	14
15	1.4483	0.6905	0.05577	0.08077	17.932	12.381	15
16	1.4845	0.6736	0.05160	0.07660	19.380	13.055	16
17	1.5216	0.6572	0.04793	0.07293	20.865	13.712	17
18	1.5597	0.6412	0.04467	0.06967	22.386	14.353	18
19	1.5987	0.6255	0.04176	0.06676	23.946	14.979	19
20	1 6386	0.6103	0.03915	0.06415	25.545	15.589	20
21	1.6796	0.5954	0.03679	0.06179	27.183	16.185	21
22	1.7216	0.5809	0.03465	0.05965	28.863	16.765	22
23	1.7646	0.5667	0.03270	0.05770	30.584	17.332	23
24	1.8087	0.5529	0.03091	0.05591	32.349	17.885	24
25	1.8539	0.5394	0.02928	0.05428	34.158	18.424	25
26	1.9003	0.5262	0.02777	0.05277	36.012	18.951	26
27	1.9478	0.5134	0.02638	0.05138	37.912	19.464	27
28	1.9965	0.5009	0.02509	0.05009	39.860	19.965	28
29	2.0464	0.4887	0.02389	0.04889	41.856	20.454	29
30	2.0976	0.4767	0.02278	0.04778	43.903	20.930	30
31	2.1500	0.4651	0.02174	0.04674	46.000	21.395	31
32	2.2038	0.4538	0.02077	0.04577	48.150	21.849	32
33	2.2589	0.4427	0.01986	0.04486	50.354	22.292	33
34	2.3153	0.4319	0.01901	0.04401	52.613	22.724	34
35	2.3732	0.4214	0.01821	0.04321	54.928	23.145	35
40	2.6851	0.3724	0.01484	0.03984	67.403	25.103	40
45	3.0379	0.3292	0.01227	0.03727	81.516	26.833	45
50	3.4371	0.2909	0.01026	0.03526	97.484	28.362	50
55	3.8888	0.2572	0.00865	0.03365	115.551	29.714	55
60	4.3998	0.2273	0.00735	0.03235	135.992	30.909	60
65	4.9780	0.2009	0.00628	0.03128	159.118	31.965	65
70	5.6321	0.1776	0.00540	0.03040	185.284	32.898	70
75	6.3722	0.1569	0.00465	0.02965	214.888	33.723	75
80	7.2100	0.1387	0.00403	0.02903	248.383	34.452	80
85	8.1570	0.1226	0.00349	0.02849	286.279	35.096	85
90	9.2289	0.1084	0.00304	0.02804	329.154	35.666	90
95	10.4416	0.0958	0.00265	0.02765	377.664	36.169	95
100	11.8137	0.0846	0.00231	0.02731	432.549	36.614	100

TABLE A–7. 3% Compound Interest Factors

	SINGLE PAYMENT		UNIFORM SERIES				
	Compound Amount Factor	Present Worth Factor	Sinking Fund Factor	Capital Recovery Factor	Compound Amount Factor	Present Worth Factor	
n	F → P	P → F	A → F	A → P	F → A	P → A	n
1	1.0300	0.9709	1.00000	1.03000	1.000	0.971	1
2	1.0609	0.9426	0.49261	0.52261	2.030	1.913	2
3	1.0927	0.9151	0.32353	0.35353	3.091	2.829	3
4	1.1255	0.8885	0.23903	0.26903	4.184	3.717	4
5	1.1593	0.8626	0.18835	0.21835	5.309	4.580	5
6	1.1941	0.8375	0.15460	0.18460	6.468	5.417	6
7	1.2299	0.8131	0.13051	0.16051	7.662	6.230	7
8	1.2668	0.7894	0.11246	0.14246	8.892	7.020	8
9	1.3048	0.7664	0.09843	0.12843	10.159	7.786	9
10	1.3439	0.7441	0.08723	0.11723	11.464	8.530	10
11	1.3842	0.7224	0.07808	0.10808	12.808	9.253	11
12	1.4258	0.7014	0.07046	0.10046	14.192	9.954	12
13	1.4685	0.6810	0.06403	0.09403	15.618	10.635	13
14	1.5126	0.6611	0.05853	0.08853	17.086	11.296	14
15	1.5580	0.6419	0.05377	0.08377	18.599	11.938	15
16	1.6047	0.6232	0.04961	0.07961	20.157	12.561	16
17	1.6528	0.6050	0.04595	0.07595	21.762	13.166	17
18	1.7024	0.5874	0.04271	0.07271	23.414	13.754	18
19	1.7535	0.5703	0.03981	0.06981	25.117	14.324	19
20	1.8061	0.5537	0.03722	0.06722	26.870	14.877	20
21	1.8603	0.5375	0.03487	0.06487	28.676	15.415	21
22	1.9161	0.5219	0.03275	0.06275	30.537	15.937	22
23	1.9736	0.5067	0.03081	0.06081	32.453	16.444	23
24	2.0328	0.4919	0.02905	0.05905	34.426	16.936	24
25	2.0938	0.4776	0.02743	0.05743	36.459	17.413	25
26	2.1566	0.4637	0.02594	0.05594	38.553	17.877	26
27	2.2213	0.4502	0.02456	0.05456	40.710	18.327	27
28	2.2879	0.4371	0.02329	0.05329	42.931	18.764	28
29	2.3566	0.4243	0.02211	0.05211	45.219	19.188	29
30	2.4273	0.4120	0.02102	0.05102	47.575	19.600	30
31	2.5001	0.4000	0.02000	0.05000	50.003	20.000	31
32	2.5751	0.3883	0.01905	0.04905	52.503	20.389	32
33	2.6523	0.3770	0.01816	0.04816	55.078	20.766	33
34	2.7319	0.3660	0.01732	0.04732	57.730	21.132	34
35	2.8139	0.3554	0.01654	0.04654	60.462	21.487	35
40	3.2620	0.3066	0.01326	0.04326	75.401	23.115	40
45	3.7816	0.2644	0.01079	0.04079	92.720	24.519	45
50	4.3839	0.2281	0.00887	0.03887	112.797	25.730	50
55	5.0821	0.1968	0.00735	0.03735	136.072	26.774	55
60	5.8916	0.1697	0.00613	0.03613	163.053	27.676	60
65	6.8300	0.1464	0 00515	0.03515	194.333	28.453	65
70	7.9178	0.1263	0.00434	0.03434	230.594	29.123	70
75	9.1789	0.1089	0.00367	0.03367	272.631	29.702	75
80	10.6409	0.0940	0.00311	0.03311	321.363	30.201	80
85	12.3357	0.0811	0.00265	0.03265	377.857	30.631	85
90	14.3005	0.0699	0.00226	0.03226	443.349	31.002	90
95	16.5782	0.0603	0.00193	0.03193	519.272	31.323	95
100	19.2186	0.0520	0.00165	0.03165	607.288	31.599	100

TABLE A–8. 3½% Compound Interest Factors

	SINGLE PAYMENT		UNIFORM SERIES				
	Compound Amount Factor	Present Worth Factor	Sinking Fund Factor	Capital Recovery Factor	Compound Amount Factor	Present Worth Factor	
n	F → P	P → F	A → F	A → P	F → A	P → A	n
1	1.0350	0.9662	1.00000	1.03500	1.000	0.966	1
2	1.0712	0.9335	0.49140	0.52640	2.035	1.900	2
3	1.1087	0.9019	0.32193	0.35693	3.106	2.802	3
4	1.1475	0.8714	0.23725	0.27225	4.215	3.673	4
5	1.1877	0.8420	0.18648	0.22148	5.362	4.515	5
6	1.2293	0.8135	0.15267	0.18767	6.550	5.329	6
7	1.2723	0.7860	0.12854	0.16354	7.779	6.115	7
8	1.3168	0.7594	0.11048	0.14548	9.052	6.874	8
9	1.3629	0.7337	0.09645	0.13145	10.368	7.608	9
10	1.4106	0.7089	0.08524	0.12024	11.731	8.317	10
11	1.4600	0.6849	0.07609	0.11109	13.142	9.002	11
12	1.5111	0.6618	0.06848	0.10348	14.602	9.663	12
13	1.5640	0.6394	0.06206	0.09706	16.113	10.303	13
14	1.6187	0.6178	0.05657	0.09157	17.677	10.921	14
15	1.6753	0.5969	0.05183	0.08683	19.296	11.517	15
16	1.7340	0.5767	0.04768	0.08268	20.971	12.094	16
17	1.7947	0.5572	0.04404	0.07904	22.705	12.651	17
18	1.8575	0.5384	0.04082	0.07582	24.500	13.190	18
19	1.9225	0.5202	0.03794	0.07294	26.357	13.710	19
20	1.9898	0.5026	0.03536	0.07036	28.280	14.212	20
21	2.0594	0.4856	0.03304	0.06804	30.269	14.698	21
22	2.1315	0.4692	0.03093	0.06593	32.329	15.167	22
23	2.2061	0.4533	0.02902	0.06402	34.460	15.620	23
24	2.2833	0.4380	0.02727	0.06227	36.667	16.058	24
25	2.3632	0.4231	0.02567	0.06067	38.950	16.482	25
26	2.4460	0.4088	0.02421	0.05921	41.313	16.890	26
27	2.5316	0.3950	0.02285	0.05785	43.759	17.285	27
28	2.6202	0.3817	0.02160	0.05660	46.291	17.667	28
29	2.7119	0.3687	0.02045	0.05545	48.911	18.036	29
30	2.8068	0.3563	0.01937	0.05437	51.623	18.392	30
31	2.9050	0.3442	0.01837	0.05337	54.429	18.736	31
32	3.0067	0.3326	0.01744	0.05244	57.335	19.069	32
33	3.1119	0.3213	0.01657	0.05157	60.341	19.390	33
34	3.2209	0.3105	0.01576	0.05076	63.453	19.701	34
35	3.3336	0.3000	0.01500	0.05000	66.674	20.001	35
40	3.9593	0.2526	0.01183	0.04683	84.550	21.355	40
45	4.7024	0.2127	0.00945	0.04445	106.782	22.495	45
50	5.5849	0.1791	0.00763	0.04263	130.998	23.456	50
55	6.6331	0.1508	0.00621	0.04121	160.947	24.264	55
60	7.8781	0.1269	0.00509	0.04009	196.517	24.945	60
65	9.3567	0.1069	0.00419	0.03919	238.763	25.518	65
70	11.1128	0.0900	0.00346	0.03846	288.938	26.000	70
75	13.1986	0.0758	0.00287	0.03787	348.530	26.407	75
80	15.6757	0.0638	0.00238	0.03738	419.307	26.749	80
85	18.6179	0.0537	0.00199	0.03699	503.367	27.037	85
90	22.1122	0.0452	0.00166	0.03666	603.205	27.279	90
95	26.2623	0.0381	0.00139	0.03639	721.781	27.484	95
100	31.1914	0.0321	0.00116	0.03616	862.612	27.655	100

TABLE A–9. 4% Compound Interest Factors

	SINGLE PAYMENT		UNIFORM SERIES				
	Compound Amount Factor	Present Worth Factor	Sinking Fund Factor	Capital Recovery Factor	Compound Amount Factor	Present Worth Factor	
n	$F \longrightarrow P$	$P \longrightarrow F$	$A \longrightarrow F$	$A \longrightarrow P$	$F \longrightarrow A$	$P \longrightarrow A$	n
1	1.0400	0.9615	1.00000	1.04000	1.000	0.962	1
2	1.0816	0.9246	0.49020	0.53020	2.040	1.886	2
3	1.1249	0.8890	0.32035	0.36035	3.122	2.775	3
4	1.1699	0.8548	0.23549	0.27549	4.246	3.630	4
5	1.2167	0.8219	0.18463	0.22463	5.416	4.452	5
6	1.2653	0.7903	0.15076	0.19076	6.633	5.242	6
7	1.3159	0.7599	0.12661	0.16661	7.898	6.002	7
8	1.3686	0.7307	0.10853	0.14853	9.214	6.733	8
9	1.4233	0.7026	0.09449	0.13449	10.583	7.435	9
10	1.4802	0.6756	0.08329	0.12329	12.006	8.111	10
11	1.5395	0.6496	0.07415	0.11415	13.486	8.760	11
12	1.6010	0.6246	0.06655	0.10655	15.026	9.385	12
13	1.6651	0.6006	0.06014	0.10014	16.627	9.986	13
14	1.7317	0.5775	0.05467	0.09467	18.292	10.563	14
15	1.8009	0.5553	0.04994	0.08994	20.024	11.118	15
16	1.8730	0.5339	0.04582	0.08582	21.825	11.652	16
17	1.9479	0.5134	0.04220	0.08220	23.698	12.166	17
18	2.0258	0.4936	0.03899	0.07899	25.645	12.659	18
19	2.1068	0.4746	0.03614	0.07614	27.671	13.134	19
20	2.1911	0.4564	0.03358	0.07358	29.778	13.590	20
21	2.2788	0.4388	0.03128	0.07128	31.969	14.029	21
22	2.3699	0.4220	0.02920	0.06920	34.248	14.451	22
23	2.4647	0.4057	0.02731	0.06731	36.618	14.857	23
24	2.5633	0.3901	0.02559	0.06559	39.083	15.247	24
25	2.6658	0.3751	0.02401	0.06401	41.646	15.622	25
26	2.7725	0.3607	0.02257	0.06257	44.312	15.983	26
27	2.8834	0.3468	0.02124	0.06124	47.084	16.330	27
28	2.9987	0.3335	0.02001	0.06001	49.968	16.663	28
29	3.1187	0.3207	0.01888	0.05888	52.966	16.984	29
30	3.2434	0.3083	0.01783	0.05783	56.085	17.292	30
31	3.3731	0.2965	0.01686	0.05686	59.328	17.588	31
32	3.5081	0.2851	0.01595	0.05595	62.701	17.874	32
33	3.6484	0.2741	0.01510	0.05510	66.210	18.148	33
34	3.7943	0.2636	0.01431	0.05431	69.858	18.411	34
35	3.9461	0.2534	0.01358	0.05358	73.652	18.665	35
40	4.8010	0.2083	0.01052	0.05052	95.026	19.793	40
45	5.8412	0.1712	0.00826	0.04826	121.029	20.720	45
50	7.1067	0.1407	0.00655	0.04655	152.667	21.482	50
55	8.6464	0.1157	0.00523	0.04523	191.159	22.109	55
60	10.5196	0.0951	0.00420	0.04420	237.991	22.623	60
65	12.7987	0.0781	0.00339	0.04339	294.968	23.047	65
70	15.5716	0.0642	0.00275	0.04275	364.290	23.395	70
75	18.9453	0.0528	0.00223	0.04223	448.631	23.680	75
80	23.0500	0.0434	0.00181	0.04181	551.245	23.915	80
85	28.0436	0.0357	0.00148	0.04148	676.090	24.109	85
90	34.1193	0.0293	0.00121	0.04121	827.983	24.267	90
95	41.5114	0.0241	0.00099	0.04099	1,012.785	24.398	95
100	50.5049	0.0198	0.00081	0.04081	1,237.624	24.505	100

TABLE A–10. 4½% Compound Interest Factors

	SINGLE PAYMENT		UNIFORM SERIES				
	Compound Amount Factor	Present Worth Factor	Sinking Fund Factor	Capital Recovery Factor	Compound Amount Factor	Present Worth Factor	
n	F → P	P → F	A → F	A → P	F → A	P → A	n
1	1.0450	0.9569	1.00000	1.04500	1.000	0.957	1
2	1.0920	0.9157	0.48900	0.53400	2.045	1.873	2
3	1.1412	0.8763	0.31877	0.36377	3.137	2.749	3
4	1.1925	0.8386	0.23374	0.27874	4.278	3.588	4
5	1.2462	0.8025	0.18279	0.22779	5.471	4.390	5
6	1.3023	0.7679	0.14888	0.19388	6.717	5.158	6
7	1.3609	0.7348	0.12470	0.16970	8.019	5.893	7
8	1.4221	0.7032	0.10661	0.15161	9.380	6.596	8
9	1.4861	0.6729	0.09257	0.13757	10.802	7.269	9
10	1.5530	0.6439	0.08138	0.12638	12.288	7.913	10
11	1.6229	0.6162	0.07225	0.11725	13.841	8.529	11
12	1.6959	0.5897	0.06467	0.10967	15.464	9.119	12
13	1.7722	0.5643	0.05828	0.10328	17.160	9.683	13
14	1.8519	0.5400	0.05282	0.09782	18.932	10.223	14
15	1.9353	0.5167	0.04811	0.09311	20.784	10.740	15
16	2.0224	0.4945	0.04402	0.08902	22.719	11.234	16
17	2.1134	0.4732	0.04042	0.08542	24.742	11.707	17
18	2.2085	0.4528	0.03724	0.08224	26.855	12.160	18
19	2.3079	0.4333	0.03441	0.07941	29.064	12.593	19
20	2.4117	0.4146	0.03188	0.07688	31.371	13.008	20
21	2.5202	0.3968	0.02960	0.07460	33.783	13.405	21
22	2.6337	0.3797	0.02755	0.07255	36.303	13.784	22
23	2.7522	0.3634	0.02568	0.07068	38.937	14.148	23
24	2.8760	0.3477	0.02399	0.06899	41.689	14.495	24
25	3.0054	0.3327	0.02244	0.06744	44.565	14.828	25
26	3.1407	0.3184	0.02102	0.06602	47.571	15.147	26
27	3.2820	0.3047	0.01972	0.06472	50.711	15.451	27
28	3.4397	0.2916	0.01852	0.06352	53.993	15.743	28
29	3.5840	0.2790	0.01741	0.06241	57.423	16.022	29
30	3.7453	0.2670	0.01639	0.06139	61.007	16.289	30
31	3.9139	0.2555	0.01544	0.06044	64.752	16.544	31
32	4.0900	0.2445	0.01456	0.05956	68.666	16.789	32
33	4.2740	0.2340	0.01374	0.05874	72.756	17.023	33
34	4.4664	0.2239	0.01298	0.05798	77.030	17.247	34
35	4.6673	0.2143	0.01227	0.05727	81.497	17.461	35
40	5.8164	0.1719	0.00934	0.05434	107.030	18.402	40
45	7.2482	0.1380	0.00720	0.05220	138.850	19.156	45
50	9.0326	0.1107	0.00560	0.05060	178.503	19.762	50
55	11.2563	0.0888	0.00439	0.04939	227.918	20.248	55
60	14.0274	0.0713	0.00345	0.04845	289.498	20.638	60
65	17,4807	0.0572	0.00273	0.04773	366.238	20.951	65
70	21.7841	0.0459	0.00217	0.04717	461.870	21.202	70
75	27.1470	0.0368	0.00172	0.04672	581.044	21.404	75
80	33.8301	0.0296	0.00137	0.04637	729.558	21.565	80
85	42.1585	0.0237	0.00109	0.04609	914.632	21.695	85
90	52.5371	0.0190	0.00087	0.04587	1,145.269	21.799	90
95	65.4708	0.0153	0.00070	0.04570	1,432.684	21.883	95
100	81.5885	0.0123	0.00056	0.04556	1.790.856	21.950	100

TABLE A–11. 5%Compound Interest Factors

	SINGLE PAYMENT		UNIFORM SERIES				
	Compound Amount Factor	Present Worth Factor	Sinking Fund Factor	Capital Recovery Factor	Compound Amount Factor	Present Worth Factor	
n	F → P	P → F	A → F	A → P	F → A	P → A	n
1	1.0500	0.9524	0.00000	1.05000	1.000	0.952	1
2	1.1025	0.9070	0.48780	0.53780	2.050	1.859	2
3	1.1576	0.8638	0.31721	0.36721	3.153	2.723	3
4	1.2155	0.8227	0.23201	0.28201	4.310	3.546	4
5	1.2763	0.7835	0.18097	0.23097	5.526	4.329	5
6	1.3401	0.7462	0.14702	0.19702	6.802	5.076	6
7	1.4071	0.7107	0.12282	0.17282	8.142	5.786	7
8	1.4775	0.6768	0.10472	0.15472	9.549	6.463	8
9	1.5513	0.6446	0.09069	0.14069	11.027	7.108	9
10	1.6289	0.6139	0.07950	0.12950	12.578	7.722	10
11	1.7103	0.5847	0.07039	0.12039	14.207	8.306	11
12	1.7959	0.5568	0.06283	0.11283	15.917	8.863	12
13	1.8856	0.5303	0.05646	0.10646	17.713	9.394	13
14	1.9800	0.5051	0.05102	0.10102	19.599	9.899	14
15	2.0789	0.4810	0.04634	0.09634	21.579	10.380	15
16	2.1829	0.4581	0.04227	0.09227	23.657	10.838	16
17	2.2920	0.4363	0.03870	0.08870	25.840	11.274	17
18	2.4066	0.4155	0.03555	0.08555	28.132	11.690	18
19	2.5270	0.3957	0.03275	0.08275	30.539	12.085	19
20	2.6533	0.3769	0.03024	0.08024	33.066	12.462	20
21	2.7860	0.3589	0.02800	0.07800	35.719	12.821	21
22	2.9253	0.3418	0.02597	0.07597	38.505	13.163	22
23	3.0715	0.3256	0.02414	0.07414	41.430	13.489	23
24	3.2251	0.3101	0.02247	0.07247	44.502	13.799	24
25	3.3864	0.2953	0.02095	0.07095	47.727	14.094	25
26	3.5557	0.2812	0.01956	0.06956	51.113	14.375	26
27	3.7335	0.2678	0.01829	0.06829	54.669	14.643	27
28	3.9201	0.2551	0.01712	0.06712	58.403	14.898	28
29	4.1161	0.2429	0.01605	0.06605	62.323	15.141	29
30	4.3219	0.2314	0.01505	0.06505	66.439	15.372	30
31	4.5380	0.2204	0.01413	0.06413	70.761	15.593	31
32	4.7649	0.2099	0.01328	0.06328	75.299	15.803	32
33	5.0032	0.1999	0.01249	0.06249	80.064	16.003	33
34	5.2533	0.1904	0.01176	0.06176	85.067	16.193	34
35	5.5160	0.1813	0.01107	0.06107	90.320	16.374	35
40	7.0400	0.1420	0.00828	0.05828	120.800	17.159	40
45	8.9850	0.1113	0.00626	0.05626	159.700	17.774	45
50	11.4674	0.0872	0.00478	0.05478	209.348	18.256	50
55	14.6356	0.0683	0.00367	0.05367	272.713	18.633	55
60	18.6792	0.0535	0.00283	0.05283	353.584	18.929	60
65	23.8399	0.0419	0.00219	0.05219	456.798	19.161	65
70	30.4264	0.0329	0.00170	0.05170	588.529	19.343	70
75	38.8327	0.0258	0.00132	0.05132	756.654	19.485	75
80	49.5614	0.0202	0.00103	0.05103	971.229	19.596	80
85	63.2544	0.0158	0.00080	0.05080	1,245.087	19.684	85
90	80.7304	0.0124	0.00063	0.05063	1,594.607	19.752	90
95	103.0357	0.0097	0.00049	0.05049	2,040.694	19.806	95
100	131.5013	0.0076	0.00038	0.05038	2,610.025	19.848	100

TABLE A–12. $5\frac{1}{2}$% Compound Interest Factors

| | SINGLE PAYMENT | | UNIFORM SERIES | | | |
| | Compound Amount Factor | Present Worth Factor | Sinking Fund Factor | Capital Recovery Factor | Compound Amount Factor | Present Worth Factor | |
n	$F \rightarrow P$	$P \rightarrow F$	$A \rightarrow F$	$A \rightarrow P$	$F \rightarrow A$	$P \rightarrow A$	n
1	1.0550	0.9479	1.00000	1.05500	1.000	0.948	1
2	1.1130	0.8985	0.48662	0.54162	2.055	1.846	2
3	1.1742	0.8516	0.31565	0.37065	3.168	2.698	3
4	1.2388	0.8072	0.23029	0.28529	4.342	3.505	4
5	1.3070	0.7651	0.17918	0.23418	5.581	4.270	5
6	1.3788	0.7252	0.14518	0.20018	6.888	4.996	6
7	1.4547	0.6874	0.12096	0.17596	8.267	5.683	7
8	1.5347	0.6516	0.10286	0.15786	9.722	6.335	8
9	1.6191	0.6176	0.08884	0.14384	11.256	6.952	9
10	1.7081	0.5854	0.07767	0.13267	12.875	7.538	10
11	1.8021	0.5549	0.06857	0.12357	14.583	8.093	11
12	1.9012	0.5260	0.06103	0.11603	16.386	8.619	12
13	2.0058	0.4986	0.05468	0.10968	18.287	9.117	13
14	2.1161	0.4726	0.04928	0.10428	20.293	9.590	14
15	2.2325	0.4479	0.04463	0.09963	22.409	10.038	15
16	2.3553	0.4246	0.04058	0.09558	24.641	10.462	16
17	2.4848	0.4024	0.03704	0.09204	26.996	10.865	17
18	2.6215	0.3815	0.03392	0.08892	29.481	11.246	18
19	2.7656	0.3616	0.03115	0.08615	32.103	11.608	19
20	2.9178	0.3427	0.02868	0.08368	34.868	11.950	20
21	3.0782	0.3249	0.02646	0.08146	37.786	12.275	21
22	3.2475	0.3079	0.02447	0.07947	40.864	12.583	22
23	3.4262	0.2919	0.02267	0.07767	44.112	12.875	23
24	3.6146	0.2767	0.02104	0.07604	47.538	13.152	24
25	3.8134	0.2622	0.01955	0.07455	51.153	13.414	25
26	4.0231	0.2486	0.01819	0.07319	54.966	13.662	26
27	4.2444	0.2356	0.01695	0.07195	58.989	13.898	27
28	4.4778	0.2233	0.01581	0.07081	63.234	14.121	28
29	4.7241	0.2117	0.01477	0.06977	67.711	14.333	29
30	4.9840	0.2006	0.01381	0.06881	72.435	14.534	30
31	5.2581	0.1902	0.01292	0.06792	77.419	14.724	31
32	5.5473	0.1803	0.01210	0.06710	82.677	14.904	32
33	5.8524	0.1709	0.01133	0.06633	88.225	15.075	33
34	6.1742	0.1620	0.01063	0.06563	94.077	15.237	34
35	6.5138	0.1535	0.00997	0.06497	100.251	15.391	35
40	8.5133	0.1175	0.00732	0.06232	136.606	16.046	40
45	11.1266	0.0899	0.00543	0.06043	184.119	16.548	45
50	14.5420	0.0688	0.00406	0.05906	246.217	16.932	50
55	19.0058	0.0526	0.00305	0.05805	327.377	17.225	55
60	24.8398	0.0403	0.00231	0.05731	433.450	17.450	60
65	32.4646	0.0308	0.00175	0.05675	572.083	17.622	65
70	42.4299	0.0236	0.00133	0.05633	753.271	17.753	70
75	55.4542	0.0180	0.00101	0.05601	990.076	17.854	75
80	72.4764	0.0138	0.00077	0.05577	1,299.571	17.931	80
85	94.7238	0.0106	0.00059	0.05559	1,704.069	17.990	85
90	123.8002	0.0081	0.00045	0.05545	2,232.731	18.035	90
95	161.8019	0.0062	0.00034	0.05534	2,923.671	18.069	95
100	211.4686	0.0047	0.00026	0.05526	3,826.702	18.096	100

TABLE A–13. **6% Compound Interest Factors**

	SINGLE PAYMENT		UNIFORM SERIES				
	Compound Amount Factor	Present Worth Factor	Sinking Fund Factor	Capital Recovery Factor	Compound Amount Factor	Present Worth Factor	
n	F → P	P → F	A → F	A → P	F → A	P → A	n
1	1.0600	0.9434	1.00000	1.06000	1.000	0.943	1
2	1.1236	0.8900	0.48544	0.54544	2.060	1.833	2
3	1.1910	0.8396	0.31411	0.37411	3.184	2.673	3
4	1.2625	0.7921	0.22859	0.28859	4.375	3.465	4
5	1.3382	0.7473	0.17740	0.23740	5.637	4.212	5
6	1.4185	0.7050	0.14336	0.20336	6.975	4.917	6
7	1.5036	0.6651	0.11914	0.17914	8.394	5.582	7
8	1.5938	0.6274	0.10104	0.16104	9.897	6.210	8
9	1.6895	0.5919	0.08702	0.14702	11.491	6.802	9
10	1.7908	0.5584	0.07587	0.13587	13.181	7.360	10
11	1.8983	0.5268	0.06679	0.12679	14.972	7.887	11
12	2.0122	0.4970	0.05928	0.11928	16.870	8.384	12
13	2.1329	0.4688	0.05296	0.11296	18.882	8.853	13
14	2.2609	0.4423	0.04758	0.10758	21.015	9.295	14
15	2.3966	0.4173	0.04296	0.10296	23.276	9.712	15
16	2.5404	0.3936	0.03895	0.09895	25.673	10.106	16
17	2.6928	0.3714	0.03544	0.09544	28.213	10.477	17
18	2.8543	0.3503	0.03236	0.09236	30.906	10.828	18
19	3.0256	0.3305	0.02962	0.08962	33.760	11.158	19
20	3.2071	0.3118	0.02718	0.08718	36.786	11.470	20
21	3.3996	0.2942	0.02500	0.08500	39.993	11.764	21
22	3.6035	0.2775	0.02305	0.08305	43.392	12.042	22
23	3.8197	0.2618	0.02128	0.08128	46.996	12.303	23
24	4.0489	0.2470	0.01968	0.07968	50.816	12.550	24
25	4.2919	0.2330	0.01823	0.07823	54.865	12.783	25
26	4.5494	0.2198	0.01690	0.07690	59.156	13.003	26
27	4.8223	0.2074	0.01570	0.07570	63.706	13.211	27
28	5.1117	0.1956	0.01459	0.07459	68.528	13.406	28
29	5.4184	0.1846	0.01358	0.07358	73.640	13.591	29
30	5.7435	0.1741	0.01265	0.07265	79.058	13.765	30
31	6.0881	0.1643	0.01179	0.07179	84.802	13.929	31
32	6.4534	0.1550	0.01100	0.07100	90.890	14.084	32
33	6.8406	0.1462	0.01027	0.07027	97.343	14.230	33
34	7.2510	0.1379	0.00960	0.06960	104.184	14.368	34
35	7.6861	0.1301	0.00897	0.06897	111.435	14.498	35
40	10.2857	0.0972	0.00646	0.06646	154.762	15.046	40
45	13.7646	0.0727	0.00470	0.06470	212.744	15.456	45
50	18.4202	0.0543	0.00344	0.06344	290.336	15.762	50
55	24.6503	0.0406	0.00254	0.06254	394.172	15.991	55
60	32.9877	0.0303	0.00188	0.06188	533.128	16.161	60
65	44.1450	0.0227	0.00139	0.06139	719.083	16.289	65
70	59.0759	0.0169	0.00103	0.06103	967.932	16.385	70
75	79.0569	0.0126	0.00077	0.06077	1,300.949	16.456	75
80	105.7960	0.0095	0.00057	0.06057	1,746.600	16.509	80
85	141.5789	0.0071	0.00043	0.06043	2,342.982	16.549	85
90	189.4645	0.0053	0.00032	0.06032	3,141.075	16.579	90
95	253.5463	0.0039	0.00024	0.06024	4,209.104	16.601	95
100	339.3021	0.0029	0.00018	0.06018	5,638.368	16.618	100

TABLE A–14. 7% Compound Interest Factors

	SINGLE PAYMENT		UNIFORM SERIES				
n	Compound Amount Factor $F \rightarrow P$	Present Worth Factor $P \rightarrow F$	Sinking Fund Factor $A \rightarrow F$	Capital Recovery Factor $A \rightarrow P$	Compound Amount Factor $F \rightarrow A$	Present Worth Factor $P \rightarrow A$	n
1	1.0700	0.9346	1.00000	1.07000	1.000	0.935	1
2	1.1449	0.8734	0.48309	0.55309	2.070	1.808	2
3	1.2250	0.8163	0.31105	0.38105	3.215	2.624	3
4	1.3108	0.7629	0.22523	0.29523	4.440	3.387	4
5	1.4026	0.7130	0.17389	0.24389	5.751	4.100	5
6	1.5007	0.6663	0.13980	0.20980	7.153	4.767	6
7	1.6058	0.6227	0.11555	0.18555	8.654	5.389	7
8	1.7182	0.5820	0.09747	0.16747	10.260	5.971	8
9	1.8385	0.5439	0.08349	0.15349	11.978	6.515	9
10	1.9672	0.5083	0.07238	0.14238	13.816	7.024	10
11	2.1049	0.4751	0.06336	0.13336	15.784	7.499	11
12	2.2522	0.4440	0.05590	0.12590	17.888	7.943	12
13	2.4098	0.4150	0.04965	0.11965	20.141	8.358	13
14	2.5785	0.3878	0.04434	0.11434	22.550	8.745	14
15	2.7590	0.3624	0.03979	0.10979	25.129	9.108	15
16	2.9522	0.3387	0.03586	0.10586	27.888	9.447	16
17	3.1588	0.3166	0.03243	0.10243	30.840	9.763	17
18	3.3799	0.2959	0.02941	0.09941	33.999	10.059	18
19	3.6165	0.2765	0.02675	0.09675	37.379	10.336	19
20	3.8697	0.2584	0.02439	0.09439	40.995	10.594	20
21	4.1406	0.2415	0.02229	0.09229	44.865	10.836	21
22	4.4304	0.2257	0.02041	0.09041	49.006	11.061	22
23	4.7405	0.2109	0.01871	0.08871	53.436	11.272	23
24	5.0724	0.1971	0.01719	0.08719	58.177	11.469	24
25	5.4274	0.1842	0.01581	0.08581	63.249	11.654	25
26	5.8074	0.1722	0.01456	0.08456	68.676	11.826	26
27	6.2139	0.1609	0.01343	0.08343	74.484	11.987	27
28	6.6488	0.1504	0.01239	0.08239	80.698	12.137	28
29	7.1143	0.1406	0.01145	0.08145	87.347	12.278	29
30	7.6123	0.1314	0.01059	0.08059	94.461	12.409	30
31	8.1451	0.1228	0.00980	0.07980	102.073	12.532	31
32	8.7153	0.1147	0.00907	0.07907	110.218	12.647	32
33	9.3253	0.1072	0.00841	0.07841	118.933	12.754	33
34	9.9781	0.1002	0.00780	0.07780	128.259	12.854	34
35	10.6766	0.0937	0.00723	0.07723	138.237	12.948	35
40	14.9745	0.0668	0.00501	0.07501	199.635	13.332	40
45	21.0025	0.0476	0.00350	0.07350	285.749	13.606	45
50	29.4570	0.0339	0.00246	0.07246	406.529	13.801	50
55	41.3150	0.0242	0.00174	0.07174	575.929	13.940	55
60	57.9464	0.0173	0.00123	0.07123	813.520	14.039	60
65	81.2729	0.0123	0.00087	0.07087	1,146.755	14.110	65
70	113.9894	0.0088	0.00062	0.07062	1,614.134	14.160	70
75	159.8760	0.0063	0.00044	0.07044	2,269.657	14.196	75
80	224.2344	0.0045	0.00031	0.00031	3,189.063	14.222	80
85	314.5003	0.0032	0.00022	0.07022	4,478.576	14.240	85
90	441.1030	0.0023	0.00016	0.07016	6,287.185	14.253	90
95	618.6697	0.0016	0.00011	0.07011	8,823.854	14.263	95
100	867.7163	0.0012	0.00008	0.07008	12,381.662	14.269	100

TABLE A-15. 8% Compound Interest Factors

	SINGLE PAYMENT		UNIFORM SERIES				
	Compound Amount Factor	Present Worth Factor	Sinking Fund Factor	Capital Recovery Factor	Compound Amount Factor	Present Worth Factor	
n	F → P	P → F	A → F	A → P	F → A	P → A	n
1	1.0800	0.9259	1.00000	1.08000	1.000	0.926	1
2	1.1664	0.8573	0.48077	0.56077	2.080	1.783	2
3	1.2597	0.7938	0.30803	0.38803	3.246	2.577	3
4	1.3605	0.7350	0.22192	0.30192	4.506	3.312	4
5	1.4693	0.6806	0.17046	0.25046	5.867	3.993	5
6	1.5869	0.6302	0.13632	0.21632	7.336	4.623	6
7	1.7138	0.5835	0.11207	0.19207	8.923	5.206	7
8	1.8509	0.5403	0.09401	0.17401	10.637	5.747	8
9	1.9990	0.5002	0.08008	0.16008	12.488	6.247	9
10	2.1589	0.4632	0.06903	0.14903	14.487	6.710	10
11	2.3316	0.4289	0.06008	0.14008	16.645	7.139	11
12	2.5182	0.3971	0.05270	0.13270	18.977	7.536	12
13	2.7196	0.3677	0.04652	0.12652	21.495	7.904	13
14	2.9372	0.3405	0.04130	0.12130	24.215	8.244	14
15	3.1722	0.3152	0.03683	0.11683	27.152	8.559	15
16	3.4259	0.2919	0.03298	0.11298	30.324	8.851	16
17	3.7000	0.2703	0.02963	0.10963	33.750	9.122	17
18	3.9960	0.2502	0.02670	0.10670	37.450	9.372	18
19	4.3157	0.2317	0.02413	0.10413	41.446	9.604	19
20	4.6610	0.2145	0.02185	0.10185	45.762	9.818	20
21	5.0338	0.1987	0.01983	0.09983	50.423	10.017	21
22	5.4365	0.1839	0.01803	0.09803	55.457	10.201	22
23	5.8715	0.1703	0.01642	0.09642	60.893	10.371	23
24	6.3412	0.1577	0.01498	0.09498	66.765	10.529	24
25	6.8485	0.1460	0.01368	0.09368	73.106	10.675	25
26	7.3964	0.1352	0.01251	0.09251	79.954	10.810	26
27	7.9881	0.1252	0.01145	0.09145	87.351	10.935	27
28	8.6271	0.1159	0.01049	0.09049	95.339	11.051	28
29	9.3173	0.1073	0.00962	0.08962	103.966	11.158	29
30	10.0627	0.0994	0.00883	0.08883	113.283	11.258	30
31	10.8677	0.0920	0.00811	0.08811	123.346	11.350	31
32	11.7371	0.0852	0.00745	0.08745	134.214	11.435	32
33	12.6760	0.0789	0.00685	0.08685	145.951	11.514	33
34	13.6901	0.0730	0.00630	0.08630	158.627	11.587	34
35	14.7853	0.0676	0.00580	0.08580	172.317	11.655	35
40	21.7245	0.0460	0.00386	0.08386	259.057	11.925	40
45	31.9204	0.0313	0.00259	0.08259	386.506	12.108	45
50	46.9016	0.0213	0.00174	0.08174	573.770	12.233	50
55	68.9139	0.0145	0.00118	0.08118	848.923	12.319	55
60	101.2571	0.0099	0.00080	0.08080	1,253.213	12.377	60
65	148.7798	0.0067	0.00054	0.08054	1,847.248	12.416	65
70	218.6064	0.0046	0.00037	0.08037	2,720.080	12.443	70
75	321.2045	0.0031	0.00025	0.08025	4,002.557	12.461	75
80	471.9548	0.0021	0.00017	0.08017	5,886.935	12.474	80
85	693.4565	0.0014	0.00012	0.08012	8,655.706	12.482	85
90	1,018.9151	0.0010	0.00008	0.08008	12,723.939	12.488	90
95	1,497.1205	0.0007	0.00005	0.08005	18,701.507	12.492	95
100	2,199.7613	0.0005	0.00004	0.08004	27,484.516	12.494	100

TABLE A–16. 10% Compound Interest Factors

	SINGLE PAYMENT		UNIFORM SERIES				
n	Compound Amount Factor $F \rightarrow P$	Present Worth Factor $P \rightarrow F$	Sinking Fund Factor $A \rightarrow F$	Capital Recovery Factor $A \rightarrow P$	Compound Amount Factor $F \rightarrow A$	Present Worth Factor $P \rightarrow A$	n
1	1.1000	0.9091	1.00000	1.10000	1.000	0.909	1
2	1.2100	0.8264	0.47619	0.57619	2.100	1.736	2
3	1.3310	0.7513	0.30211	0.40211	3.310	2.487	3
4	1.4641	0.6830	0.21547	0.31547	4.641	3.170	4
5	1.6105	0.6209	0.16380	0.26380	6.105	3.791	5
6	1.7716	0.5645	0.12961	0.22961	7.716	4.355	6
7	1.9487	0.5132	0.10541	0.20541	9.487	4.868	7
8	2.1436	0.4665	0.08744	0.18744	11.436	5.335	8
9	2.3579	0.4241	0.07364	0.17364	13.579	5.759	9
10	2.5937	0.3855	0.06275	0.16275	15.937	6.144	10
11	2.8531	0.3505	0.05396	0.15396	18.531	6.495	11
12	3.1384	0.3186	0.04676	0.14676	21.384	6.814	12
13	3.4523	0.2897	0.04078	0.14078	24.523	7.103	13
14	3.7975	0.2633	0.03575	0.13575	27.975	7.367	14
15	4.1772	0.2394	0.03147	0.13147	31.772	7.606	15
16	4.5950	0.2176	0.02782	0.12782	35.950	7.824	16
17	5.0545	0.1978	0.02466	0.12466	40.545	8.022	17
18	5.5599	0.1799	0.02193	0.12193	45.599	8.201	18
19	6.1159	0.1635	0.01955	0.11955	51.159	8.365	19
20	6.7275	0.1486	0.01746	0.11746	57.275	8.514	20
21	7.4002	0.1351	0.01562	0.11562	64.002	8.649	21
22	8.1403	0.1228	0.01401	0.11401	71.403	8.772	22
23	8.9543	0.1117	0.01257	0.11257	79.543	8.883	23
24	9.8497	0.1015	0.01130	0.11130	88.497	8.985	24
25	10.8347	0.0923	0.01017	0.11017	98.347	9.077	25
26	11.9182	0.0839	0.00916	0.10916	109.182	9.161	26
27	13.1100	0.0763	0.00826	0.10826	121.100	9.237	27
28	14.4210	0.0693	0.00745	0.10745	134.210	9.307	28
29	15.8631	0.0630	0.00673	0.10673	148.631	9.370	29
30	17.4494	0.0573	0.00608	0.10608	164.494	9.427	30
31	19.1943	0.0521	0.00550	0.10550	181.943	9.479	31
32	21.1138	0.0474	0.00497	0.10497	201.138	9.526	32
33	23.2252	0.0431	0.00450	0.10450	222.252	9.569	33
34	25.5477	0.0391	0.00407	0.10407	245.477	9.609	34
35	28.1024	0.0356	0.00369	0.10369	271.024	9.644	35
40	45.2593	0.0221	0.00226	0.10226	442.593	9.779	40
45	72.8905	0.0137	0.00139	0.10139	718.905	9.863	45
50	117.3909	0.0085	0.00086	0.10086	1,163.909	9.915	50
55	189.0591	0.0053	0.00053	0.10053	1,880.591	9.947	55
60	304.4816	0.0033	0.00033	0.10033	3,034.816	9.967	60
65	490.3707	0.0020	0.00020	0.10020	4,893.707	9.980	65
70	789.7470	0.0013	0.00013	0.10013	7,887.470	9.987	70
75	1,271.8952	0.0008	0.00008	0.10008	12,708.954	9.992	75
80	2,048.4002	0.0005	0.00005	0.10005	20,474.002	9.995	80
85	3,298.9690	0.0003	0.00003	0.10003	32,979.690	9.997	85
90	5,313.0226	0.0002	0.00002	0.10002	53,120.226	9.998	90
95	8,556.6760	0.0001	0.00001	0.10001	85,556.760	9.999	95
100	13,780.6123	0.0001	0.00001	0.10001	137,796.123	9.999	100

TABLE A–17.   12% Compound Interest Factors

	SINGLE PAYMENT		UNIFORM SERIES				
	Compound Amount Factor	Present Worth Factor	Sinking Fund Factor	Capital Recovery Factor	Compound Amount Factor	Present Worth Factor	
n	F → P	P → F	A → F	A → P	F → A	P → A	n
1	1.1200	0.8929	1.00000	1.12000	1.000	0.893	1
2	1.2544	0.7972	0.47170	0.59170	2.120	1.690	2
3	1.4049	0.7118	0.29635	0.41635	3.374	2.402	3
4	1.5735	0.6355	0.20923	0.32923	4.779	3.037	4
5	1.7623	0.5674	0.15741	0.27741	6.353	3.605	5
6	1.9738	0.5066	0.12323	0.24323	8.115	4.111	6
7	2.2107	0.4523	0.09912	0.21912	10.089	4.564	7
8	2.4760	0.4039	0.08130	0.20130	12.300	4.968	8
9	2.7731	0.3606	0.06768	0.18768	14.776	5.328	9
10	3.1058	0.3220	0.05698	0.17698	17.549	5.650	10
11	3.4785	0.2875	0.04842	0.16842	20.655	5.938	11
12	3.8960	0.2567	0.04144	0.16144	24.133	6.194	12
13	4.3635	0.2292	0.03568	0.15568	28.029	6.424	13
14	4.8871	0.2046	0.03087	0.15087	32.393	6.628	14
15	5.4736	0.1827	0.02682	0.14682	37.280	6.811	15
16	6.1304	0.1631	0.02339	0.14339	42.753	6.974	16
17	6.8660	0.1456	0.02046	0.14046	48.884	7.120	17
18	7.6900	0.1300	0.01794	0.13794	55.750	7.250	18
19	8.6128	0.1161	0.01576	0.13576	63.440	7.366	19
20	9.6463	0.1037	0.01388	0.13388	72.052	7.469	20
21	10.8038	0.0926	0.01224	0.13224	81.699	7.562	21
22	12.1003	0.0826	0.01081	0.13081	92.503	7.645	22
23	13.5523	0.0738	0.00956	0.12956	104.603	7.718	23
24	15.1786	0.0659	0.00846	0.12846	118.155	7.784	24
25	17.0001	0.0588	0.00750	0.12750	133.334	7.843	25
26	19.0401	0.0525	0.00665	0.12665	150.334	7.896	26
27	21.3249	0.0469	0.00590	0.12590	169.374	7.943	27
28	23.8839	0.0419	0.00524	0.12524	190.699	7.984	28
29	26.7499	0.0374	0.00466	0.12466	214.583	8.022	29
30	29.9599	0.0334	0.00414	0.12414	241.333	8.055	30
31	33.5551	0.0298	0.00369	0.12369	271.292	8.085	31
32	37.5817	0.0266	0.00328	0.12328	304.847	8.112	32
33	42.0915	0.0238	0.00292	0.12292	342.429	8.135	33
34	47.1425	0.0212	0.00260	0.12260	384.520	8.157	34
35	52.7996	0.0189	0.00232	0.12232	431.663	8.176	35
40	93.0510	0.0107	0.00130	0.12130	767.091	8.244	40
45	163.9876	0.0061	0.00074	0.12074	1,358.230	8.283	45
50	289.0022	0.0035	0.00042	0.12042	2,400.018	8.305	50
∞				0.12000		8.333	∞

TABLE A–18. 15% Compound Interest Factors

	SINGLE PAYMENT		UNIFORM SERIES				
	Compound Amount Factor	Present Worth Factor	Sinking Fund Factor	Capital Recovery Factor	Compound Amount Factor	Present Worth Factor	
n	$F \rightarrow P$	$P \rightarrow F$	$A \rightarrow F$	$A \rightarrow P$	$F \rightarrow A$	$P \rightarrow A$	n
1	1.1500	0.8696	1.00000	1.15000	1.000	0.870	1
2	1.3225	0.7561	0.46512	0.61512	2.150	1.626	2
3	1.5209	0.6575	0.28798	0.43798	3.472	2.283	3
4	1.7490	0.5718	0.20026	0.35027	4.993	2.855	4
5	2.0114	0.4972	0.14822	0.29832	6.742	3.352	5
6	2.3131	0.4323	0.11424	0.26424	8.754	3.784	6
7	2.6600	0.3759	0.09036	0.24036	11.067	4.160	7
8	3.0590	0.3269	0.07285	0.22285	13.727	4.487	8
9	3.5179	0.2843	0.05957	0.20957	16.786	4.772	9
10	4.0456	0.2472	0.04925	0.19925	20.304	5.019	10
11	4.6524	0.2149	0.04107	0.19107	24.349	5.234	11
12	5.3503	0.1869	0.03448	0.18448	29.002	5.421	12
13	6.1528	0.1625	0.02911	0.17911	34.352	5.583	13
14	7.0757	0.1413	0.02469	0.17469	40.505	5.724	14
15	8.1371	0.1229	0.02102	0.17102	47.580	5.847	15
16	9.3576	0.1069	0.01795	0.16795	55.717	5.954	16
17	10.7613	0.0929	0.01537	0.16537	65.075	6.047	17
18	12.3755	0.0808	0.01319	0.16319	75.836	6.128	18
19	14.2318	0.0703	0.01134	0.16134	88.212	6.198	19
20	16.3665	0.0611	0.00976	0.15976	102.444	6.259	20
21	18.8215	0.0531	0.00842	0.15842	118.810	6.312	21
22	21.6447	0.0462	0.00727	0.15727	137.632	6.359	22
23	24.8915	0.0402	0.00628	0.15628	159.276	6.399	23
24	28.6252	0.0349	0.00543	0.15543	184.168	6.434	24
25	32.9190	0.0304	0.00470	0.15470	212.793	6.464	25
26	37.8568	0.0264	0.00407	0.15407	245.712	6.491	26
27	43.5353	0.0230	0.00353	0.15353	283.569	6.514	27
28	50.0656	0.0200	0.00306	0.15306	327.104	6.534	28
29	57.5755	0.0174	0.00265	0.15265	377.170	6.551	29
30	66.2118	0.0151	0.00230	0.15230	434.745	6.566	30
31	76.1435	0.0131	0.00200	0.15200	500.957	6.579	31
32	87.5651	0.0114	0.00173	0.15173	577.100	6.591	32
33	100.6998	0.0099	0.00150	0.15150	664.666	6.600	33
34	115.8048	0.0086	0.00131	0.15131	765.365	6.609	34
35	133.1755	0.0075	0.00113	0.15113	881.170	6.617	35
40	267.8635	0.0037	0.00056	0.15056	1,779.090	6.642	40
45	538.7693	0.0019	0.00028	0.15028	3,585.128	6.654	45
50	1,083.6574	0.0009	0.00014	0.15014	7,217.716	6.661	50
∞				0.15000		6.667	∞

TABLE A–19. 20% Compound Interest Factors

	SINGLE PAYMENT		UNIFORM SERIES				
	Compound Amount Factor	Present Worth Factor	Sinking Fund Factor	Capital Recovery Factor	Compound Amount Factor	Present Worth Factor	
n	$F \rightarrow P$	$P \rightarrow F$	$A \rightarrow F$	$A \rightarrow P$	$F \rightarrow A$	$P \rightarrow A$	n
1	1.2000	0.8333	1.00000	1.20000	1.000	0.833	1
2	1.4400	0.6944	0.45455	0.65455	2.200	1.528	2
3	1.7280	0.5787	0.27473	0.47473	3.640	2.106	3
4	2.0736	0.4823	0.18629	0.38629	5.368	2.589	4
5	2.4883	0.4019	0.13438	0.33438	7.442	2.991	5
6	2.9860	0.3349	0.10071	0.30071	9.930	3.326	6
7	3.5832	0.2791	0.07742	0.27742	12.916	3.605	7
8	4.2998	0.2326	0.06061	0.26061	16.499	3.837	8
9	5.1598	0.1938	0.04808	0.24808	20.799	4.031	9
10	6.1917	0.1615	0.03852	0.23852	25.959	4.192	10
11	7.4301	0.1346	0.03110	0.23110	32.150	4.327	11
12	8.9161	0.1122	0.02526	0.22526	39.581	4.439	12
13	10.6993	0.0935	0.02062	0.22062	48.497	4.533	13
14	12.8392	0.0779	0.01689	0.21689	59.196	4.611	14
15	15.4070	0.0649	0.01388	0.21388	72.035	4.675	15
16	18.4884	0.0541	0.01144	0.21144	87.442	4.730	16
17	22.1861	0.0451	0.00944	0.20944	105.931	4.775	17
18	26.6233	0.0376	0.00781	0.20781	128.117	4.812	18
19	31.9480	0.0313	0.00646	0.20646	154.740	4.844	19
20	38.3376	0.0261	0.00536	0.20536	186.688	4.870	20
21	46.0051	0.0217	0.00444	0.20444	225.026	4.891	21
22	55.2061	0.0181	0.00369	0.20369	271.031	4.909	22
23	66.2474	0.0151	0.00307	0.20307	326.237	4.925	23
24	79.4968	0.0126	0.00255	0.20255	392.484	4.937	24
25	95.3962	0.0105	0.00212	0.20212	471.981	4.948	25
26	114.4755	0.0087	0.00176	0.20176	567.377	4.956	26
27	137.3706	0.0073	0.00147	0.20147	681.853	4.964	27
28	164.8447	0.0061	0.00122	0.20122	819.223	4.970	28
29	197.8136	0.0051	0.00102	0.20102	984.068	4.975	29
30	237.3763	0.0042	0.00085	0.20085	1,181.882	4.979	30
31	284.8516	0.0035	0.00070	0.20070	1,419.258	4.982	31
32	341.8219	0.0029	0.00059	0.20059	1,704.109	4.985	32
33	410.1863	0.0024	0.00049	0.20049	2,045.931	4.988	33
34	492.2235	0.0020	0.00041	0.20041	2,456.118	4.990	34
35	590.6682	0.0017	0.00034	0.20034	2,948.341	4.992	35
40	1,469.7716	0.0007	0.00014	0.20014	7,343.858	4.997	40
45	3,657.2620	0.0003	0.00005	0.20005	18,281.310	4.999	45
50	9,100.4382	0.0001	0.00002	0.20002	45,497.191	4.999	50
∞				0.20000		5.000	∞

TABLE A–20. 25% Compound Interest Factors

	SINGLE PAYMENT		UNIFORM SERIES			
	Compound Amount Factor	Present Worth Factor	Sinking Fund Factor	Capital Recovery Factor	Compound Amount Factor	Present Worth Factor
n	F → P	P → F	A → F	A → P	F → A	P → A	n
1	1.2500	0.8000	1.00000	1.25000	1.000	0.800	1
2	1.5625	0.6400	0.44444	0.69444	2.250	1.440	2
3	1.9531	0.5120	0.26230	0.51230	3.813	1.952	3
4	2.4414	0.4096	0.17344	0.42344	5.766	2.362	4
5	3.0518	0.3277	0.12185	0.37185	8.207	2.689	5
6	3.8147	0.2621	0.08882	0.33882	11.259	2.951	6
7	4.7684	0.2097	0.06634	0.31634	15.073	3.161	7
8	5.9605	0.1678	0.05040	0.30040	19.842	3.329	8
9	7.4506	0.1342	0.03876	0.28876	25.802	3.463	9
10	9.3132	0.1074	0.03007	0.28007	33.253	3.571	10
11	11.6415	0.0859	0.02349	0.27349	42.566	3.656	11
12	14.5519	0.0687	0.01845	0.26845	54.208	3.725	12
13	18.1899	0.0550	0.01454	0.26454	68.760	3.780	13
14	22.7374	0.0440	0.01150	0.26150	86.949	3.824	14
15	28.4217	0.0352	0.00912	0.25912	109.687	3.859	15
16	35.5271	0.0281	0.00724	0.25724	138.109	3.887	16
17	44.4089	0.0225	0.00576	0.25576	173.636	3.910	17
18	55.5112	0.0180	0.00459	0.25459	218.045	3.928	18
19	69.3889	0.0144	0.00366	0.25366	273.556	3.942	19
20	86.7362	0.0115	0.00292	0.25292	342.945	3.954	20
21	108.4202	0.0092	0.00233	0.25233	429.681	3.963	21
22	135.5253	0.0074	0.00186	0.25186	538.101	3.970	22
23	169.4066	0.0059	0.00148	0.25148	673.626	3.976	23
24	211.7582	0.0047	0.00119	0.25119	843.033	3.981	24
25	264.6978	0.0038	0.00095	0.25095	1,054.791	3.985	25
26	330.8722	0.0030	0.00076	0.25076	1,319.489	3.988	26
27	413.5903	0.0024	0.00061	0.25061	1,650.361	3.990	27
28	516.9879	0.0019	0.00048	0.25048	2,063.952	3.992	28
29	646.2349	0.0015	0.00039	0.25039	2,580.939	3.994	29
30	807.7936	0.0012	0.00031	0.25031	3,227.174	3.995	30
31	1,009.7420	0.0010	0.00025	0.25025	4,034.968	3.996	31
32	1,262.1774	0.0008	0.00020	0.25020	5,044.710	3.997	32
33	1,577.7218	0.0006	0.00016	0.25016	6,306.887	3.997	33
34	1,972.1523	0.0005	0.00013	0.25013	7,884.609	3.998	34
35	2,465.1903	0.0004	0.00010	0.25010	9,856.761	3.998	35
40	7,523.1638	0.0001	0.00003	0.25003	30,088.655	3.999	40
45	22,958.8740	0.0001	0.00001	0.25001	91,831.496	4.000	45
50	70,064.9232	0.0000	0.00000	0.25000	280,255.693	4.000	50
∞				0.25000		4.000	∞

TABLE A-21. 30% Compound Interest Factors

	SINGLE PAYMENT		UNIFORM SERIES				
	Compound Amount Factor	Present Worth Factor	Sinking Fund Factor	Capital Recovery Factor	Compound Amount Factor	Present Worth Factor	
n	F → P	P → F	A → F	A → P	F → A	P → A	n
1	1.3000	0.7692	1.00000	1.30000	1.000	0.769	1
2	1.6900	0.5917	0.43478	0.73478	2.300	1.361	2
3	2.1970	0.4552	0.25063	0.55063	3.990	1.816	3
4	2.8561	0.3501	0.16163	0.46163	6.187	2.166	4
5	3.7129	0.2693	0.11058	0.41058	9.043	2.436	5
6	4.8268	0.2072	0.07839	0.37839	12.756	2.643	6
7	6.2749	0.1594	0.05687	0.35687	17.583	2.802	7
8	8.1573	0.1226	0.04192	0.34192	23.858	2.925	8
9	10.6045	0.0943	0.03124	0.33124	32.015	3.019	9
10	13.7858	0.0725	0.02346	0.32346	42.619	3.092	10
11	17.9216	0.0558	0.01773	0.31773	56.405	3.147	11
12	23.2981	0.0429	0.01345	0.31345	74.327	3.190	12
13	30.2875	0.0330	0.01024	0.31024	97.625	3.223	13
14	39.3738	0.0254	0.00782	0.30782	127.913	3.249	14
15	51.1859	0.0195	0.00598	0.30598	167.286	3.268	15
16	66.5417	0.0150	0.00458	0.30458	218.472	3.283	16
17	86.5042	0.0116	0.00351	0.30351	285.014	3.295	17
18	112.4554	0.0089	0.00269	0.30269	371.518	3.304	18
19	146.1920	0.0068	0.00207	0.30207	483.973	3.311	19
20	190.0496	0.0053	0.00159	0.30159	630.165	3.316	20
21	247.0645	0.0040	0.00122	0.30122	820.215	3.320	21
22	321.1839	0.0031	0.00094	0.30094	1,067.280	3.323	22
23	417.5391	0.0024	0.00072	0.30072	1,388.464	3.325	23
24	542.8008	0.0018	0.00055	0.30055	1,806.003	3.327	24
25	705.6410	0.0014	0.00043	0.30043	2,348.803	3.329	25
26	917.3333	0.0011	0.00033	0.30033	3,054.444	3.330	26
27	1,192.5333	0.0008	0.00025	0.30025	3,971.778	3.331	27
28	1,550.2933	0.0006	0.00019	0.30019	5,164.311	3.331	28
29	2,015.3813	0.0005	0.00015	0.30015	6,714.604	3.332	29
30	2,619.9956	0.0004	0.00011	0.30011	8,729.985	3.332	30
31	3,405.9943	0.0003	0.00009	0.30009	11,349.981	3.332	31
32	4,427.7926	0.0002	0.00007	0.30007	14,755.975	3.333	32
33	5,756.1304	0.0002	0.00005	0.30005	19,183.768	3.333	33
34	7,482.9696	0.0001	0.00004	0.30004	24,939.899	3.333	34
35	9,727.8604	0.0001	0.00003	0.30003	32,422.868	3.333	35
∞				0.30000		3.333	∞

TABLE A–22. 35% Compound Interest Factors

	SINGLE PAYMENT		UNIFORM SERIES				
	Compound Amount Factor	Present Worth Factor	Sinking Fund Factor	Capital Recovery Factor	Compound Amount Factor	Present Worth Factor	
n	$F \rightarrow P$	$P \rightarrow F$	$A \rightarrow F$	$A \rightarrow P$	$F \rightarrow A$	$P \rightarrow A$	n
1	1.3500	0.7407	1.00000	1.35000	1.000	0.741	1
2	1.8225	0.5487	0.42553	0.77553	2.350	1.289	2
3	2.4604	0.4064	0.23966	0.58966	4.172	1.696	3
4	3.3215	0.3011	0.15076	0.50076	6.633	1.997	4
5	4.4840	0.2230	0.10046	0.45046	9.954	2.220	5
6	6.0534	0.1652	0.06926	0.41926	14.438	2.385	6
7	8.1722	0.1224	0.04880	0.39880	20.492	2.507	7
8	11.0324	0.0906	0.03489	0.38489	28.664	2.598	8
9	14.8937	0.0671	0.02519	0.37519	39.696	2.665	9
10	20.1066	0.0497	0.01832	0.36832	54.590	2.715	10
11	27.1439	0.0368	0.01339	0.36339	74.697	2.752	11
12	36.6442	0.0273	0.00982	0.35982	101.841	2.779	12
13	49.4697	0.0202	0.00722	0.35722	138.485	2.799	13
14	66.7841	0.0150	0.00532	0.35532	187.954	2.814	14
15	90.1585	0.0111	0.00393	0.35393	254.738	2.825	15
16	121.7139	0.0082	0.00290	0.35290	344.897	2.834	16
17	164.3138	0.0061	0.00214	0.35214	466.611	2.840	17
18	221.8236	0.0045	0.00159	0.35158	630.925	2.844	18
19	299.4619	0.0033	0.00117	0.35117	852.748	2.848	19
20	404.2736	0.0025	0.00087	0.35087	1,152.210	2.850	20
21	545.7693	0.0018	0.00064	0.35064	1,556.484	2.852	21
22	736.7886	0.0014	0.00048	0.35048	2,102.253	2.853	22
23	994.6646	0.0010	0.00035	0.35035	2,839.042	2.854	23
24	1,342.7973	0.0007	0.00026	0.35026	3,833.706	2.855	24
25	1,812.7763	0.0006	0.00019	0.35019	5,176.504	2.856	25
26	2,447.2480	0.0004	0.00014	0.35014	6,989.280	2.856	26
27	3,303.7848	0.0003	0.00011	0.35011	9,436.528	2.856	27
28	4,460.1095	0.0002	0.00008	0.35008	12,740.313	2.857	28
29	6,021.1478	0.0002	0.00006	0.35006	17,200.422	2.857	29
30	8,128.5495	0.0001	0.00004	0.35004	23,221.570	2.857	30
31	10,973.5418	0.0001	0.00003	0.35003	31,350.120	2.857	31
32	14,814.2815	0.0001	0.00002	0.35002	42,323.661	2.857	32
33	19,999.2800	0.0001	0.00002	0.35002	57,137.943	2.857	33
34	26,999.0280	0.0000	0.00001	0.35001	77,137.223	2.857	34
35	36,448.6878		0.00001	0.35001	104,136.251	2.857	35
∞				0.35000		2.857	∞

TABLE A–23. 40% Compound Interest Factors

	SINGLE PAYMENT		UNIFORM SERIES				
	Compound Amount Factor	Present Worth Factor	Sinking Fund Factor	Capital Recovery Factor	Compound Amount Factor	Present Worth Factor	
n	F → P	P → F	A → F	A → P	F → A	P → A	n
1	1.4000	0.7143	1.00000	1.40000	1.000	0.714	1
2	1.9600	0.5102	0.41667	0.81667	2.400	1.224	2
3	2.7440	0.3644	0.22936	0.62936	4.360	1.589	3
4	3.8416	0.2603	0.14077	0.54077	7.104	1.849	4
5	5.3782	0.1859	0.09136	0.49136	10.946	2.035	5
6	7.5295	0.1328	0.06126	0.46126	16.324	2.168	6
7	10.5414	0.0949	0.04192	0.44192	23.853	2.263	7
8	14.7579	0.0678	0.02907	0.42907	34.395	2.331	8
9	20.6610	0.0484	0.02034	0.42034	49.153	2.379	9
10	28.9255	0.0346	0.01432	0.41432	69.814	2.414	10
11	40.4957	0.0247	0.01013	0.41013	98.739	2.438	11
12	56.6939	0.0176	0.00718	0.40718	139.235	2.456	12
13	79.3715	0.0126	0.00510	0.40510	195.929	2.469	13
14	111.1201	0.0090	0.00363	0.40363	275.300	2.478	14
15	155.5681	0.0064	0.00259	0.40259	386.420	2.484	15
16	217.7953	0.0046	0.00185	0.40185	541.988	2.489	16
17	304.9135	0.0033	0.00132	0.40132	759.784	2.492	17
18	426.8789	0.0023	0.00094	0.40094	1,064.697	2.494	18
19	597.6304	0.0017	0.00067	0.40067	1,491.576	2.496	19
20	836.6826	0.0012	0.00048	0.40048	2,089.206	2.497	20
21	1,171.3554	0.0009	0.00034	0.40034	2,925.889	2.498	21
22	1,639.8976	0.0006	0.00024	0.40024	4,097.245	2.498	22
23	2,295.8569	0.0004	0.00317	0.40017	5,737.142	2.499	23
24	3,214.1997	0.0003	0.00012	0.40012	8,032.999	2.499	24
25	4,499.8796	0.0002	0.00009	0.40009	11,247.199	2.499	25
26	6,299.8314	0.0002	0.00006	0.40006	15,747.079	2.500	26
27	8,819.7640	0.0001	0.00005	0.40005	22,046.910	2.500	27
28	12,347.6696	0.0001	0.00003	0.40003	30,866.674	2.500	28
29	17,286.7374	0.0001	0.00002	0.40002	43,214.343	2.500	29
30	24,201.4324	0.0000	0.00001	0.40002	60,501.081	2.500	30
31	33,882.0053		0.00001	0.40001	84,702.513	2.500	31
32	47,434.8074		0.00001	0.40001	118,584.519	2.500	32
33	66,408.7304		0.00001	0.40001	166,019.326	2.500	33
34	92,972.2225		0.00000	0.40000	232,428.056	2.500	34
35	130,161.1116			0.40000	325,400.279	2.500	35
∞				0.40000		2.500	∞

TABLE A–24. 45% Compound Interest Factors

| | SINGLE PAYMENT | | UNIFORM SERIES | | | | |
| | Compound Amount Factor | Present Worth Factor | Sinking Fund Factor | Capital Recovery Factor | Compound Amount Factor | Present Worth Factor | |
n	$F \rightarrow P$	$P \rightarrow F$	$A \rightarrow F$	$A \rightarrow P$	$F \rightarrow A$	$P \rightarrow A$	n
1	1.4500	0.6897	1.00000	1.45000	1.000	0.690	1
2	2.1025	0.4756	0.40816	0.85816	2.450	1.165	2
3	3.0486	0.3280	0.21966	0.66966	4.552	1.493	3
4	4.4205	0.2262	0.13156	0.58156	7.601	1.720	4
5	6.4097	0.1560	0.08318	0.53318	12.022	1.876	5
6	9.2941	0.1076	0.05426	0.50426	18.431	1.983	6
7	13.4765	0.0742	0.03607	0.48607	27.725	2.057	7
8	19.5409	0.0512	0.02427	0.47427	41.202	2.109	8
9	28.3343	0.0353	0.01646	0.46646	60.743	2.144	9
10	41.0847	0.0243	0.01123	0.46123	89.077	2.168	10
11	59.5728	0.0168	0.00768	0.45768	130.162	2.185	11
12	86.3806	0.0116	0.00527	0.45527	189.735	2.196	12
13	125.2518	0.0080	0.00362	0.45362	276.115	2.204	13
14	181.6151	0.0055	0.00249	0.45249	401.367	2.210	14
15	263.3419	0.0038	0.00172	0.45172	582.982	2.214	15
16	381.8458	0.0026	0.00118	0.45118	846.324	2.216	16
17	553.6764	0.0018	0.00081	0.45081	1,228.170	2.218	17
18	802.8308	0.0012	0.00056	0.45056	1,781.846	2.219	18
19	1,164.1047	0.0009	0.00039	0.45039	2,584.677	2.220	19
20	1,687.9518	0.0006	0.00027	0.45027	3,748.782	2.221	20
21	2,447.5301	0.0004	0.00018	0.45018	5,436.734	2.221	21
22	3,548.9187	0.0003	0.00013	0.45013	7,884.264	2.222	22
23	5,145.9321	0.0002	0.00009	0.45009	11,433.182	2 222	23
24	7,461.6015	0.0001	0.00006	0.45006	16,579.115	2.222	24
25	10,819.3222	0.0001	0.00004	0.45004	24,040.716	2.222	25
26	15,688.0173	0.0001	0.00003	0.45003	34,860.038	2.222	26
27	22,747.6250	0.0000	0.00002	0.45002	50,548.056	2.222	27
28	32,984.0563		0.00001	0.45001	73,295.681	2.222	28
29	47,826.8816		0.00001	0.45001	106,279.737	2.222	29
30	69,348.9783		0.00001	0.45001	154,106.618	2.222	30
∞				0.45000		2.222	∞

TABLE A–25. **50% Compound Interest Factors**

	SINGLE PAYMENT		UNIFORM SERIES				
	Compound Amount Factor	Present Worth Factor	Sinking Fund Factor	Capital Recovery Factor	Compound Amount Factor	Present Worth Factor	
n	$F \longrightarrow P$	$P \longrightarrow F$	$A \longrightarrow F$	$A \longrightarrow P$	$F \longrightarrow A$	$P \longrightarrow A$	n
1	1.5000	0.6667	1.00000	1.50000	1.000	0.667	1
2	2.2500	0.4444	0.40000	0.90000	2.500	1.111	2
3	3.3750	0.2963	0.21053	0.71053	4.750	1.407	3
4	5.0625	0.1975	0.12308	0.62308	8.125	1.605	4
5	7.5938	0.1317	0.07583	0.57583	13.188	1.737	5
6	11.3906	0.0878	0.04812	0.54812	20.781	1.824	6
7	17.0859	0.0585	0.03108	0.53108	32.172	1.883	7
8	25.6289	0.0390	0.02030	0.52030	49.258	1.922	8
9	38.4434	0.0260	0.01335	0.51335	74.887	1.948	9
10	57.6650	0.0173	0.00882	0.50882	113.330	1.965	10
11	86.4976	0.0116	0.00585	0.50585	170.995	1.977	11
12	129.7463	0.0077	0.00388	0.50388	257.493	1.985	12
13	194.6195	0.0051	0.00258	0.50258	387.239	1.990	13
14	291.9293	0.0034	0.00172	0.50172	581.859	1.993	14
15	437.8939	0.0023	0.00114	0.50114	873.788	1.995	15
16	656.8408	0.0015	0.00076	0.50076	1,311.682	1.997	16
17	985.2613	0.0010	0.00051	0.50051	1,968.523	1.998	17
18	1,477.8919	0.0007	0.00034	0.50034	2,953.784	1.999	18
19	2,216.8378	0.0005	0.00023	0.50023	4,431.676	1.999	19
20	3,325.2567	0.0003	0.00015	0.50015	6,648.513	1.999	20
21	4,987.8851	0.0002	0.00010	0.50010	9,973.770	2.000	21
22	7,481.8276	0.0001	0.00007	0.50007	14,961.655	2.000	22
23	11,222.7415	0.0001	0.00004	0.50004	22,443.483	2.000	23
24	16,834.1122	0.0001	0.00003	0.50003	33,666.224	2.000	24
25	25,251.1683	0.0000	0.00002	0.50002	50,500.337	2.000	25
∞				0.50000		2.000	∞

APPENDIX # B

Areas Under
the Normal Curve

TABLE B–1.   Areas Under the Normal Curve[*].

(Proportion of total area under the curve from $-\infty$ to designated Z value)

Z	0.09	0.08	0.07	0.06	0.05	0.04	0.03	0.02	0.01	0.00
−3.5	0.00017	0.00017	0.00018	0.00019	0.00019	0.00020	0.00021	0.00022	0.00022	0.0000
−3.4	0.00024	0.00025	0.00026	0.00027	0.00028	0.00029	0.00030	0.00031	0.00033	0.0000
−3.3	0.00035	0.00036	0.00038	0.00039	0.00040	0.00042	0.00043	0.00045	0.00017	0.0000
−3.2	0.00050	0.00052	0.00054	0.00056	0.00058	0.00060	0.00062	0.00064	0.00066	0.0000
−3.1	0.00071	0.00074	0.00076	0.00079	0.00082	0.00085	0.00087	0.00090	0.00094	0.0000
−3.0	0.00100	0.00104	0.00107	0.00111	0.00114	0.00118	0.00122	0.00126	0.00131	0.0000
−2.9	0.0014	0.0014	0.0015	0.0015	0.0016	0.0016	0.0017	0.0017	0.0018	0.0000
−2.8	0.0019	0.0020	0.0020	0.0021	0.0022	0.0023	0.0023	0.0024	0.0025	0.0000
−2.7	0.0026	0.0027	0.0028	0.0029	0.0030	0.0031	0.0032	0.0033	0.0034	0.0000
−2.6	0.0036	0.0037	0.0038	0.0039	0.0040	0.0041	0.0043	0.0044	0.0045	0.0000
−2.5	0.0048	0.0049	0.0051	0.0052	0.0054	0.0055	0.0057	0.0059	0.0060	0.0000
−2.4	0.0064	0.0066	0.0068	0.0069	0.0071	0.0073	0.0075	0.0078	0.0080	0.0000
−2.3	0.0084	0.0087	0.0089	0.0091	0.0094	0.0096	0.0099	0.0102	0.0104	0.0000
−2.2	0.0110	0.0113	0.0116	0.0119	0.0122	0.0125	0.0129	0.0132	0.0136	0.0000
−2.1	0.0143	0.0146	0.0150	0.0154	0.0158	0.0162	0.0166	0.0170	0.0174	0.0000
−2.0	0.0183	0.0188	0.0192	0.0197	0.0202	0.0207	0.0212	0.0217	0.0222	0.0000
−1.9	0.0233	0.0239	0.0244	0.0250	0.0256	0.0262	0.0268	0.0274	0.0281	0.0000
−1.8	0.0294	0.0301	0.0307	0.0314	0.0322	0.0329	0.0336	0.0344	0.0351	0.0000
−1.7	0.0367	0.0375	0.0384	0.0392	0.0401	0.0409	0.0418	0.0427	0.0436	0.0000
−1.6	0.0455	0.0465	0.0475	0.0485	0.0495	0.0505	0.0516	0.0526	0.0537	0.0000
−1.5	0.0559	0.0571	0.0582	0.0594	0.0606	0.0618	0.0630	0.0643	0.0655	0.0000
−1.4	0.0681	0.0694	0.0703	0.0721	0.0735	0.0749	0.0764	0.0778	0.0793	0.0000
−1.3	0.0823	0.0838	0.0853	0.0869	0.0885	0.0901	0.0918	0.0934	0.0951	0.0000
−1.2	0.0985	0.1003	0.1020	0.1038	0.1057	0.1075	0.1093	0.1112	0.1131	0.0000
−1.1	0.1170	0.1190	0.1210	0.1230	0.1251	0.1271	0.1292	0.1314	0.1335	0.000
−1.0	0.1379	0.1401	0.1423	0.1446	0.1469	0.1492	0.1515	0.1539	0.1562	0.000
−0.9	0.1611	0.1635	0.1660	0.1685	0.1711	0.1736	0.1762	0.1788	0.1814	0.000
−0.8	0.1867	0.1894	0.1922	0.1949	0.1977	0.2005	0.2033	0.2061	0.2000	0.000
−0.7	0.2148	0.2177	0.2207	0.2236	0.2266	0.2297	0.2327	0.2358	0.2389	0.000
−0.6	0.2451	0.2483	0.2514	0.2546	0.2578	0.2611	0.2643	0.2676	0.2709	0.000
−0.5	0.2776	0.2810	0.2843	0.2877	0.2912	0.2946	0.2981	0.3015	0.3050	0.000
−0.4	0.3121	0.3156	0.3192	0.3228	0.3264	0.3300	0.3336	0.3372	0.3409	0.000
−0.3	0.3483	0.3520	0.3557	0.3594	0.3632	0.3669	0.3707	0.3745	0.3783	0.000
−0.2	0.3859	0.3897	0.3936	0.3974	0.4013	0.4052	0.4090	0.4129	0.4168	0.000
−0.1	0.4247	0.4286	0.4325	0.4364	0.4404	0.4443	0.4483	0.4522	0.4562	0.000
−0.0	0.4641	0.4681	0.4721	0.4761	0.4801	0.4840	0.4880	0.4920	0.4960	0.000

[*] $Z = \dfrac{[x_i - E(x_i)]}{\sigma}$.

TABLE B–1.   Areas Under the Normal Curve.

(Proportion of total area under the curve from $-\infty$ to designated Z value)

	0.00	0.01	0.02	0.03	0.04	0.05	0.06	0.07	0.08	0.09
0	0.5000	0.5040	0.5080	0.5120	0.5160	0.5199	0.5239	0.5279	0.5319	0.5359
.1	0.5398	0.5438	0.5478	0.5517	0.5557	0.5596	0.5636	0.5675	0.5714	0.5753
.2	0.5793	0.5832	0.5871	0.5910	0.5948	0.5987	0.6026	0.6064	0.6103	0.6141
.3	0.6179	0.6217	0.6255	0.6293	0.6331	0.6368	0.6406	0.6443	0.6480	0.6517
.4	0.6554	0.6591	0.6628	0.6664	0.6700	0.6736	0.6772	0.6808	0.6844	0.6879
.5	0.6915	0.6950	0.6985	0.7019	0.7054	0.7088	0.7123	0.7157	0.7190	0.7224
.6	0.7257	0.7291	0.7324	0.7357	0.7389	0.7422	0.7454	0.7486	0.7517	0.7549
.7	0.7580	0.7611	0.7642	0.7673	0.7704	0.7734	0.7764	0.7794	0.7823	0.7852
.8	0.7881	0.7910	0.7939	0.7967	0.7995	0.8023	0.8051	0.8079	0.8106	0.8133
.9	0.8159	0.8186	0.8212	0.8238	0.8264	0.8289	0.8315	0.8340	0.8365	0.8389
.0	0.8413	0.8438	0.8461	0.8485	0.8508	0.8531	0.8554	0.8577	0.8599	0.8621
.1	0.8643	0.8665	0.8686	0.8708	0.8729	0.8749	0.8770	0.8790	0.8810	0.8830
.2	0.8849	0.8869	0.8888	0.8907	0.8925	0.8944	0.8962	0.8980	0.8997	0.9015
.3	0.9032	0.9049	0.9066	0.9082	0.9099	0.9115	0.9131	0.9147	0.9162	0.9177
.4	0.9192	0.9207	0.9222	0.9236	0.9251	0.9265	0.9279	0.9292	0.9306	0.9319
.5	0.9332	0.9345	0.9357	0.9370	0.9382	0.9394	0.9406	0.9418	0.9429	0.9441
.6	0.9452	0.9463	0.9474	0.9484	0.9495	0.9505	0.9515	0.9525	0.9535	0.9545
.7	0.9554	0.9564	0.9573	0.9582	0.9591	0.9599	0.9608	0.9616	0.9625	0.9633
.8	0.9641	0.9649	0.9656	0.9664	0.9671	0.9678	0.9686	0.9693	0.9699	0.9706
.9	0.9713	0.9719	0.9726	0.9732	0.9738	0.9744	0.9750	0.9756	0.9761	0.9767
.0	0.9773	0.9778	0.9783	0.9788	0.9793	0.9798	0.9803	0.9808	0.9812	0.9817
.1	0.9821	0.9826	0.9830	0.9834	0.9838	0.9842	0.9846	0.9850	0.9854	0.9857
.2	0.9861	0.9804	0.9868	0.9871	0.9875	0.9878	0.9881	0.9884	0.9887	0.9890
.3	0.9893	0.9896	0.9898	0.9901	0.9904	0.9906	0.9909	0.9911	0.9913	0.9916
.4	0.9918	0.9920	0.9922	0.9925	0.9927	0.9929	0.9931	0.9932	0.9934	0.9936
.5	0.9938	0.9940	0.9941	0.9943	0.9945	0.9946	0.9948	0.9949	0.9951	0.9952
.6	0.9953	0.9955	0.9956	0.9957	0.9959	0.9960	0.9961	0.9962	0.9963	0.9964
.7	0.9965	0.9966	0.9967	0.9968	0.9969	0.9970	0.9971	0.9972	0.9973	0.9974
.8	0.9974	0.9975	0.9976	0.9977	0.9977	0.9978	0.9979	0.9979	0.9980	0.9981
.9	0.9981	0.9982	0.9983	0.9983	0.9984	0.9984	0.9985	0.9985	0.9986	0.9986
.0	0.99865	0.99869	0.99874	0.99878	0.99882	0.99886	0.99889	0.99893	0.99896	0.99900
.1	0.99903	0.99906	0.99910	0.99913	0.99915	0.99918	0.99921	0.99924	0.99926	0.99929
.2	0.99931	0.99934	0.99936	0.99938	0.99940	0.99942	0.99944	0.99946	0.99948	0.99950
.3	0.99952	0.99953	0.99955	0.99957	0.99958	0.99960	0.99961	0.99962	0.99964	0.99965
.4	0.99966	0.99967	0.99969	0.99970	0.99971	0.99972	0.99973	0.99974	0.99975	0.99976
.5	0.99977	0.99978	0.99978	0.99979	0.99980	0.99981	0.99981	0.99982	0.99983	0.99983

Index